Ground Vehicle Dynamics

Originally published in German. Popp, K. and Schiehlen,W. (1993).
Fahrzeugdynamik. B.G. Teubner, Stuttgart. ISBN 3-519-02373-3

Karl Popp and Werner Schiehlen

Ground Vehicle Dynamics

In cooperation with Matthias Kröger and Lars Panning

Springer

Prof. Karl Popp (†)

Prof. Dr.-Ing. Werner Schiehlen
Institute of Engineering and
Computational Mechanics
University of Stuttgart
Pfaffenwaldring 9
70569 Stuttgart, Germany
E-mail: schiehlen@itm.uni-stuttgart.de

In cooperation with:

Prof. Dr.-Ing. Matthias Kröger
Institute of Machine Elements,
Design and Manufacturing
Technical University
Bergakademie Freiberg
Agricolastrasse 1
09596 Freiberg, Germany
E-mail: kroeger@imkf.tu-freiberg.de

Dr.-Ing. Lars Panning
Institute of Dynamics and
Vibration Research
Leibniz Universitaet Hannover
Appelstrasse 11
30167 Hanover, Germany
E-mail: panning@ids.uni-hannover.de

ISBN 978-3-642-42580-6

ISBN 978-3-540-68553-1 (eBook)

DOI 10.1007/978-3-540-68553-1

Typesetting: Data supplied by the authors

Production: Scientific Publishing Services Pvt. Ltd., Chennai, India

Cover Design: WMX Design, Heidelberg, Germany

Printed in acid-free paper

9 8 7 6 5 4 3 2 1

springer.com

To our wives

Gitty Popp
Christine Schiehlen
Melanie Kröger

and to

Kristina von Scheidt

Preface

The book *Ground Vehicle Dynamics* is the revised English edition of the book *Fahrzeugdynamik* originally published in German back in 1993. During the preparation of this English edition the first author Karl Popp passed away far too early. In his spirit, and for his memories, two members of Karl Popp's research group at the Leibniz University Hanover, Matthias Kröger and Lars Panning, agreed to contribute to the ongoing work on this edition of the book. However, it took more time than originally planned and Matthias Kröger moved to the Technical University Bergakademie Freiberg as head of the Institute of Machine Elements, Design and Manufacturing.

Vehicle dynamics deals with the mechanical modeling as well as the mathematical description and analysis of vehicle systems. The aim of this book is a methodologically based introduction to the dynamics of ground vehicle systems. The different kinds of vehicles like automobiles, rail cars or magnetically levitated vehicles are not considered one by one but the dynamically common problems of all these vehicle systems are treated from a uniform point of view. This is achieved by a system oriented approach. The evaluation of meaningful mathematical models allows simulations of motion and parameter studies well in advance of setting up a first prototype. The trend to shorter periods for the development and to larger numbers of vehicle versions demands from the engineer comprehensive computations, the fundamentals of which are presented in this book.

The fundamental concept of this book is based on a modularization into vehicle subsystems with standardized interfaces. In the first vital part the models of vehicles, guidance and suspension systems as well as guideway systems are presented, they are mathematically described in detail and they are assembled to complete vehicle-guideway systems. The second methodologically oriented part is devoted to the performance criteria driving stability, driving safety and durability. Then, it follows a review on the computational methods for linear and nonlinear vehicle systems. The sophisticated theoretical methods related to the demanding problems in vehicle dynamics are applied in the third part to longitudinally, laterally and vertically decoupled motions providing the basics of vehicle dynamics. An appendix with some results from the theory of optimal multivariable control systems presents methods for the control design of mechatronic vehicle components. The many problems included in the book show mainly simple applications of the theory presented, and their solutions will support the reader in better understanding the theory and the fundamentals of railway and road vehicles.

The book is devoted on the one hand to students of applied mechanics and system theory as well as mechanical engineering and automotive engineering. The book will support lectures on vehicle systems and provide a view on the general behavior of ground vehicles. On the other hand the book illustrates to engineers joining or working with a vehicle company, or one of their suppliers, advanced methods which are the basis of software tools widely used today in industry. Thus, the book may contribute to continuing education. Moreover, the systematic methodologically based approach is a good example for many divisions of mechanical engineering and mechatronics.

Compared to the successful German edition which has been out of print for some years, the English edition has been only slightly revised. New references are added throughout the book, in Chapter 2 the recursive formalisms for multibody dynamics are discussed, in Chapter 6 the revised ISO Standard 2631 is considered, in Chapter 7 the standard time integration codes provided by Matlab are evaluated and in Chapter 10 a planar half-car model has been included to fill the gap between the quarter-car model and the complex vehicle model. Since the book is now written in English, some German keywords are added in the appendix. This may help the German reader to identify more easily the technical terms for subjects in which she or he is interested.

The authors and contributors of the book acknowledge the continuous support of the Institute of Engineering and Computational Mechanics at the University of Stuttgart headed by Peter Eberhard. We thank our co-workers from Hanover and the many students from Stuttgart, for typing formulas, tables and text as well as for drawing the figures. Moreover, thanks are due to members of the Institute of Engineering and Computational Mechanics for proofreading the manuscript. In particular, Daniel García Vallejo, a post-doc from the University of Seville, Spain contributed to the final editorial work on the book during his stay at the Institute. Finally it has to be pointed out that the cooperation with Petra Jantzen, Dieter Merkle and Christoph Baumann from Springer-Verlag was excellent.

Freiberg
Hanover
Stuttgart
October 2009

Matthias Kröger
Lars Panning
Werner Schiehlen

Preface to the German Edition

Das vorliegende Buch entstand durch die bereits einige Zeit zurückliegende Anregung unseres verehrten Lehrers, Herrn Prof. Dr. Dr. - Ing. E.h. K. Magnus. Berücksichtigung fanden die Ergebnisse zahlreicher neuerer, zum Teil gemeinsamer Forschungsarbeiten. Vor allem aber ist die mehr als zehnjährige Lehrerfahrung der Verfasser aus Vorlesungen über Fahrzeugdynamik an der Technischen Universität München, der Universität Hannover und der Universität Stuttgart eingeflossen. Hilfreich zur Aufbereitung des umfangreichen Stoffes waren ferner die bei der Durchführung des Kurses "Dynamics of High-Speed Verhicles" am Internationalen Zentrum für Mechanik (CISM) in Udine gesammelten Erfahrungen.

Die Fahrzeugdynamik befaßt sich mit der mechanischen Modellierung sowie der maithematischen Beschreibung und Analyse von Fahrzeugsystemen. Ziel dieses Buches ist es, eine methodenorientierte Einführung in die Dynamik landgestützter Fahrzeugsysteme zu geben. Dabei werden nicht die einzelnen Fahrzeugarten wie Kraftfahrzeuge, Schienenfahrzeuge oder Magnetschwebebahnen nebeneinander betrachtet, sondern die allen Fahrzeugsystemen gemeinsamen dynamischen Probleme unter einheitlichen Gesichtspunkten behandelt. Dies ist durch eine systemtheoretische Betrachtungsweise möglich. Die Bereitstellung aussagekräftiger mathematischer Modelle erlaubt Bewegungssimulationen und Parameterstudien lange bevor der erste Prototyp gebaut wird. Der Trend zu kürzeren Entwicklungszeiten und eine große Variantenvielfalt verlangen heute vom Ingenieur umfassende Berechnungen, für die dieses Buch die Grundlagen vermitteln soll.

Das Grundkonzept des vorliegenden Buches beruht auf einer Modularisierung der Fahrzeugteilsysteme mit standardisierten Schnittstellen. Im ersten zentralen Teil werden die Modelle für Fahrzeuge, Trag- und Führsysteme sowie Fahrwege im einzelnen begründet, mathematisch ausführlich beschrieben und zu Gesamtmodellen für Fahrzeug-Fahrweg-Systeme zusammengefaßt. Der zweite, methodenorientierte Teil wird durch die Beurteilungskriterien Fahrstabilität, Fahrsicherheit, Fahrkomfort und Bauteil-Lebensdauer eingeleitet. Anschließend folgt die Darstellung der Berechnungsmethoden für lineare und nichtlineare Fahrzeugsysteme. Die der anspruchsvollen Aufgabenstellung ent-sprechenden theoretischen Verfahren werden im dritten Teil am Beispiel einfacher Longitudinal-, Lateral- und Vertikalbewegungen verdeutlicht. Ein Anhang mit Ergebnissen aus der Theorie optimaler Mehrgrößenregelsysteme trägt dem Trend zu aktiven Fahrzeugkomponenten Rechnung. Eine Vielzahl aufeinander abgestimmter und in die einzelnen Kapitel eingestreuter Beispiele mit ausführlichen Lösungen sollen das Verständnis der Theorie erleichtern und die Anschauung fördern.

Das Buch wendet sich einerseits an die Studierenden der Angewandten Mechanik und Systemtheorie sowie der Fahrzeugtechnik. Es soll insbesondere Vorlesungen über spezielle Fahrzeugsysteme unterstützen und den Blick für allgemeine Zusammenhänge schärfen. Andererseits zeigt es dem in der

Praxis stehenden Ingenieur die Fortschritte bei der Untersuchung komplexer Fahrzeugmodelle auf und dient so der Weiterbildung. Darüber hinaus ist das systematische Vorgehen beispielhaft für viele Bereiche des Maschinenbaus. Die Verfasser danken Herrn Dipl.-Ing. R. Austermann und Herrn Dipl.-Ing. P. Eberhard für die sorgfältige Durchsicht der Druckfahnen sowie Herrn W. Pietsch für die Erstellung der Reinzeichnungen und Bilder. Dank gebührt ferner den vielen Helfern beim Schreiben des Manuskripts. Schließlich gilt unser Dank dem Verlag B. G. Teubner für die erwiesene Geduld und die stets erfreuliche Zusammenarbeit.

Hannover
Stuttgart
Sommer 1992

K. Popp
W. Schiehlen

Contents

List of Problems and their Solutions

1

System Definition and Modeling

Ground vehicle systems are composed by the vehicle body, the propulsion, guidance and suspension devices, and the guideway, see Fig. 1.1. These components are interacting dynamically with each other. As the vehicle is traveling on the guideway, internal propulsion and suspension forces as well as external disturbances are acting on the vehicle body. Furthermore, the motion of the vehicle affects passengers and goods carried on the vehicle. The dynamical analysis of this interplay of forces and motions is the subject of vehicle dynamics. This analysis requires an integrated treatment of all system components interacting with each other.

The basis of a theoretical analysis of vehicle systems is an appropriate mathematical model adapted to the given engineering task. The quality of the results achievable on a system's dynamical behavior depends on the underlying model. Therefore, the mathematical model has to be as detailed as possible to represent accurately and completely all the essential properties of the vehicle system. On the other hand, the model has to be as simple as possible to allow efficient and fast simulations of the vehicle motions essential for the manufacturers competing on a global market. These conflicting requirements show that modeling is a difficult engineering task, in particular, for complex systems like today's vehicles.

For the modeling it is advisable to decompose the total system in subsystems, e.g. the components shown in Fig. 1.1, with interfaces for forces and motions clearly defined. Then, the subsystems may be modeled separately and composed by modular assembly to the mathematical model of the overall system. The modular concept allows different modeling approaches for the individual subsystems, it offers the required flexibility for design variations and it is essential for the lucidity of large complex systems.

On principle, there are two procedures known for modeling, see Fig. 1.2,

- the empirical approach, and
- the axiomatic approach.

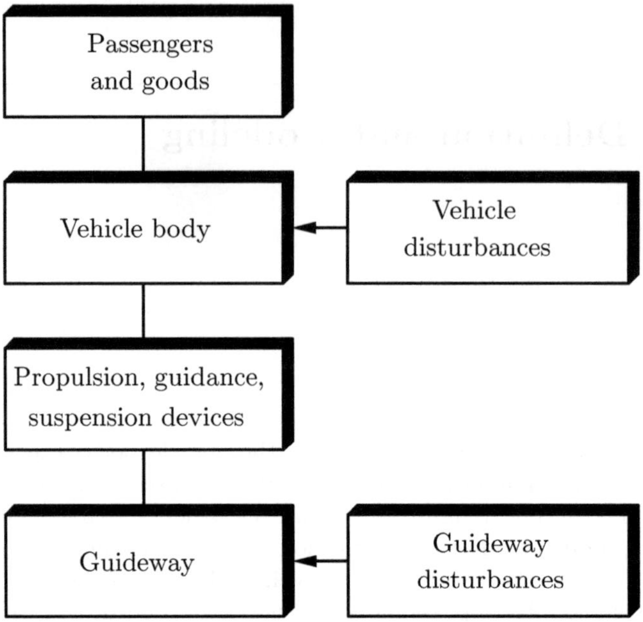

Fig. 1.1. Vehicle system components

The *empirical approach* is based on measurements with prototypes of components or vehicles, respectively, which are processed by identification methods resulting in a mathematical model. For this purpose the system under consideration is excited with known test signals and the responses are recorded. From the input-output relation the parameters of the mathematical model are identified considering the a-prior-knowledge on the structure of the system. This method is called parametric identification. If there is no information on the structure of the system available, then a black box problem is given. In this case the structure of the system and its parameters have to be found, too, and the method is called non-parametric identification. For linear time-invariant systems reliable identification methods are available, Ljung (1999) and Pintelon and Schoukens (2001). In vehicle dynamics the frequency response method proved to be successful but the covariance method can be also applied.

The *axiomatic approach* results directly in a mathematical model by application of fundamental, already mathematically described physical principles. This model emphasizes the structure of the system as well as relations between the system parameters. But some of the parameter values like damping coefficients may remain unknown. For the application of the fundamental principles on complex systems it is often necessary to deal with more simple systems of idealized elements which are called physical models. The already discussed

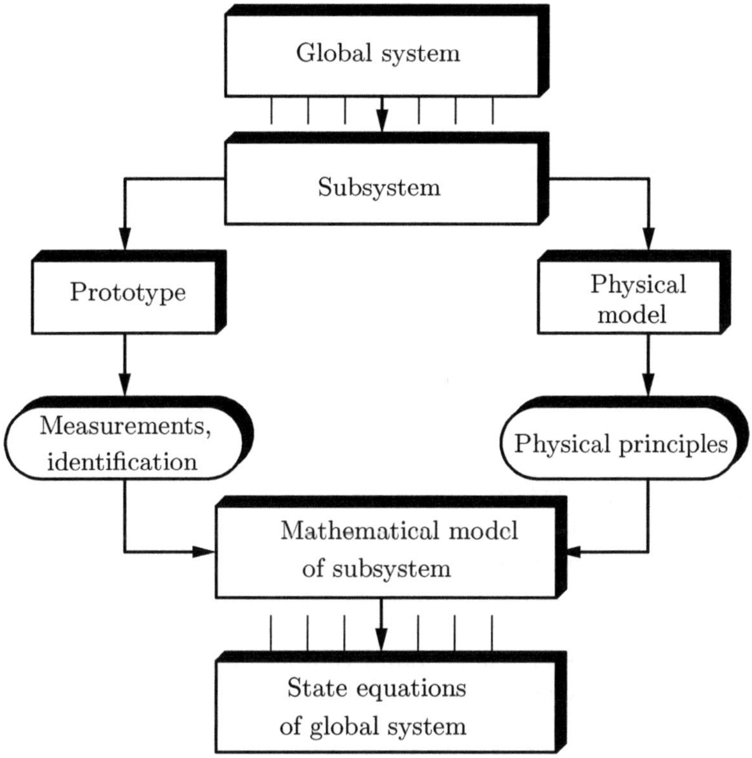

Fig. 1.2. Approaches of modeling: empirical (left) and axiomatic (right)

conflicts related to mathematical modeling apply to physical modeling, too. The models have to be as simple as possible and as detailed as required.

The mathematical description of complex dynamical systems requires in general a combination of both approaches. Whenever possible, the less expensive axiomatic approach will be applied. The missing parameter values have to be found by experiments, and the final results should be validated empirically. The modeling is considered to be satisfactory if the theoretical predictions coincide with the experimental findings.

In the following the axiomatic approach to modeling will be presented in detail for all vehicle systems under consideration. To cover as many different designs as possible, firstly the components shown in Fig. 1.1 are treated separately by realistic models. Then, the components are assembled systematically resulting in the mathematical model of the global system easy to assess and efficient to be analyzed computationally.

The first step is the choice of a *physical model* which is determined by the engineering task. Vehicles consist mainly of mechanical components which can be modeled by

- MultiBody Systems (MBS),
- Finite Element Systems (FES),
- COntinuous Systems (COS).

The range of application of these structural different modeling approaches depends on the geometry and the stiffness of the components to be modeled, see Table 1.1. However, the boundaries of these ranges of application are open, and the engineer has some freedom for his choice. The corresponding mathematical models have also different structures. An important feature is the number of degrees of freedom characterizing the linear independent possibilities of motion of the physical model mathematically described by a corresponding number of time-varying generalized coordinates.

Multibody Systems (MBS) consist of a finite number of rigid or flexible bodies, respectively, which may also degenerate to particles. All of them are characterized by matter and the corresponding inertia. The bodies and particles are interconnected with each other and with the environment by springs, dampers and actuators as well as by bearings, joints and servomechanisms, all of them without mass. An important hypothesis in vehicle dynamics is the assumption of rigid bodies for the modeling of low frequency motions. Rigid bodies are characterized by constant distances between all material points resulting in six degrees of freedom of each body. However, the number of degrees of freedom of a multibody system is reduced by the bearings. Due to the compliant elements acting at a discrete number of node points on the rigid bodies the stiffness distribution within a multibody system is inhomogeneous. Furthermore, there are no restrictions with respect to the geometry resulting in the high adaptability of multibody systems. The motion behavior is completely described by generalized coordinates the number of which is equal to the number f of degrees of freedom. The linear equations of motion of a multibody system read as

$$\boldsymbol{M}\,\ddot{\boldsymbol{y}}(t) + \boldsymbol{D}\,\dot{\boldsymbol{y}}(t) + \boldsymbol{K}\,\boldsymbol{y}(t) = \boldsymbol{f}(t)\,. \tag{1.1}$$

These equations represent a system of ordinary differential equations of second order where the symmetric $f \times f$ -matrices $\boldsymbol{M}$, $\boldsymbol{D}$, $\boldsymbol{K}$ characterize the inertia, the damping and the stiffness behavior of the vehicle component under consideration. The $f \times 1$ -column vector $\boldsymbol{y}(t)$ summarizes the generalized coordinates and the $f \times 1$ -column vector $\boldsymbol{f}(t)$ describes the time-dependent excitation forces, see e.g. Table 1.2. Equations (1.1) have to be supplemented by the initial conditions. Then, in terms of mathematics, an initial value problem with an existing and unique solution is given.

Finite element systems (FES) consist of simple, non-rigid elements like rods, beams, plates etc. with a finite number as degrees of freedom. At node

Table 1.1. Physical models for vehicle components

Model	Geometry	Stiffness distribution	Mathematical model	Number of degrees of freedom
Multibody system (MBS)	complex	homogeneous	ordinary differential equations	finite (small)
Finite element system (FES)	complex	homogeneous	ordinary differential equations	finite (large)
Continuous system (COS)	simple	inhomogeneous	partial differential equations	infinite

Table 1.2. Examples of physical models and their corresponding equations of motion

Multibody system (MBS)	Finite element system (FES)	Continuous system (COS)
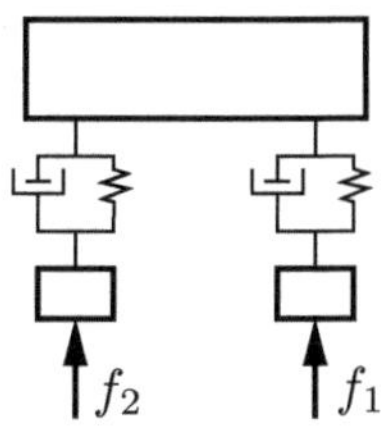	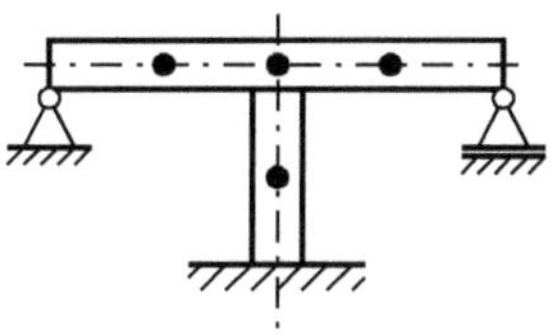	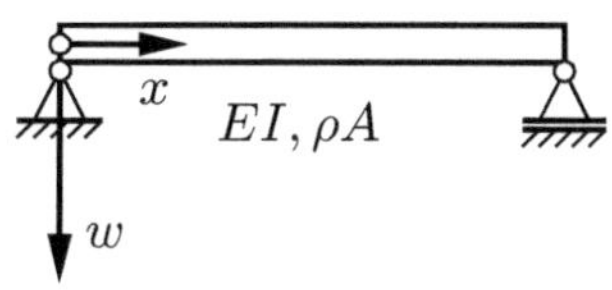
$\boldsymbol{M}\ddot{\boldsymbol{y}}(t) + \boldsymbol{D}\dot{\boldsymbol{y}}(t) + \boldsymbol{K}\boldsymbol{y}(t) = \boldsymbol{f}(t)$	$\boldsymbol{M}\ddot{\boldsymbol{y}}(t) + \boldsymbol{K}\boldsymbol{y}(t) = \boldsymbol{0}$	$EI\frac{\delta^4 w(x,t)}{\delta x^4} + \rho A \frac{\delta^2 w(x,t)}{\delta t^2} = 0$

points there are acting discrete loads, and there may be attached particles, springs, bearings etc. Further, the boundary conditions are taken into account by motions or forces, respectively, at the node points. The linear independent motions of the node points represent the degrees of freedom of the total FES. The number of degrees of freedom is finite due to the finite number of node points, however, the numbers are much higher than with comparable multibody systems due to the non-rigid elements. The deformations within the finite elements between the node points are identified by shape functions. The modeling procedure with FES is very flexible, there don't exist any restrictions with respect to the geometric system design. Modeling is very efficient by using uniform or similar elements, and by collecting groups of elements in super-elements or sub-structures, respectively. Free undamped vibrations occur in beam structures, see e.g. Table 1.2, resulting in equations of motion of the form

$$\boldsymbol{M}\,\ddot{\boldsymbol{y}}(t) + \boldsymbol{K}\,\boldsymbol{y}(t) = \boldsymbol{0} \tag{1.2}$$

where the symmetric $f \times f$ -matrices $\boldsymbol{M}$ and $\boldsymbol{K}$ characterizes the inertia and stiffness properties, again. In addition, initial conditions have to be specified. Thus, an initial value problem is given as in the case of MBS.

Continuous systems (COS) consist of non-rigid bodies with distributed mass and stiffness parameters. As an example one-dimensional continua with homogeneous mass and stiffness properties are considered, in particular flexible beams as shown in Table 1.2. Due to the time-varying distances between the material points, beams have an infinite number of degrees of freedom. Mathematically, continuous beams are represented by partial differential equations. Thus, the equation of motion of an infinitesimal beam element with plane cross sections reads as

$$EI\frac{\partial^4 w(x,t)}{\partial x^4} + \rho\,A\frac{\partial^2 w(x,t)}{\partial t^2} = 0\,, \tag{1.3}$$

where EI means the flexural stiffness, ρA the related inertia, and $w(x,t)$ is the deflection. Equation (1.3) has to be completed by initial and boundary conditions resulting in an initial and boundary value problem. In many cases such problems can be transferred by separation of the variables with the ansatz

$$w(x,t) = \varphi(x)\,y(t) \tag{1.4}$$

into a boundary value eigenvalue problem for $\varphi(x)$ and an initial value problem for $y(t)$ which may be solved at least approximately.

Nowadays, the modeling of complex systems is strongly supported by computational approaches. For MBS, and in particular for FES, many efficient computer codes are available on the market, the characteristic features of which are discussed by Kortuem and Sharp (1993).

The analysis of the different models is strongly depending on the kind and the time behavior of the acting forces. The following *kinds of forces* are distinguished:

- external forces and internal forces,
- applied forces and constraint forces,
- surface forces and volume forces.

The first category depends on the system boundary chosen. Forces originating from outside the boundary are called external forces while forces completely defined within the system boundary are called internal forces. Internal forces appear always twice according to the counteraction principal (actio = reactio) with the same amount but opposite direction within the free body diagram. Furthermore, internal forces may become external forces by changing the system boundary upon decision of the analyst.

The second distinction is related to the cause of the forces. Applied forces are a priori described by a physical force law. Constraint or reaction forces, respectively, follow a posteriori from the principle of mechanics. With vehicle systems, e.g., gravity forces represent applied forces while the forces between a rigid rolling wheel and a rigid guideway belong to the constraint forces.

The third distinction is characterized by the distribution of forces. Surface forces are area-related while volume forces are spatially distributed and, therefore, they depend on the volume of the body under consideration. E.g., friction forces are surface forces while gravitational and magnetic forces are counted as volume forces.

Regarding the time history of forces there are distinguished

- deterministic forces,
- stochastic forces.

Deterministic forces are uniquely determined by a given function of time, e.g. periodic trajectories or impulsive functions. Stochastic forces, on the other hand, are random in time, they can be characterized, however, using statistical approaches and models. Some force trajectories typical for vehicle systems are shown in Table 1.3.

A mathematical representation of generally periodic forces is obtained by Fourier series as $\Omega = 2\pi/T$

$$f(t) = f(t+T) = f_0 + \sum_{j=1}^{\infty} (f_{cj} \cos j\Omega t + f_{sj} \sin j\Omega t) \ , \tag{1.5}$$

where f_0, f_{cj} and f_{sj} are the Fourier coefficients and Ω means the frequency with the period T. In many applications there is only the basic harmonic ($j = 1$) needed.

Short-time acting impulsive forces may be described by the Dirac δ - function:

$$f(t) = p\delta(t - t_0) \ . \tag{1.6}$$

where p means the impulse intensity. Important properties of the δ - function (or more accurately δ - distribution) read as

$$\delta(t-t_0) = \begin{cases} 0 & \text{for } t \neq t_0, \\ \infty & \phantom{\text{for }} t = t_0, \end{cases} \quad \lim_{\varepsilon \to 0} \int_{t_0-\varepsilon}^{t_0+\varepsilon} \delta(\tau - t_0) \mathrm{d}\tau = 1 \ . \tag{1.7}$$

Table 1.3. Time history of forces acting on vehicles

Time history	Mathematical representation	Examples from vehicle systems
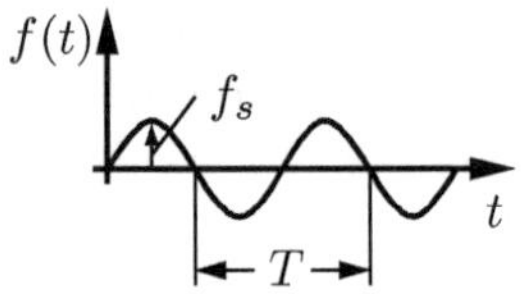	$f(t) = f_s \sin \Omega t, \Omega = \frac{2\pi}{T}$	Unbalance forces of drive-train or wheels
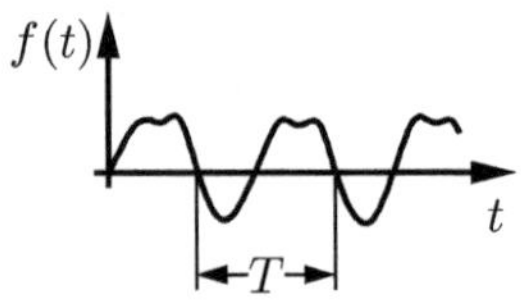	$f(t) = f_0 + \sum_{j=1}^{\infty}(f_{cj} \cos j\Omega t + f_{sj} \sin j\Omega t)$	Forces from periodically constructed guideways Disturbances from internal combustion engines
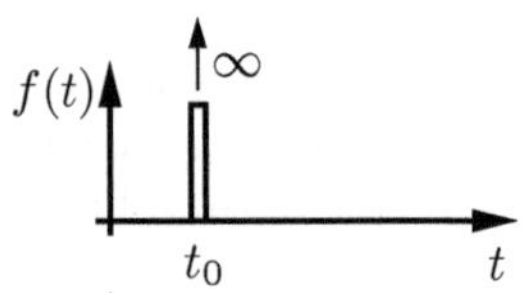	$f(t) = p\,\delta(t - t_0)$	Forces by driving over an obstacle
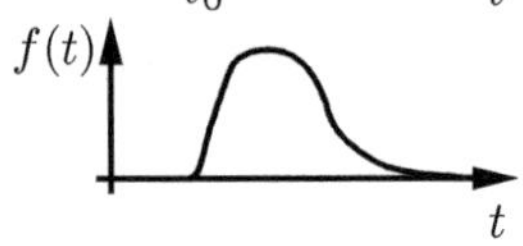	$f(t) = \sum_{j=0}^{\infty} f_j \{t - t_K\}^j$	Aerodynamic forces due to gust of wind
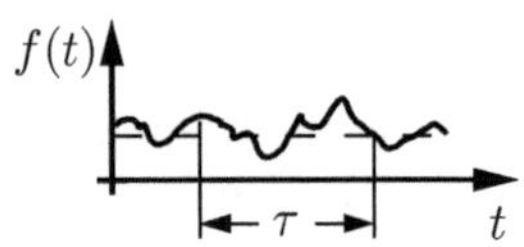	$f(t) \sim [m_f, R_f(\tau)]$ Mean value m_f Correlation function $R_f(\tau)$	Disturbances from random guideway unevenness

Due to the infinitesimal short duration of the δ - function the force function (1.6) is an idealization of real impulsive forces. The quantity p has the unit force times time or mass times velocity, respectively. Thus, it can be interpreted as a variation of the mechanical impulse due to a force impact. A more realistic description of short-time forces starting at time t_k is provided by the Heaviside functions $\{t - t_k\}^j$:

$$f(t) = \sum_{j=0}^{\infty} f_j \{t - t_k\}^j \tag{1.8}$$

where it yields

$$\{t - t_k\}^j = \begin{cases} (t - t_k)^j & \text{for} \quad t \geq t_k \ , \\ 0 & \phantom{\text{for}} \quad t < t_k \ . \end{cases} \tag{1.9}$$

As a special case it follows from (1.9) with $j = 0$ the unit step function describing curbstone obstacles. The coefficients f_j have different units depending on the index j, and they have to be adjusted or identified, respectively, to measured force functions. This completes the mathematical representation of transiently acting forces. Moreover, the Heaviside functions allow also the description of multiple obstacles when a series of starting times t_k, $k = 0$, 1, 2, ... is used.

The description of stochastic processes will be presented in Section 4.2.1 in more detail. It turns out that for many applications in vehicle dynamics Gaussian processes can be used which are uniquely defined by their mean value m_f and their correlation function $R_f(\tau)$:

$$f(t) \sim [m_f, R_f(\tau)] \ . \tag{1.10}$$

Under the assumption of ergodic processes both characteristics can also be found from one realization $f(t)$ by integration over an infinite period of time,

$$m_f = \lim_{T \to \infty} \frac{1}{2T} \int_{-T}^{T} f(t) \mathrm{d}t \ , \tag{1.11}$$

$$R_f(\tau) = \lim_{T \to \infty} \frac{1}{2T} \int_{-T}^{T} f(t + \tau) f(t) \mathrm{d}t = R_f(-\tau) \ . \tag{1.12}$$

In engineering, these definitions are used for large finite time intervals, too, resulting in acceptable approximations. The correlation function characterizes the relationship of the random variables at two times distinguished by the correlation time τ. For $\tau \to \infty$ any relationship for the considered random process disappears, one gets with $m_f = 0$ therefore $R(\tau \to \infty) = 0$. For $\tau \to 0$ the random variables are completely identical resulting in a maximum of the correlation function. This maximum agrees with the quadratic mean as shown by (1.12), too. The correlation function is an even function, $R(\tau) = R(-\tau)$.

2

Vehicle Models

For the modeling of vehicle motions with frequencies up to 50 Hz the method of multibody systems is most appropriate. The composition of the mechanical model is obtained immediately from the design of the vehicle due to great differences in the stiffness distribution, Fig. 2.1. In the following the generation of the equations of motion for multibody systems is presented step by step up to the basics of computer-aided approaches.

2.1 Elements of Multibody Systems

The elements of multibody systems for vehicle modeling, see Fig. 2.1, include *rigid bodies* which may also degenerate to particles, *coupling elements* like springs, dampers or force controlled actuators as well as ideal, i.e. inflexible, kinematical *connecting elements* like joints, bearings, rails and motion controlled actuators. The coupling and connection elements, respectively, are generating internal forces and torques between the bodies of the system or external forces with respect to the environment. Both of them are considered as massless elements.

The constraints resulting from the connecting elements may be holonomic or nonholonomic, scleronomic or rheonomic, respectively. Holonomic constraints reduce the motion space of the system while nonholonomic constraints reduce the velocity space. The constraint equations are called rheonomic if they depend explicitly on time, and scleronomic otherwise. Real vehicle systems are subject to holonomic constraints only which may be given by geometrical or integrable kinematical conditions. However, in more simplified models, e.g. rolling of a rigid wheel or wheelset on a rigid plane, nonholonomic constraints may occur. Some configurations of holonomic connecting elements are listed in Table 2.1 depending on the number of degrees of freedom characterizing the remaining possibilities of motion. Now the motion of vehicle parts will be described mathematically depending on space and time. This is the task of kinematics.

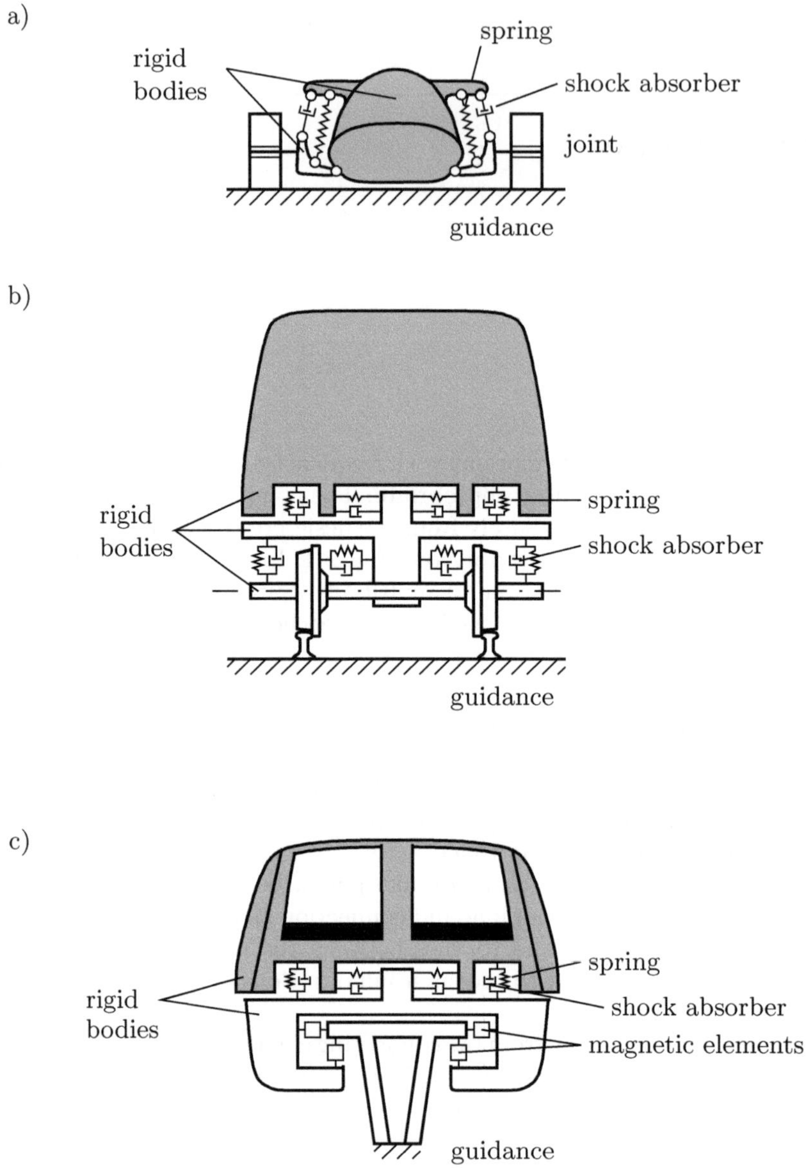

Fig. 2.1. Multibody models of vehicles: a) road vehicle; b) rail vehicle; c) magnetically levitated vehicle

Table 2.1. Configurations of holonomic connecting elements

Motion	Degrees of freedom		
	1	2	3
Rotary	Revolute joint	Universal joint	Spherical joint
Linear	Prismatic joint	Planar joint	
Mixed	Screw joint	Cylindric joint	General planar joint

2.2 Kinematics

The kinematics of a rigid body are presented in an inertial frame and a moving reference frame, and followed by the kinematics of a multibody system.

2.2.1 Frames of Reference for Vehicle Kinematics

A prerequisite for the mathematical description of position, velocity and acceleration of a mechanical system is the definition of appropriate frames of reference. The frames required in vehicle dynamics are shown in Fig. 2.2 with the details summarized in Table 2.2. There will be used only right-handed Cartesian frames with the unit base vectors $\boldsymbol{e}_\nu$, $|\boldsymbol{e}_\nu| = 1$ where the Greek indices generally take the integers 1, 2, 3. A basis or frame $\{O, \boldsymbol{e}_\nu\}$, respectively, is completely defined by its origin O and its base vectors $\boldsymbol{e}_\nu$. For distinction between different frames the upper right index is used if necessary. The inertial frame $\{O^I, \boldsymbol{e}_\nu^I\}$ serves as the general reference frame, in particular for the evaluation of the acceleration. The given trajectory of the vehicle is assumed to be a spatial curve with the moving frame $\{O^B, \boldsymbol{e}_\nu^B\}$ also known as Frenet frame or moving trihedron. The origin O^B is moving with some speed tangential to the trajectory. The reference frame $\{O^R, \boldsymbol{e}_\nu^R\}$ is closely related to the moving frame. Its origin and the first unit vector coincide with the moving frame $O^R = O^B$, $\boldsymbol{e}_1^R = \boldsymbol{e}_1^B$. The second base vector $\boldsymbol{e}_2^R$, however, is parallel to the guideway surface considering the bank of the road or the track, respectively, pointing to the right with respect to the direction of motion.

The body-fixed frame $\{O^i, \boldsymbol{e}_\nu^i\}$ is the principal axis frame of the rigid body K_i located in its center of mass C_i. This frame describes uniquely the position of the body in space. Finally, there is defined a local frame $\{O^j, \boldsymbol{e}_\nu^j\}$ to describe constraint elements between bodies. It is oriented according to the local specifications like the direction of a joint axis. In the following a frame is simply identified by its name (upper right index) only.

2.2.2 Kinematics of a Rigid Body in an Inertial Frame

First of all some definitions and remarks on the nomenclature are presented. The position of a particle P in space is uniquely defined by the position vector

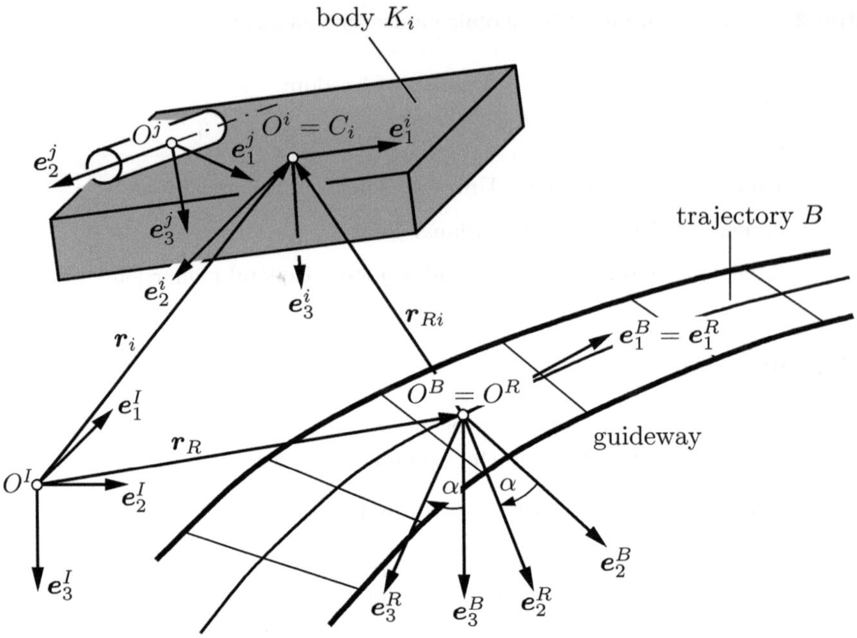

Fig. 2.2. Frames of reference

$\boldsymbol{x}$ represented in the inertial frame $\{O^I, \boldsymbol{e}_\nu^I\}$ by its coordinates x_ν as

$$\boldsymbol{x} = x_1\boldsymbol{e}_1 + x_2\boldsymbol{e}_2 + x_3\boldsymbol{e}_3 \ . \tag{2.1}$$

This set of coordinates may be summarized in a column matrix $\boldsymbol{x}^I$ often simply called a vector, i.e.,

$$\boldsymbol{x}^I = \begin{bmatrix} x_1 \\ x_2 \\ x_3 \end{bmatrix} \equiv \begin{bmatrix} x_1 \ x_2 \ x_3 \end{bmatrix}^{\mathrm{T}} \tag{2.2}$$

where upper right index defines the frame in which the coordinates are measured. This index will often be omitted if there isn't any possibility for a mix-up of frames or if there is used only one frame identified in the text. A goal may be to present all vector and tensor quantities in one common frame, e.g. the inertial frame I. Then, it is possible to integrate subsystems easily into the complete system.

For a particle P moving in time its coordinates are time-dependent, too, and they define a trajectory in space. The mathematical representation results in the vector equation $\boldsymbol{x} = \boldsymbol{x}(t)$ equivalent to three scalar equations according to the three degrees of freedom of the particle in the three-dimensional space. The velocity $\boldsymbol{v}(t)$ and the acceleration $\boldsymbol{a}(t)$ of the particle follow by differentiation with respect to time as

Table 2.2. Frames of reference

Frame of reference	Origin of frame	Orientation of axis
Inertial frame $\{O^I\,,\,\boldsymbol{e}_v^I\}$	O^I space-fixed	$\boldsymbol{e}_1^I$, $\boldsymbol{e}_2^I$ in horizontal plane $\boldsymbol{e}_3^I$
Moving frame $\{O^B\,,\,\boldsymbol{e}_v^B\}$	O^B trajectory-fixed	$\boldsymbol{e}_1^B \equiv \boldsymbol{e}_t$ tangential to trajectory $\boldsymbol{e}_2^B \equiv \boldsymbol{e}_n$ normal to trajectory $\boldsymbol{e}_3^B \equiv \boldsymbol{e}_b$ bi-normal to trajectory
Reference frame $\{O^R\,,\,\boldsymbol{e}_v^R\}$	O^R trajectory-fixed	$\boldsymbol{e}_1^R \equiv \boldsymbol{e}_1^B$ $\boldsymbol{e}_2^R$ in guideway plane $\boldsymbol{e}_3^R$ normal to guideway plane
Body-fixed frame $\{O^i\,,\,\boldsymbol{e}_v^i\}$	$O^i \equiv C_i$ body-fixed in center of mass	$\boldsymbol{e}_v^i$ principal inertia axes
Local frame $\{O^j\,,\,\boldsymbol{e}_v^j\}$	O^i arbitrary	$\boldsymbol{e}_v^j$ locally specified axes

$$\boldsymbol{v}(t) = \frac{\mathrm{d}^I\boldsymbol{x}(t)}{\mathrm{d}t}\,, \quad \boldsymbol{v}^I(t) = \dot{\boldsymbol{x}}^I(t) = \left[\,\dot{x}_1\;\dot{x}_2\;\dot{x}_3\,\right]^{\mathrm{T}}\,, \tag{2.3}$$

$$\boldsymbol{a}(t) = \frac{\mathrm{d}^I\boldsymbol{v}(t)}{\mathrm{d}t}\,, \quad \boldsymbol{a}^I(t) = \dot{\boldsymbol{v}}^I(t) = \ddot{\boldsymbol{x}}^I(t) = \left[\,\ddot{x}_1\;\ddot{x}_2\;\ddot{x}_3\,\right]^{\mathrm{T}}\,. \tag{2.4}$$

The upper right index refers to the frame of reference in which the operations, in particular the differentiation, have to be executed. In the inertial frame I the differentiation of vectors is just performed by differentiation of the scalar coordinates.

The motion of a particle P on a curvilinear trajectory in space may be presented in the moving frame B, too. The position of the point is uniquely identified by the arc length $s(t)$ as a generalized coordinate, Fig. 2.3. Then, the position vector $\boldsymbol{x}$ is a function of the arc length, $\boldsymbol{x} = \boldsymbol{x}(s)$. For the velocity and acceleration vector it follows, see e.g. Magnus and Mueller-Slany (2005),

$$\boldsymbol{v}(t) = \frac{\mathrm{d}^I\boldsymbol{x}(t)}{\mathrm{d}t} = \frac{\mathrm{d}^I\boldsymbol{x}(s)}{\mathrm{d}s}\frac{\mathrm{d}s}{\mathrm{d}t} = v\,\boldsymbol{e}_t\,, \quad v = \dot{s}\,,$$

$$\boldsymbol{v}^B(t) = \left[\,\dot{s}\;0\;0\,\right]^{\mathrm{T}}, \tag{2.5}$$

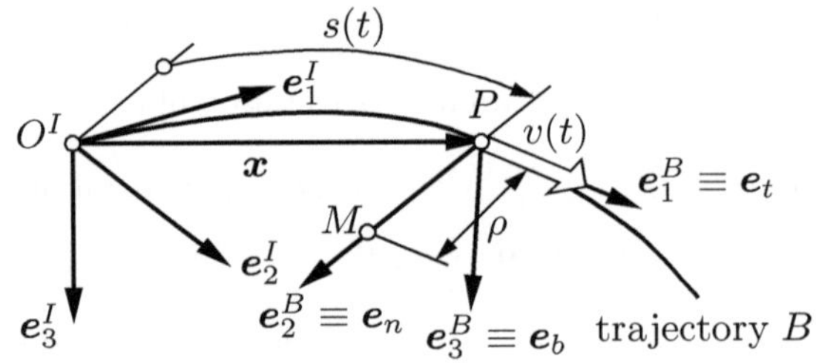

Fig. 2.3. Trajectory of particle P

$$\boldsymbol{a}(t) = \frac{\mathrm{d}^I \boldsymbol{v}(t)}{\mathrm{d}t} = a_t \boldsymbol{e}_t + a_n \boldsymbol{e}_n = \dot{v}\, \boldsymbol{e}_t + \frac{v^2}{\rho} \boldsymbol{e}_n \,,$$
$$\boldsymbol{a}^B(t) = \begin{bmatrix} \ddot{s} & \dot{s}^2/\rho & 0 \end{bmatrix}^{\mathrm{T}} , \tag{2.6}$$

where ρ is the curvature of the trajectory in point P. Further, it is

- $v = \dot{s}$ the tangent velocity,
- $a_t = \dot{v} = \ddot{s}$ the tangent acceleration, and
- $a_n = v^2/\rho = \dot{s}^2/\rho$ the normal or centripetal acceleration.

Special cases of the general motion in space are the motion in a straight line ($\rho \to \infty$), and the motion on a circle ($\rho =$ const). Often the functions $s(t)$, $\dot{s}(t)$, $\ddot{s}(t)$, $\dot{s}(s)$, $\ddot{s}(s)$, $\ddot{s}(\dot{s})$ are depicted in kinematical diagrams for graphical visualization of the motion along a track.

■ Problem 2.1 Longitudinal motion of an automobile

The tangent acceleration a_t of an automobile depends during speed up on the velocity, $a_t = a_t(v)$, and during slow down on the time, $a_t = a_t(t)$, according to the driver's input, see Fig. 2.4.

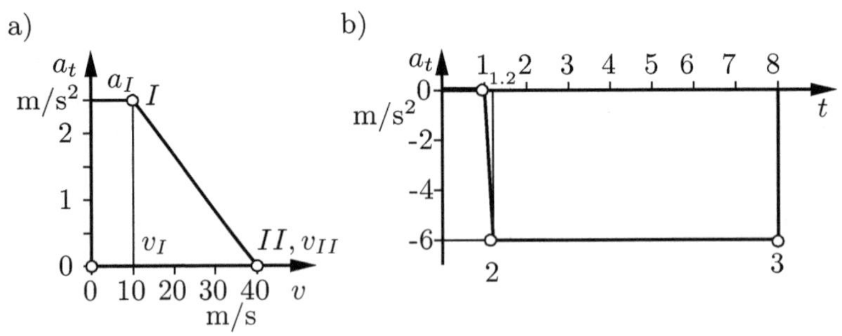

Fig. 2.4. Acceleration functions: a) speed up $a_t = a_t(v)$; b) slow down $a_t = a_t(t)$

a) Compute time t required by the vehicle to reach a speed $v > v_I$ as well as the distance s, and draw graphs of the functions $v(s)$, $a_t(s)$, $s(t)$, $v(t)$, $a_t(t)$.
b) Compute the braking distance s_B and the braking time t_B for an initial speed $v_0 = 40\,\mathrm{m/s}$, and show the corresponding graphs.

Solution
a) From the fundamental relation

$$a_t(v) = \frac{\mathrm{d}v}{\mathrm{d}t} = \frac{\mathrm{d}v}{\mathrm{d}s}\frac{\mathrm{d}s}{\mathrm{d}t} = v\frac{\mathrm{d}v}{\mathrm{d}s} \tag{1}$$

one gets by integration with the initial conditions $s(t_0) = s_0$ and $v(t_0) = v_0$ the required quantities

$$t = t_0 + \int\limits_{v_0}^{v} \frac{\mathrm{d}\overline{v}}{a_t(\overline{v})}\,, \qquad s = s_0 + \int\limits_{v_0}^{v} \frac{\overline{v}}{a_t(\overline{v})}\mathrm{d}\overline{v}\,, \tag{2}$$

where the integration variables are characterized by a bar. The acceleration $a_t(v)$, e.g., may be given by a piecewise defined function, see Fig. 2.4 a),

$$a_t(v) = \begin{cases} a_I & \text{for} \quad 0 \le v \le v_I\,, \\ a_I\dfrac{v_{II} - v}{v_{II} - v_I} & \phantom{\text{for}} \quad v_I \le v \le v_{II}\,. \end{cases} \tag{3}$$

For $v > v_I$ the evaluation of the integrals (2) with $t_0 = 0$ and $v_0 = 0$ results in

$$\begin{aligned} t = t(v) &= \frac{1}{a_I}\int\limits_0^{v_I} \mathrm{d}v + \frac{v_{II} - v_I}{a_I}\int\limits_{v_I}^{v} \frac{\mathrm{d}\overline{v}}{v_{II} - \overline{v}} \\ &= \frac{1}{a_I}\left[v_I + (v_{II} - v_I)\ln\frac{v_{II} - v_I}{v_{II} - v}\right]\,, \end{aligned} \tag{4}$$

$$\begin{aligned} s = s(v) &= \frac{1}{a_I}\int\limits_0^{v_I} v\mathrm{d}v + \frac{v_{II} - v_I}{a_I}\int\limits_{v_I}^{v} \frac{\overline{v}\mathrm{d}\overline{v}}{v_{II} - v} \\ &= \frac{1}{2a_I}\left\{v_I^2 + 2(v_{II} - v_I)\left[v_{II}\ln\frac{v_{II} - v_I}{v_{II} - v} - (v - v_I)\right]\right\}\,. \end{aligned} \tag{5}$$

The speed $v = v_{II}$ is achieved only in the limit $t \to \infty$ or $s \to \infty$, respectively. The desired kinematic diagrams follow either numerically by evaluation of the corresponding inverse functions or graphically from the known functions (4) and (5) where the special arrangement of the diagrams as shown in Fig. 2.5 is helpful due to the two auxiliary diagrams connecting the different abscissae. Starting from the known function $a_t(v)$ and the graphs $v(s)$, $v(t)$ following from (5) and (4) the missing graphs can be found pointwise.

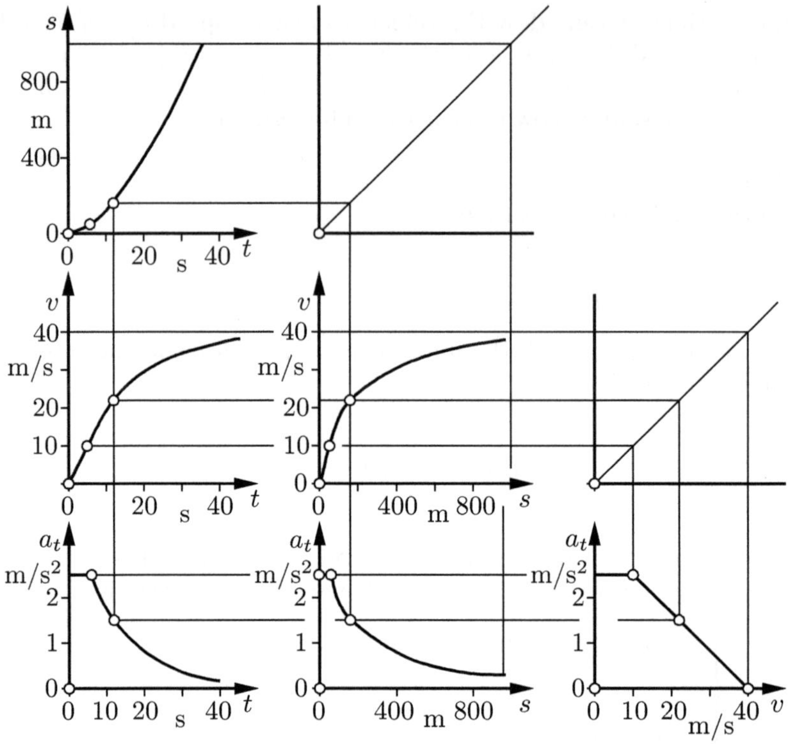

Fig. 2.5. Kinematical graphs of the speed up, given $a_t(v)$

For piecewise given functions it is often useful to introduce for each interval i a new independent variable (index i) and to compute the required quantities by continuous adding. The approach is based on the final values of interval $i-1$ which are used as initial values of interval i. To avoid any confusion, all values at the end of an interval may be characterized by an asterisk. The adding approach will be used for the solution of problem b).

b) From the fundamental relations

$$a_t(t) = \frac{\mathrm{d}v}{\mathrm{d}t} \,, \quad v(t) = \frac{\mathrm{d}s}{\mathrm{d}t} \tag{6}$$

one gets for the required quantities

$$v = v_0 + \int\limits_{t_0}^{t} a(\bar{t})\mathrm{d}\bar{t} \,, \quad s = s_0 + \int\limits_{t_0}^{t} v(\bar{t})\mathrm{d}\bar{t} \,. \tag{7}$$

With the piecewise representation of the variables

$$t = \sum_{v=0}^{i=1} t_v^* + t_i \,, \quad 0 \le t_i \le t_i^* \,, \quad t_0^* = t_0$$

$$a_t(t_i) = a_i \,, \quad v(t_i) = v_i \,, \quad s(t_i) = s_i \,, \quad i = 1, 2, \ldots \tag{8}$$

it follows from (7)

$$v_i = v_{i-1}^* + \int_0^{t_i} a_i(t)\mathrm{d}t, s_i = s_{i-1}^* + \int_0^{t_i} v_i(t)\mathrm{d}t \,,$$

$$v_0^* = v_0 \,, \quad s_0^* = s_0 \,, \quad i = 1, 2, \ldots \,. \tag{9}$$

The evaluation can be made in table form or by computer. With the intervals shown in Fig. 2.6 and the numerical values $t_0 = 0$, $s_0 = 0$, $v_0 = 40\,\mathrm{m/s}$, $|a_2^*| = |a_3^*| = a = 6\,\mathrm{m/s}^2$ one gets the numerical scheme shown in Table 2.3.

Table 2.3. Numerical scheme of the kinematics during slow down

Interval i	t_i^* [s]	a_i	$v_i = v_{i-1}^* + \int_0^{t_i} a_i(t)\mathrm{d}t$	v_i^* [m/s]
1	1	0	$v_1 = v_0$	40
2	0.2	$-a\dfrac{t_2}{t_2^*}$	$v_2 = v_1^* - \dfrac{a}{2t_2^*}t_2^2$	39.4
3	6.57	$-a$	$v_3 = v_2^* - at_3$	0
Interval i	t_i^* [s]		$s_i = s_{i-1}^* + \int_0^{t_i} v_i(t)dt$	s_i^* [m]
1	1		$s_1 = v_0 t_1$	40
2	0.2		$s_2 = s_1^* + v_1^* t_2 - \dfrac{at_2^3}{6t_2^*}$	47.84
3	6.57		$s_3 = s_2^* + v_2^* t_3 - \dfrac{a}{2}t_3^2$	177.20

■

The *translational motion* of a rigid body K_i is completely described by the general relations for a particle related to a body-fixed point, e.g. the center of mass C_i of the rigid body, and the corresponding position vector $\boldsymbol{r}_i$, see Fig. 2.7. The *rotational motion* of a rigid body K_i follows from the relative position of two frames where one of them is a body-fixed frame. For coinciding origins the position of the body-fixed frame i relative to the inertial frame I is defined by three rotation angles according to the three rotational degrees of freedom of a rigid body in space. Both frames are related to each other

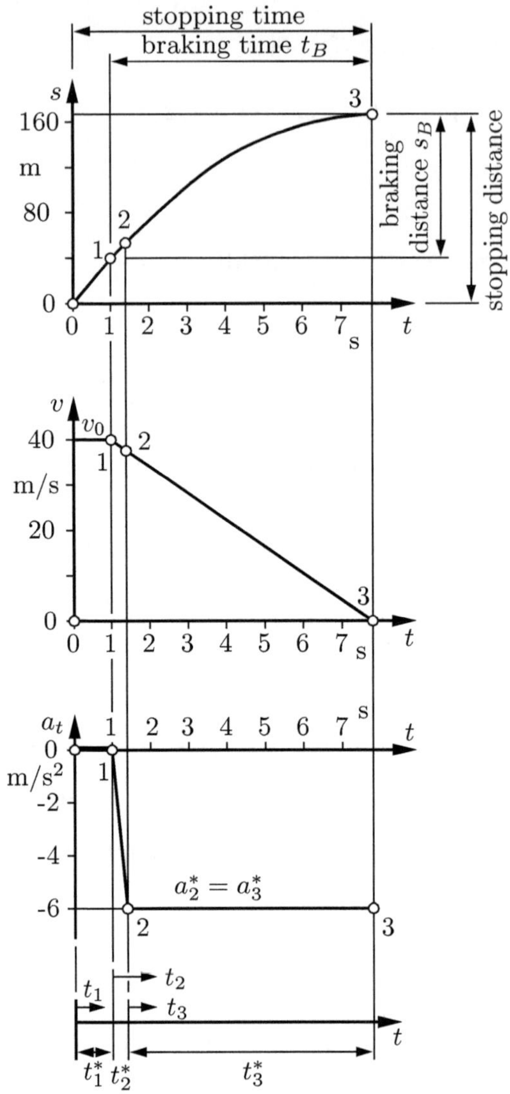

Fig. 2.6. Kinematical graphs of the slow down, given $a_t(t)$

by three elementary rotations performed successively around different base vectors using three rotation angles. If, for example, the frame i is revolved around the coinciding 3-axes of frame I and i by the angle γ, the relation of the corresponding base vectors is given by the matrix $\boldsymbol{S}^{Ii}$ as shown in Fig. 2.8 and reads as

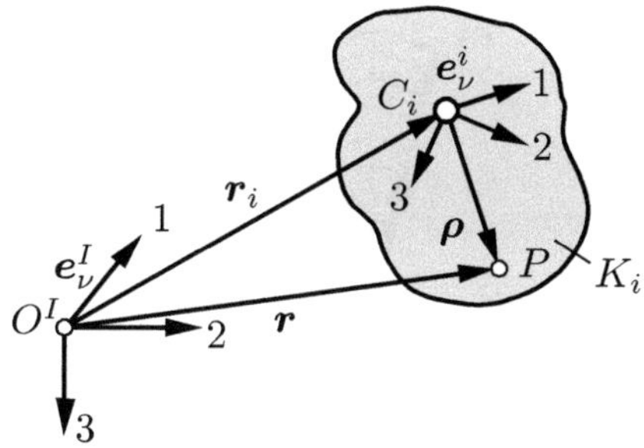

Fig. 2.7. Position of a rigid body K_i in the inertial frame I

$$\begin{bmatrix} \boldsymbol{e}_1^I \\ \boldsymbol{e}_2^I \\ \boldsymbol{e}_3^I \end{bmatrix} = \underbrace{\begin{bmatrix} \cos\gamma & -\sin\gamma & 0 \\ \sin\gamma & \cos\gamma & 0 \\ 0 & 0 & 1 \end{bmatrix}}_{\boldsymbol{S}^{Ii} = \boldsymbol{\gamma}_3} \begin{bmatrix} \boldsymbol{e}_1^i \\ \boldsymbol{e}_2^i \\ \boldsymbol{e}_3^i \end{bmatrix} \tag{2.7}$$

The row ν of the elementary rotation matrix $\boldsymbol{\gamma}_3$ is composed of the coordinates of the base vector $\boldsymbol{e}_\nu^I$ in frame i. The corresponding matrices for positive rotations around the remaining axes read as

$$\boldsymbol{\alpha}_1 = \begin{bmatrix} 1 & 0 & 0 \\ 0 & \cos\alpha & -\sin\alpha \\ 0 & \sin\alpha & \cos\alpha \end{bmatrix}, \quad \boldsymbol{\beta}_2 = \begin{bmatrix} \cos\beta & 0 & \sin\beta \\ 0 & 1 & 0 \\ -\sin\beta & 0 & \cos\beta \end{bmatrix} \tag{2.8}$$

where the elementary rotation matrices are characterized by the name of the rotation angle and the index related to the axis of rotation. There are numerous possibilities to choose the name of the angle, however, the sequence of the rotation axes used is not commutative. In vehicle dynamics the Cardano angles α, β, γ are often used, see Fig. 2.9, which are different from the well-known Euler angles ψ, ϑ, φ.

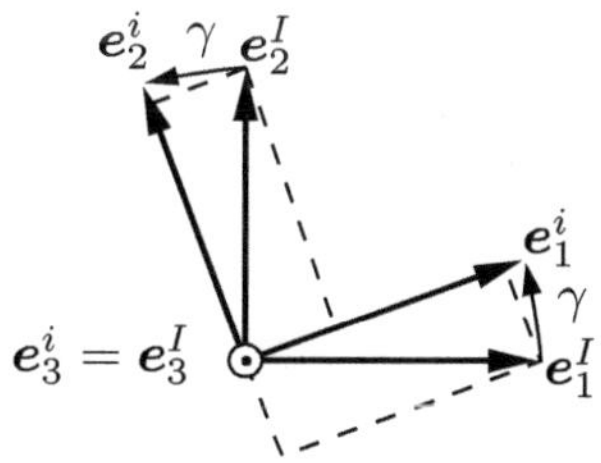

Fig. 2.8. Elementary rotation with angle γ around 3-axes

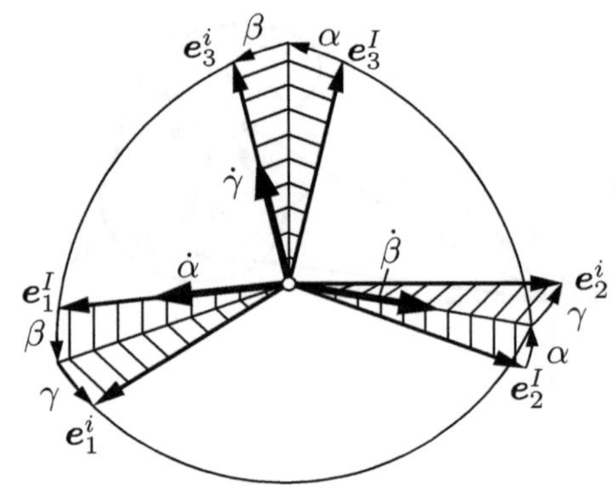

Fig. 2.9. Spatial rotation with Cardano angles α, β, γ

The resulting rotation matrices $\boldsymbol{S}^{Ii}$ which present the relation between frames I and i are obtained by the corresponding matrix multiplications

$$\boldsymbol{S}^{Ii}(\alpha,\beta,\gamma) = \boldsymbol{\alpha}_1\,\boldsymbol{\beta}_2\,\boldsymbol{\gamma}_3\,, \quad \boldsymbol{S}^{Ii}(\psi,\vartheta,\varphi) = \boldsymbol{\psi}_3\,\boldsymbol{\vartheta}_1\,\boldsymbol{\varphi}_3\,. \tag{2.9}$$

Since matrix products are not commutative, the sequence of the elementary rotations has to be observed strongly. The Cardano and Euler angles, respectively, are defined by successive rotations around the 1-, 2-, 3-axis and 3-, 1-, 3-axis, respectively, starting from the inertial frame I. The sequence of the elementary rotations is uniquely identified by the sequence of the indices of the elementary rotation matrices as shown in (2.9).

The rotation matrices are orthogonal matrices

$$\boldsymbol{S}^{iI} = \left(\boldsymbol{S}^{Ii}\right)^{-1} = \left(\boldsymbol{S}^{Ii}\right)^{\mathrm{T}} = \boldsymbol{S}^{iI}\,, \quad \det \boldsymbol{S} = +1 \tag{2.10}$$

where the inversion is also represented by the exchange of the upper indices. Thus, the inverse rotation matrix is simply found by transposition of the original rotation matrix $\boldsymbol{S}^{Ii}$. Using Cardano angles the rotation matrix reads explicitly as

$$\boldsymbol{S}^{Ii}(\alpha,\beta,\gamma) = \begin{bmatrix} \mathrm{c}\beta\mathrm{c}\gamma & -\mathrm{c}\beta\mathrm{s}\gamma & \mathrm{s}\beta \\ \mathrm{c}\alpha\mathrm{s}\gamma + \mathrm{s}\alpha\mathrm{s}\beta\mathrm{c}\gamma & \mathrm{c}\alpha\mathrm{c}\gamma - \mathrm{s}\alpha\mathrm{s}\beta\mathrm{s}\gamma & -\mathrm{s}\alpha\mathrm{c}\beta \\ \mathrm{s}\alpha\mathrm{s}\gamma - \mathrm{c}\alpha\mathrm{s}\beta\mathrm{c}\gamma & \mathrm{s}\alpha\mathrm{c}\gamma + \mathrm{c}\alpha\mathrm{s}\beta\mathrm{s}\gamma & \mathrm{c}\alpha\mathrm{c}\beta \end{bmatrix}, \tag{2.11}$$

where the abbreviations c and s stands for cos and sin, respectively. In applications often small rotations are found, $\alpha, \beta, \gamma \ll 1$, resulting in the linearized rotation matrix

$$\boldsymbol{S}^{Ii}(\alpha,\beta,\gamma) = \begin{bmatrix} 1 & -\gamma & \beta \\ \gamma & 1 & -\alpha \\ -\beta & \alpha & 1 \end{bmatrix}. \tag{2.12}$$

This result is also obtained if the elementary rotation matrices are linearized and multiplied with each other. Due to the vector property of small rotations the sequence of the multiplications has no longer to be considered. For small rotations the Cardano angles may be assigned directly to the rotational motions around the body-fixed axes, Fig. 2.10. In vehicle engineering the following notations are used,

- α roll motion,
- β pitch motion,
- γ yaw motion.

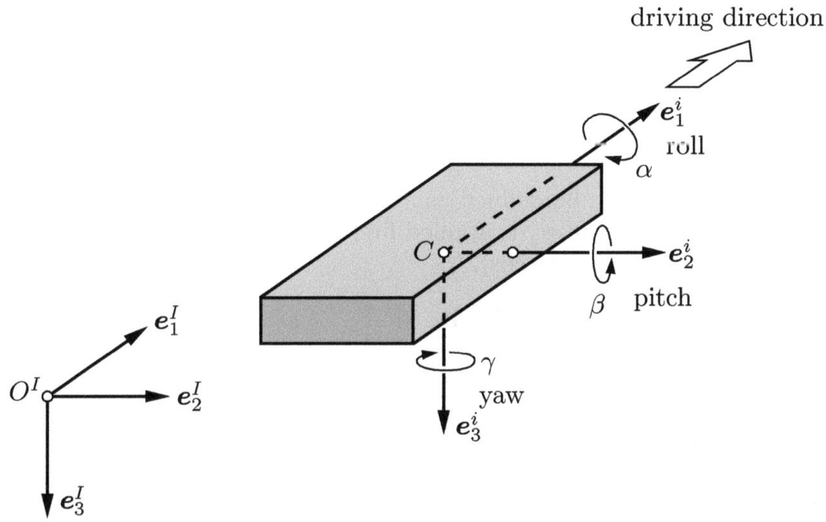

Fig. 2.10. Notation of rotary vehicle motions

■ Problem 2.2 Rotation matrix for a railway wheelset

A railway wheelset is running on a straight track. Evaluate the rotation matrix $\boldsymbol{S}^{RZ}$ between the track related reference frame R and an intermediate frame Z with origin in the wheelset's center of mass C and the 2-axis along the wheelset axle, see Fig. 2.11. The frame Z describes the wheelset motion without the large rotation around the wheelset axle. The frame R is transformed into the frame Z by two elementary rotations, the first one by angle γ around the 3-axis and the second one by angle α around the 1-axis. Both rotation angles may be considered as small, $\alpha, \gamma \ll 1$.

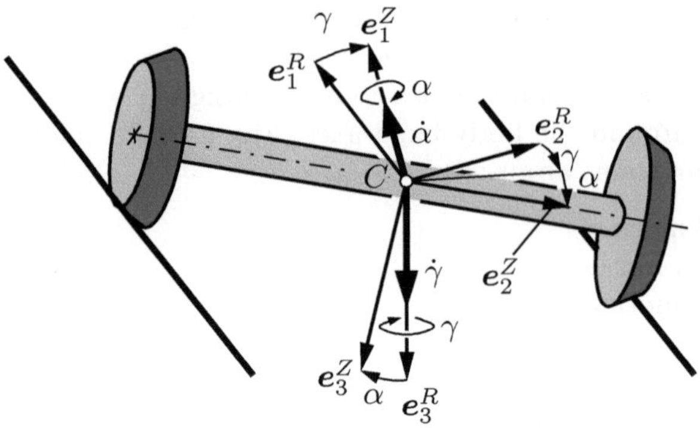

Fig. 2.11. Wheelset frames and rotations

Solution

The rotation matrix can be defined as a product of the elementary rotation matrices $\boldsymbol{\gamma}_3$ and $\boldsymbol{\alpha}_1$ where $\boldsymbol{\gamma}_3$ is applied firstly with respect to frame R and then it follows $\boldsymbol{\alpha}_1$ (s$\hat{=}$sin, c$\hat{=}$cos),

$$\boldsymbol{S}^{RZ} = \boldsymbol{\gamma}_3\boldsymbol{\alpha}_1 = \begin{bmatrix} c\gamma & -s\gamma & 0 \\ s\gamma & c\gamma & 0 \\ 0 & 0 & 1 \end{bmatrix} \begin{bmatrix} 1 & 0 & 0 \\ 0 & c\alpha & -s\alpha \\ 0 & s\alpha & c\alpha \end{bmatrix} = \begin{bmatrix} c\gamma & -c\alpha s\gamma & s\alpha s\gamma \\ s\gamma & c\alpha c\gamma & -s\alpha c\gamma \\ 0 & s\alpha & c\alpha \end{bmatrix} . \tag{1}$$

Obviously, Eq. (1) is different from the definition (2.11) of the Cardano angle even if $\beta = 0$ is considered. The reason is the interchange of the sequence of the elementary rotations. For small angle one obtains from (1) by linearization

$$\boldsymbol{S}^{RZ} \approx \begin{bmatrix} 1 & -\gamma & 0 \\ \gamma & 1 & -\alpha \\ 0 & \alpha & 1 \end{bmatrix} . \tag{2}$$

Comparing (2) with (2.12) for $\beta = 0$ shows complete agreement. Thus, it turns out again that the sequence of small elementary rotations has not to be considered. ■

The coordinates of a vector $\boldsymbol{x}$ read differently for different frames. The relation between the coordinates $\boldsymbol{x}^i$ in frame i and the coordinates $\boldsymbol{x}^I$ in frame I is given by the transformation law for vector coordinates as

$$\boldsymbol{x}^i = \boldsymbol{S}^{iI}\boldsymbol{x}^I \quad \text{and} \quad \boldsymbol{x}^I = \boldsymbol{S}^{Ii}\boldsymbol{x}^i , \tag{2.13}$$

respectively, what is easily proven by (2.10). Please observe that the same indices appear in both forms in a neighboring sequence. This property is often helpful in applications. Further, it has to be pointed out that the rotation matrix $\boldsymbol{S}$ is a function of time, $\boldsymbol{S} = \boldsymbol{S}(t)$, what has to be considered for time derivations.

The position of a rigid body K_i in the inertial frame is uniquely described by the position quantities $\{\boldsymbol{r}_i, \boldsymbol{S}^{Ii}\}$ which characterize the body-fixed frame $\{C^i, \boldsymbol{e}^i_\nu\}$. During motion the position quantities are functions of time. Thus, the position coordinates of an arbitrary particle P of the rigid body read in the inertial frame I as

$$\boldsymbol{r}^I(t) = \boldsymbol{r}^I_i(t) + \boldsymbol{\rho}^I(t)\ , \quad \boldsymbol{\rho}^I(t) = \boldsymbol{S}^{Ii}(t)\boldsymbol{\rho}^i\ , \tag{2.14}$$

where in the body-fixed frame it yields $\boldsymbol{\rho}^i = \mathbf{const}$, see also Fig. 2.7.

The kinematics of a rigid body K_i will be now presented in the inertial frame I. The change of the position of its particle P with respect to time relative to frame I is found by differentiation of (2.14) as

$$\dot{\boldsymbol{r}}^I(t) = \dot{\boldsymbol{r}}^I_i(t) + \dot{\boldsymbol{S}}^{Ii}(t)\boldsymbol{\rho}^i = \dot{\boldsymbol{r}}^I_i(t) + \dot{\boldsymbol{S}}^{Ii}(t)\boldsymbol{S}^{iI}(t)\boldsymbol{\rho}^I(t)\ . \tag{2.15}$$

The first term on the right-hand side represents the translational velocity of the origin C_i of the body-fixed frame i. The second term is obviously related to the rotation of the body-fixed frame and represents the body's rotation. This term will now be discussed in more detail. The matrix product $[\dot{\boldsymbol{S}}(t)\boldsymbol{S}^{\mathrm{T}}(t)]$ is skew symmetric, i.e. $[\bullet] = -[\bullet]^{\mathrm{T}}$, what follows immediately from the differentiation of the orthogonality condition $\boldsymbol{S}(t)\boldsymbol{S}^{\mathrm{T}}(t) = \boldsymbol{E}$ according to (2.10):

$$\begin{aligned}\frac{\mathrm{d}}{\mathrm{d}t}\left[\boldsymbol{S}(t)\boldsymbol{S}^{\mathrm{T}}(t)\right] &= \dot{\boldsymbol{S}}(t)\boldsymbol{S}^{\mathrm{T}}(t) + \boldsymbol{S}(t)\dot{\boldsymbol{S}}^{\mathrm{T}}(t)\\ &= \left[\dot{\boldsymbol{S}}(t)\boldsymbol{S}^{\mathrm{T}}(t)\right] + \left[\dot{\boldsymbol{S}}(t)\boldsymbol{S}^{\mathrm{T}}(t)\right]^{\mathrm{T}} = \boldsymbol{0}\ .\end{aligned} \tag{2.16}$$

The matrix product $[\bullet]$ will be abbreviated by the symbol $\tilde{\boldsymbol{\omega}}(t)$ and identified by the corresponding three coordinates $\omega_\nu = \omega_\nu(t)$, $\nu = 1(1)3$, as follows

$$\dot{\boldsymbol{S}}^{Ii}\left(\boldsymbol{S}^{Ii}\right)^{\mathrm{T}} = \dot{\boldsymbol{S}}^{Ii}\boldsymbol{S}^{iI} = \tilde{\boldsymbol{\omega}}^I_{Ii}(t) = \begin{bmatrix} 0 & -\omega_3 & \omega_2 \\ \omega_3 & 0 & -\omega_1 \\ -\omega_2 & \omega_1 & 0 \end{bmatrix}\ , \quad \boldsymbol{\omega}^I_{Ii} = \begin{bmatrix}\omega_1\\ \omega_2\\ \omega_3\end{bmatrix}\ . \tag{2.17}$$

Both quantities, the skew symmetric tensor $\tilde{\boldsymbol{\omega}}^I_{Ii}$ and the corresponding rotational velocity vector $\boldsymbol{\omega}^I_{Ii}$, respectively, describe the rotational motion of system i or body K_i, respectively, relative to the inertial frame I. The upper indices indicate that both quantities are represented in the inertial frame I. If there is no chance for mixing up the frames the upper and lower index I is simply deleted. The skew symmetric tensor corresponding with a vector $[\bullet]$ is further identified by the symbol $[\tilde{\bullet}]$ and it replaces the vector product

$$\tilde{\boldsymbol{\omega}}\boldsymbol{\rho} \equiv \boldsymbol{\omega} \times \boldsymbol{\rho}\ . \tag{2.18}$$

In coordinates, in any frame, one gets accordingly

$$\tilde{\boldsymbol{\omega}}\,\boldsymbol{\rho} = \begin{bmatrix} 0 & -\omega_3 & \omega_2 \\ \omega_3 & 0 & -\omega_1 \\ -\omega_2 & \omega_1 & 0 \end{bmatrix} \begin{bmatrix} \rho_1 \\ \rho_2 \\ \rho_3 \end{bmatrix} = \begin{bmatrix} \omega_2\rho_3 - \omega_3\rho_2 \\ \omega_3\rho_1 - \omega_1\rho_3 \\ \omega_1\rho_2 - \omega_2\rho_1 \end{bmatrix} . \tag{2.19}$$

This notation of the vector product is most valuable for numerical computations since the vector product is not defined in matrix calculus.

The rotational velocity vector $\boldsymbol{\omega}^i_{Ii}$ in the body-fixed frame i follows from transformation or direct evaluation, respectively. The application of transformation (2.13) to (2.19) results in

$$\tilde{\boldsymbol{\omega}}^i_{Ii} = \boldsymbol{S}^{iI}\tilde{\boldsymbol{\omega}}^I_{Ii}\boldsymbol{S}^{Ii} = \boldsymbol{S}^{iI}\left(\dot{\boldsymbol{S}}^{Ii}\boldsymbol{S}^{iI}\right)\boldsymbol{S}^{Ii} = \boldsymbol{S}^{iI}\dot{\boldsymbol{S}}^{Ii} = \left(\boldsymbol{S}^{Ii}\right)^{\mathrm{T}}\dot{\boldsymbol{S}}^{Ii} . \tag{2.20}$$

The first and second term of (2.20) represents the transformation law for tensor coordinates where the same indices appear again in a neighboring sequence, the first and last term show the direct evaluation. The corresponding vector to (2.20) is $\boldsymbol{\omega}^i_{Ii} = \boldsymbol{S}^{iI}\boldsymbol{\omega}^I_{Ii}$ where the transformation law for vector coordinates has been used again.

From (2.15) and (2.17) it follows for rigid body K_i

$$\boldsymbol{v}^I(t) = \boldsymbol{v}^I_i(t) + \tilde{\boldsymbol{\omega}}^I_{Ii}(t)\boldsymbol{\rho}^I(t) . \tag{2.21}$$

Considering (2.18), this is the relation for rigid body kinematics well-known from each mechanics textbook as

$$\boldsymbol{v}(t) = \boldsymbol{v}_i(t) + \boldsymbol{\omega}_{Ii}(t) \times \boldsymbol{\rho}(t) . \tag{2.22}$$

The relations (2.21) or (2.22), respectively, represent the motion of a rigid body composed by an absolute translational velocity $\boldsymbol{v}_i$ of the body-fixed reference point $O^i = C^i$, and a rotation with the angular velocity $\boldsymbol{\omega}_{Ii}$. The fundamental kinematical quantities $\{\boldsymbol{v}_i, \boldsymbol{\omega}_{Ii}\}$ are also denoted as twist characterizing uniquely the motion of a rigid body.

■ Problem 2.3 Angular velocity of a rigid body

A rigid body is rotating in space. Evaluate the coordinates of the angular velocity vector $\boldsymbol{\omega}_{Ii}$ of a rigid body K_i in the inertial frame I and in a body-fixed frame i provided that its position is given by time-varying Cardano angles, see Fig. 2.9.

Solution

For the evaluation two different approaches are available. The analytical approach is based on the definitions (2.17) or (2.20), respectively, where the rotation matrix $\boldsymbol{S}^{Ii}$ according to (2.11) has to be used. The necessary computation are left to the reader. The geometrical approach is based on the angular velocities $\dot\alpha, \dot\beta, \dot\gamma$ of the three elementary rotations the vectors of which are simply added. From Fig. 2.9 follows

$$\boldsymbol{\omega}_{Ii} = \dot\alpha\ \boldsymbol{e}_1^I + \dot\beta\ \boldsymbol{e}_2^Z + \dot\gamma\ \boldsymbol{e}_1^i \tag{1}$$

where $\boldsymbol{e}_2^Z$ is the 2-axis of an intermediate frame Z. If the unit vectors appearing in (1) are transformed to the inertial frame I it remains

$$\boldsymbol{\omega}_{Ii}^I = \begin{bmatrix}1\\0\\0\end{bmatrix}\dot\alpha + \begin{bmatrix}0\\ \mathrm{c}\alpha\\ \mathrm{s}\alpha\end{bmatrix}\dot\beta + \begin{bmatrix}\mathrm{s}\beta\\ -\mathrm{s}\alpha\mathrm{c}\beta\\ \mathrm{c}\alpha\mathrm{c}\beta\end{bmatrix}\dot\gamma = \begin{bmatrix}1 & 0 & \mathrm{s}\beta\\ 0 & \mathrm{c}\alpha & -\mathrm{s}\alpha\mathrm{c}\beta\\ 0 & \mathrm{s}\alpha & \mathrm{c}\alpha\mathrm{c}\beta\end{bmatrix}\begin{bmatrix}\dot\alpha\\ \dot\beta\\ \dot\gamma\end{bmatrix}. \tag{2}$$

The coordinates $\boldsymbol{\omega}_{Ii}^I$ in the body-fixed frame i are obtained either by coordinate transformation of (2), $\boldsymbol{\omega}_{Ii}^i = \boldsymbol{S}^{iI}\boldsymbol{\omega}_{Ii}^I$, or by the geometrical approach according to (2) by presentation of (1) in frame i,

$$\boldsymbol{\omega}_{Ii}^i = \begin{bmatrix}\mathrm{c}\beta\mathrm{c}\gamma\\ -\mathrm{c}\beta\mathrm{s}\gamma\\ \mathrm{s}\beta\end{bmatrix}\dot\alpha + \begin{bmatrix}\mathrm{s}\gamma\\ \mathrm{c}\gamma\\ 0\end{bmatrix}\dot\beta + \begin{bmatrix}0\\0\\1\end{bmatrix}\dot\gamma = \begin{bmatrix}\mathrm{c}\beta\mathrm{c}\gamma & \mathrm{s}\gamma & 0\\ -\mathrm{c}\beta\mathrm{s}\gamma & \mathrm{c}\gamma & 0\\ \mathrm{s}\beta & 0 & 1\end{bmatrix}\begin{bmatrix}\dot\alpha\\ \dot\beta\\ \dot\gamma\end{bmatrix}. \tag{3}$$

It turns out from both representations that the coordinates of the angular velocity vector depend on the time derivatives of the Cardano angles and on the Cardano angles itself in addition. Only for small angles $\alpha, \beta, \gamma \ll 1$ the coordinates of the angular velocity vector

$$\boldsymbol{\omega}_{Ii}^I = \boldsymbol{\omega}_{Ii}^i = \begin{bmatrix}\dot\alpha\ \dot\beta\ \dot\gamma\end{bmatrix}^{\mathrm{T}} \text{ for } \alpha, \beta, \gamma \ll 1 \tag{4}$$

depend only on the time derivatives of the Cardano angles independent of the frame used, frame I or frame i, respectively. ■

■ Problem 2.4 Angular velocity of a railway wheelset

A railway wheelset is running on a straight track. Evaluate the angular velocity $\boldsymbol{\omega}_{RZ}^Z$ of the intermediate frame Z with the respect to the trajectory-fixed reference frame R, represented in frame R, for the wheelset given in Problem 2.2.

Solution

From Fig. 2.11 it follows immediately

$$\boldsymbol{\omega}_{RZ}^Z = \begin{bmatrix}\mathrm{c}\gamma\\ \mathrm{s}\gamma\\ 0\end{bmatrix}\dot\alpha + \begin{bmatrix}0\\0\\1\end{bmatrix}\dot\gamma = \begin{bmatrix}\mathrm{c}\gamma\dot\alpha\\ \mathrm{s}\gamma\dot\alpha\\ \dot\gamma\end{bmatrix}. \tag{1}$$

On the other hand one obtains by definition (2.17) using the rotational matrix $\boldsymbol{S}^{RZ}$ found in Problem 2.2 the skew symmetric matrix

$$\boldsymbol{\omega}_{RZ}^{R} = \dot{\boldsymbol{S}}^{RZ}\boldsymbol{S}^{ZR} = \begin{bmatrix} c\gamma & -c\alpha s\gamma & s\alpha s\gamma \\ s\gamma & c\alpha c\gamma & -s\alpha s\gamma \\ 0 & s\alpha & c\alpha \end{bmatrix}^{\bullet} \begin{bmatrix} c\gamma & s\gamma & 0 \\ -c\alpha s\gamma & c\alpha c\gamma & s\alpha \\ s\alpha s\gamma & -s\alpha s\gamma & c\alpha \end{bmatrix}$$

$$= \begin{bmatrix} 0 & -\dot{\gamma} & s\gamma\dot{\alpha} \\ \dot{\gamma} & 0 & -c\gamma\dot{\alpha} \\ -s\gamma\dot{\alpha} & c\gamma\dot{\alpha} & 0 \end{bmatrix} \equiv \begin{bmatrix} 0 & -\omega_3 & \omega_2 \\ \omega_3 & 0 & -\omega_1 \\ -\omega_2 & \omega_1 & 0 \end{bmatrix} . \quad (2)$$

Then, it follows by rewriting the angular velocity vector

$$\boldsymbol{\omega}_{RZ}^{R} = \begin{bmatrix} \omega_1 \\ \omega_2 \\ \omega_3 \end{bmatrix} = \begin{bmatrix} c\gamma\dot{\alpha} \\ s\gamma\dot{\alpha} \\ \dot{\gamma} \end{bmatrix} . \quad (3)$$

Comparing the computational efficiency of both approaches it turns out that the geometrical considerations (1) are more efficient. ■

2.2.3 Kinematics of a Rigid Body in a Moving Reference Frame

From a mathematical point of view the kinematical description of the motion of a rigid body is most convenient in the inertial frame I resulting in a more simple representation of the fundamental laws of mechanics. In engineering applications, however, a moving reference frame R related to the vehicle or the guideway, respectively, turns out to be more adequate. The frame R allows a problem-oriented choice of the coordinates and an efficient description of the forces and torques acting on the system. Moving reference frames are also useful in experiments since many measurement data are not related to the inertial frame. The choice of the reference frame R depends on the problem under consideration. In many cases the frame R characterizes the large nonlinear reference motion of a vehicle while the small deviations from the reference motion result in linear kinematical relations even for rotations.

In the following the motion of a rigid body is represented in a moving reference frame R the motion of which is given in the inertial frame I by the position vector $\boldsymbol{r}_R(t)$ of its origin and the rotation matrix $\boldsymbol{S}^{IR}(t)$, see Fig. 2.12. This means that the translational and rotational guidance velocities are also known, $\boldsymbol{v}_R^I = \dot{\boldsymbol{r}}_R^I, \tilde{\boldsymbol{\omega}}_{IR}^I = \dot{\boldsymbol{S}}^{IR}\boldsymbol{S}^{RI}$ according to (2.3) and (2.17). Considering Fig. 2.12, the absolute position quantities $\{\boldsymbol{r}_i, \boldsymbol{S}^{Ii}\}$ of the rigid body read as

$$\boldsymbol{r}_i^I(t) = \boldsymbol{r}_R^I(t) + \boldsymbol{S}^{IR}(t)\boldsymbol{r}_{Ri}^R(t) , \quad (2.23)$$

$$\boldsymbol{S}^{Ii}(t) = \boldsymbol{S}^{IR}(t)\boldsymbol{S}^{Ri}(t) . \quad (2.24)$$

The notation of the position vectors originating from the inertial frame I is simplified by omitting the index I. The vector $\boldsymbol{r}_{Ri}$ has the double index R_i due to his origin O_R, and it is represented in the reference frame R, too.

The absolute motion of the rigid body K_i is denoted by the twist $\{\boldsymbol{v}_i, \boldsymbol{\omega}_{Ii}\}$ which may be also evaluated in frame R. Formal differentiation of (2.23) in frame I results in

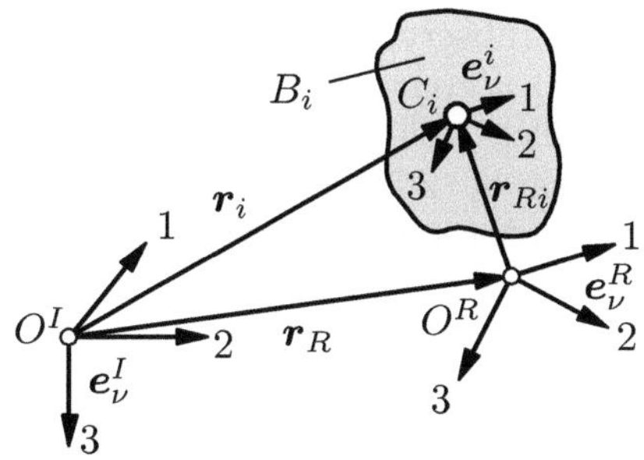

Fig. 2.12. Position of a rigid body in the reference frame R

$$\dot{\boldsymbol{r}}_i^I(t) = \dot{\boldsymbol{r}}_R^I(t) + \dot{\boldsymbol{S}}^{IR}(t)\boldsymbol{r}_{Ri}^R(t) + \boldsymbol{S}^{IR}(t)\dot{\boldsymbol{r}}_{Ri}^R(t) \; . \tag{2.25}$$

By left-multiplication with $\boldsymbol{S}^{RI}(t)$, i.e. transformation in the reference frame R, it follows considering (2.20)

$$\overset{*}{\boldsymbol{r}}{}_R^R = \boldsymbol{v}_i^R(t) = \boldsymbol{v}_R^R(t) + \tilde{\boldsymbol{\omega}}_{IR}^R(t)\,\boldsymbol{r}_{Ri}^R(t) + \dot{\boldsymbol{r}}_{Ri}^R(t) \; , \tag{2.26}$$

The symbol $(*)$ characterizes the differentiation in the inertial frame. Equation (2.26) means that the absolute velocity $\boldsymbol{v}_i$ (observed in the inertial frame I) includes the guidance velocity $\boldsymbol{v}_R + \boldsymbol{\omega}_{IR} \times \boldsymbol{r}_{Ri}$ supplemented by the relative velocity $\boldsymbol{v}_{rel} = \dot{\boldsymbol{r}}_{Ri}$ (observed in the reference frame R). The guidance velocity is the velocity of a particle which is fixed in the reference frame at position $\boldsymbol{r}_{Ri}$. Accordingly one gets by differentiation of (2.24) and subsequent right-multiplication with $\boldsymbol{S}^{iI} = \boldsymbol{S}^{iR}\boldsymbol{S}^{RI}$ considering (2.20) the corresponding angular velocities as

$$\boldsymbol{\omega}_{Ii}^I = \boldsymbol{\omega}_{IR}^I + \boldsymbol{\omega}_{Ri}^I \quad \text{or} \quad \boldsymbol{\omega}_{Ii}^R = \boldsymbol{\omega}_{IR}^R + \boldsymbol{\omega}_{Ri}^R \; . \tag{2.27}$$

Thus, the absolute angular velocity $\boldsymbol{\omega}_{Ii}$ is just the sum of guidance and relative angular velocity, $\boldsymbol{\omega}_{IR}^I$ and $\boldsymbol{\omega}_{Ri}^I$, respectively.

Due to the rotation of the frame R the guidance motion and the relative motion characterized by the indices IR and Ri, respectively, are not simply added but there appears an additional term in (2.26). This implies the well-known law of differentiation in a rotating frame

$$\begin{aligned} \frac{\mathrm{d}^I}{\mathrm{d}t}\boldsymbol{r}(t) &= \frac{\mathrm{d}^R}{\mathrm{d}t}\boldsymbol{r}(t) + \boldsymbol{\omega}_{IR}(t) \times \boldsymbol{r}(t) \; , \\ \overset{*}{\boldsymbol{r}}(t) &= \dot{\boldsymbol{r}}(t) + \boldsymbol{\omega}_{IR}(t) \times \boldsymbol{r}(t) \; . \end{aligned} \tag{2.28}$$

This equation is valid for arbitrary vectors $\boldsymbol{r}$ and can be used as a general rule for the differentiation of vectors in different frames moving relative to each other. Equation (2.28) means that the absolute variation

$\mathrm{d}^I/\mathrm{d}t\, \boldsymbol{r}(t) = \overset{*}{\boldsymbol{r}}(t)$ (observed in the inertial frame I) consists of the relative variation $\mathrm{d}^R/\mathrm{d}t\, \boldsymbol{r}(t) = \dot{\boldsymbol{r}}(t)$ (observed in the inertial frame R) supplemented by the rotation part $\boldsymbol{\omega}_{IR}(t) \times \boldsymbol{r}(t)$. Absolute and relative variation coincide only if the vector $\boldsymbol{r}$ is parallel to the angular velocity vector $\boldsymbol{\omega}_{IR}$. By formal differentiation or application of (2.28) to (2.26) and (2.27), respectively, one gets finally the absolute translational and rotational acceleration of the rigid body K_i again written in the reference frame R as

$$\boldsymbol{a}_i^R(t) = \overset{*}{\boldsymbol{v}}_i^R = \overset{*}{\boldsymbol{v}}_R^R + \left(\dot{\tilde{\boldsymbol{\omega}}}_{IR}^R + \tilde{\boldsymbol{\omega}}_{IR}^R \tilde{\boldsymbol{\omega}}_{IR}^R\right) \boldsymbol{r}_{Ri}^R + 2\tilde{\boldsymbol{\omega}}_{IR}^R \dot{\boldsymbol{r}}_{Ri}^R + \ddot{\boldsymbol{r}}_{Ri}^R \,, \tag{2.29}$$

$$\boldsymbol{\alpha}_{Ii}^R(t) = \overset{*}{\boldsymbol{\omega}}_{Ii}^R = \dot{\boldsymbol{\omega}}_{IR}^R + \tilde{\boldsymbol{\omega}}_{IR}^R \boldsymbol{\omega}_{Ri}^R + \dot{\boldsymbol{\omega}}_{Ri}^R \,. \tag{2.30}$$

Thus, in addition to the guidance and relative acceleration the Coriolis acceleration with the characteristic factor 2 is found for translations.

■ Problem 2.5 Relative motion during cornering

A vehicle is running with constant speed v_F on a circular elevated track of radius R and height h. A passenger throws horizontally an empty bottle modeled as a particle K under an angle φ opposite to the driving direction with relative velocity v_K, see Fig. 2.13 a).

a) Which relative velocity v_K and which angle φ are required to hit the ground at the center of the track?

b) How does the trajectory of the bottle look like observed by a passenger in the vehicle?

Solution

First of all an inertial frame I and a vehicle-fixed frame R are defined,

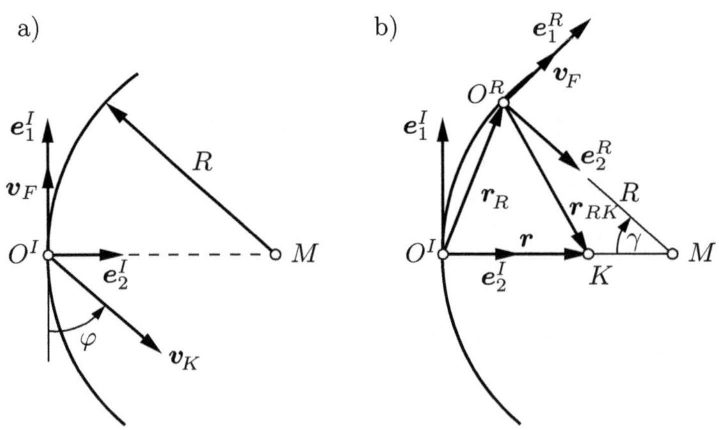

Fig. 2.13. Relative motion during cornering: a) initial position; b) general position

Fig. 2.13 b). The origin O^I of the inertial frame represents the initial condition of the bottle's trajectory.

a) The initial position $\boldsymbol{r}_0^I$, the final position $\boldsymbol{r}_M^I$ and the absolute initial velocity $\boldsymbol{v}_0^I$ of the bottle K read in the inertial frame

$$\boldsymbol{r}_0^I = \begin{bmatrix} 0 \\ 0 \\ 0 \end{bmatrix} , \quad \boldsymbol{r}_M^I = \begin{bmatrix} 0 \\ R \\ h \end{bmatrix} , \quad \boldsymbol{v}_0^I = \begin{bmatrix} v_F - v_K \cos\varphi \\ v_K \sin\varphi \\ 0 \end{bmatrix} . \tag{1}$$

The bottle K reaches the center M of the circuit only if the motion takes place in the $\boldsymbol{e}_2^I$,$\boldsymbol{e}_3^I$-plane resulting in a parabolic trajectory. This requires that the velocity in $\boldsymbol{e}_1^I$-direction is vanishing. Then, it yields

$$v_F - v_K \cos\varphi = 0 . \tag{2}$$

The absolute acceleration $\boldsymbol{a}^I$ of the bottle K in frame I reads as

$$\boldsymbol{a}^I = \begin{bmatrix} 0 \\ 0 \\ g \end{bmatrix} , \tag{3}$$

where $g = 9.81\,\mathrm{m/s^2}$ means the gravitational acceleration. A time integration considering (1) and (2) results in

$$\boldsymbol{v}^I(t) = \int_0^t \boldsymbol{a}^I \mathrm{d}\bar{t} + \boldsymbol{v}_0^I = \begin{bmatrix} 0 \\ v_K \sin\varphi \\ gt \end{bmatrix} , \tag{4}$$

$$\boldsymbol{r}^I(t) = \int_0^t \boldsymbol{v}^I \mathrm{d}\bar{t} + \boldsymbol{r}_0^I = \begin{bmatrix} 0 \\ v_K t \sin\varphi \\ \frac{1}{2} g t^2 \end{bmatrix} , \tag{5}$$

From the final condition of the trajectory $\boldsymbol{r}^I(t = t_M) = \boldsymbol{r}_M^I$ it follows

$$t_M = \sqrt{\frac{2h}{g}} , \quad v_K \sin\varphi - \frac{R}{t_M} - R\sqrt{\frac{g}{2h}} . \tag{6}$$

Equations (2) and (6) represent two equations for the required quantities v_K and φ. The evaluation yields

$$v_K = \sqrt{v_F^2 + \frac{R^2 g}{2h}} , \quad \tan\varphi = \frac{R}{v_F}\sqrt{\frac{g}{2h}} . \tag{7}$$

b) The computation of the trajectory of the bottle $\boldsymbol{r}_{RK}^R(t)$ in the vehicle-fixed frame R, Fig. 2.13 b), means a transformation of the trajectory (5) from frame I to frame R. From Fig. 2.13 b) it follows

$$\boldsymbol{r}_{RK}^{I}(t) = \boldsymbol{r}^{I}(t) - \boldsymbol{r}_{R}^{I}(t)\,, \quad \boldsymbol{r}_{R}^{I}(t) = \begin{bmatrix} R\sin\gamma \\ R(1-\cos\gamma) \\ 0 \end{bmatrix}\,, \quad \gamma = \frac{v_F t}{R}\,, \tag{8}$$

where angle γ identifies the position of the vehicle on the circuit. The transformation law reads as

$$\boldsymbol{r}_{RK}^{R}(t) = \boldsymbol{S}^{RI}\boldsymbol{r}_{RK}^{I}(t) = \boldsymbol{S}^{RI}\left[\boldsymbol{r}^{I}(t) - \boldsymbol{r}_{R}^{I}(t)\right]\,,$$
$$\boldsymbol{S}^{RI} = \boldsymbol{\gamma}_3 = \begin{bmatrix} \cos\gamma & -\sin\gamma & 0 \\ \sin\gamma & \cos\gamma & 0 \\ 0 & 0 & 1 \end{bmatrix}\,. \tag{9}$$

The evaluation results in

$$\boldsymbol{r}_{RK}^{R}(t) = \begin{bmatrix} \cos\gamma & \sin\gamma & 0 \\ -\sin\gamma & \cos\gamma & 0 \\ 0 & 0 & 1 \end{bmatrix} \begin{bmatrix} -R\sin\gamma \\ v_K t\sin\varphi - R\,(1-\cos\gamma) \\ \frac{1}{2}gt^2 \end{bmatrix}$$
$$= \begin{bmatrix} v_K t\sin\gamma\sin\varphi - R\sin\gamma \\ v_K t\cos\gamma\sin\varphi + R(1-\cos\gamma) \\ \frac{1}{2}gt^2 \end{bmatrix}\,,$$
$$\gamma = \frac{v_F}{R}t\,, \quad 0 \le t \le t_M = \sqrt{\frac{2h}{g}}\,. \tag{10}$$

Finally, substitution of the time by the relation $t = \dfrac{R}{v_F}\gamma$ yields a trajectory depending on angle γ only,

$$\boldsymbol{r}_{RK}^{R}(\gamma) \equiv \begin{bmatrix} x \\ y \\ z \end{bmatrix} = R\begin{bmatrix} -(1-\gamma/\gamma_M)\ \sin\gamma \\ 1-(1-\gamma/\gamma_M)\ \cos\gamma \\ (\gamma/\gamma_M)^2 h/R \end{bmatrix}\,, \quad 0 \le \gamma \le \gamma_M = \frac{v_F}{R}\sqrt{\frac{2h}{g}}\,. \tag{11}$$

In Fig. 2.14 some projections of trajectories on the horizontal $\boldsymbol{e}_1^R, \boldsymbol{e}_2^R$-plane are shown depending on the system parameter γ_M as seen by an observer in the vehicle. The parameter γ_M characterizes the position angle of the vehicle when the bottle K hits the ground in the center M. It turns out that with increasing speed $v_F \sim \gamma_M$ the initial throw direction deviates more and more from the direct line to the center M and that the tangent of the trajectory turns against the driving direction of the vehicle. The trajectory appears more and more bulged. At very high speeds ($\gamma_M > 180°$) the relative trajectory crosses the center line and approaches the center M helically. ■

2.2.4 Kinematics of Multibody Systems

So far only one free rigid body K_i was considered the position of which is uniquely described in the inertial frame I as

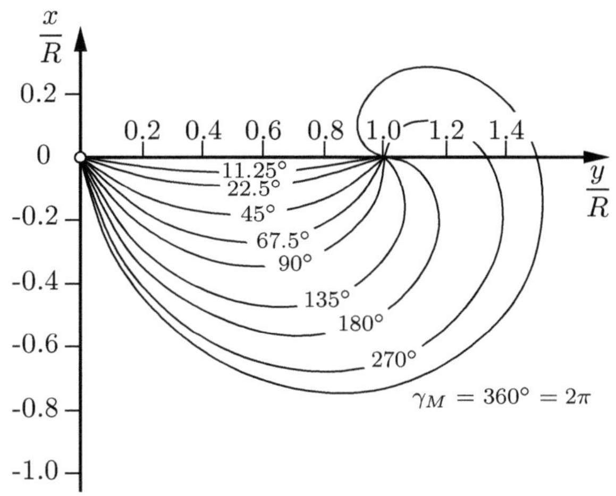

Fig. 2.14. Trajectory seen by an observer in the vehicle (projection on the horizontal plane)

$$\boldsymbol{r}_i^I = \left[r_{i1} r_{i2} r_{i3}\right]^{\mathrm{T}} , \quad \boldsymbol{S}^{Ii} \equiv \boldsymbol{S}_i = \boldsymbol{S}_i(\alpha_i, \beta_i, \gamma_i) . \tag{2.31}$$

There are six position coordinates which are summarized in a 6×1- column matrix, simply called local position vector, as

$$\boldsymbol{x}_i(t) = \left[r_{i1}\, r_{i2}\, r_{i3}\, \alpha_i\, \beta_i\, \gamma_i\right]^{\mathrm{T}} . \tag{2.32}$$

For a free multibody system consisting of p disassembled rigid bodies K_i, $i = 1(1)p$, there exist $6p$ position coordinates resulting in a $6p \times 1$ global position vector of an unconstrained system

$$\boldsymbol{x}(t) = \left[\boldsymbol{x}_1^{\mathrm{T}} \ldots\ldots \boldsymbol{x}_p^{\mathrm{T}}\right]^{\mathrm{T}} . \tag{2.33}$$

Assembling the free system there appear constraints between the position coordinates and their derivatives. In realistic models of vehicles only holonomic constraints are found restricting the motion of the position coordinates by geometric or integrable kinematic constraints. These constraints are implicitly described by algebraical equations which may be time-dependent (rheonomic), too,

$$\varphi_j(\boldsymbol{x}, t) = 0 , \quad j = 1(1)q . \tag{2.34}$$

Due to the q constraints there remain f linear independent position coordinates characterizing $f = 6p - q$ degrees of freedom. The f independent position coordinates are also called generalized coordinates and may be summarized in a $f \times 1$- column matrix as global position vector of the constraint system

$$\boldsymbol{y}(t) = [y_1 \ldots\ldots y_f]^{\mathrm{T}} \ . \tag{2.35}$$

By (2.34) and (2.35) the vector $\boldsymbol{x}$ is an explicit function of the f generalized coordinates representing the constraints explicitly,

$$\boldsymbol{x} = \boldsymbol{x}(\boldsymbol{y}, t) \ . \tag{2.36}$$

The choice of generalized coordinates is not unique. E.g., some of the local position coordinates (absolute coordinates) or differences between local coordinates (relative coordinates) may be chosen as generalized coordinates. However, there exists a unique relation between different sets of generalized coordinates represented by a regular, time-invariant $f \times f$ -matrix $\boldsymbol{T}$ resulting in the transformation

$$\boldsymbol{y}\,(t) = \boldsymbol{T}\overline{\boldsymbol{y}}\,(t) \ , \tag{2.37}$$

where $\boldsymbol{y}$ and $\overline{\boldsymbol{y}}$ are the corresponding global position vectors. The position variables (2.31) may be rewritten for the whole system as

$$\boldsymbol{r}_i(t) = \boldsymbol{r}_i(\boldsymbol{y}, t) \ , \quad \boldsymbol{S}^{Ii}(t) \equiv \boldsymbol{S}_i(t) = \boldsymbol{S}_i(\boldsymbol{y}, t) \ , \quad i = 1(1)p \ . \tag{2.38}$$

The corresponding velocity variables $\{\boldsymbol{v}_i, \boldsymbol{\omega}_i\}$, $\boldsymbol{\omega}_i = \boldsymbol{\omega}_{Ii}$, are obtained by differentiation as

$$\boldsymbol{v}_i(t) = \dot{\boldsymbol{r}}_i(t) = \frac{\partial \boldsymbol{r}_i}{\partial \boldsymbol{y}^{\mathrm{T}}}\dot{\boldsymbol{y}} + \frac{\partial \boldsymbol{r}_i}{\partial t} = \boldsymbol{J}_{Ti}(\boldsymbol{y}, t)\dot{\boldsymbol{y}} + \overline{\boldsymbol{v}}_i(\boldsymbol{y}, t) \ , \tag{2.39}$$

$$\boldsymbol{\omega}_i(t) = \dot{\boldsymbol{s}}_i(t) = \frac{\partial \boldsymbol{s}_i}{\partial \boldsymbol{y}^{\mathrm{T}}}\dot{\boldsymbol{y}} + \frac{\partial \boldsymbol{s}_i}{\partial t} = \boldsymbol{J}_{Ri}(\boldsymbol{y}, t)\dot{\boldsymbol{y}} + \overline{\boldsymbol{\omega}}_i(y, t) \ , \tag{2.40}$$

where $\partial \boldsymbol{s}_i$ describes the 3×1-vector of the infinitesimal rotation following from the rotation matrix analogously to the rotational velocity (2.17) as

$$\partial \tilde{\boldsymbol{s}}_i = \partial \boldsymbol{S}_i \boldsymbol{S}_i^{\mathrm{T}} := \begin{bmatrix} 0 & -\partial s_{i3} & \partial s_{i2} \\ \partial s_{i3} & 0 & -\partial s_{i1} \\ -\partial s_{i2} & \partial s_{i1} & 0 \end{bmatrix} , \quad \partial \boldsymbol{s}_i = \begin{bmatrix} \partial s_{i1} \\ \partial s_{i2} \\ \partial s_{i3} \end{bmatrix} . \tag{2.41}$$

The $3 \times f$-functional or Jacobian matrices $\boldsymbol{J}_{Ti}, \boldsymbol{J}_{Ri}$ of translation and rotation, respectively, identify the relation between the local and the generalized or global coordinates. The structure of these matrices may be defined using the rules of matrix multiplication as shown for the translation matrix

$$\frac{\partial \boldsymbol{r}_i}{\partial \boldsymbol{y}^{\mathrm{T}}} = \partial \boldsymbol{r}_i \left(\frac{1}{\partial \boldsymbol{y}^{\mathrm{T}}} \right) = \boldsymbol{J}_{Ti} = \begin{bmatrix} \dfrac{\partial r_{i1}}{\partial y_1} & \dfrac{\partial r_{i1}}{\partial y_2} & \cdots & \dfrac{\partial r_{i1}}{\partial y_f} \\ \dfrac{\partial r_{i2}}{\partial y_1} & \dfrac{\partial r_{i2}}{\partial y_2} & \cdots & \dfrac{\partial r_{i2}}{\partial y_f} \\ \dfrac{\partial r_{i3}}{\partial y_1} & \dfrac{\partial r_{i3}}{\partial y_2} & \cdots & \dfrac{\partial r_{i3}}{\partial y_f} \end{bmatrix} . \tag{2.42}$$

In practice, the computation of the Jacobian matrices is often more easy if firstly the position variables are characterized as functions of the vector $\boldsymbol{x}$ considering (2.36) and secondly the chain rule of differentiation is applied. Then, it yields, e.g.,

$$\boldsymbol{J}_{Ti} = \frac{\partial \boldsymbol{r}_i}{\partial \boldsymbol{y}^{\mathrm{T}}} = \frac{\partial \boldsymbol{r}_i}{\partial \boldsymbol{x}^{\mathrm{T}}} \frac{\partial \boldsymbol{x}}{\partial \boldsymbol{y}^{\mathrm{T}}} \,. \tag{2.43}$$

From (2.39) and (2.40) one obtains by a second differentiation the acceleration variables $\{\boldsymbol{a}_i, \boldsymbol{\alpha}_i\}$ depending on the position vector $\boldsymbol{y}$ and its derivatives,

$$\boldsymbol{a}_i(t) = \dot{\boldsymbol{v}}_i(t) = \boldsymbol{J}_{Ti}(\boldsymbol{y}, t)\ddot{\boldsymbol{y}} + \frac{\partial \boldsymbol{v}_i}{\partial \boldsymbol{y}^{\mathrm{T}}}\dot{\boldsymbol{y}} + \frac{\partial \boldsymbol{v}_i}{\partial t} \,, \tag{2.44}$$

$$\boldsymbol{\alpha}_i(t) = \dot{\boldsymbol{\omega}}_i(t) = \boldsymbol{J}_{Ri}(\boldsymbol{y}, t)\ddot{\boldsymbol{y}} + \frac{\partial \boldsymbol{\omega}_i}{\partial \boldsymbol{y}^{\mathrm{T}}}\dot{\boldsymbol{y}} + \frac{\partial \boldsymbol{\omega}_i}{\partial t} \,. \tag{2.45}$$

For scleronomic, time-invariant constraints the partial time derivatives in (2.39), (2.40), (2.44) and (2.45) are vanishing.

In addition to the real motions, the virtual motions are required in the next chapter dealing with dynamics. A virtual motion is defined as an arbitrary, infinitesimally small variation of the position completely compatible with the constraints at any time. Rheonomic constraints are treated as frozen at the time under consideration. The symbol δ of the virtual motion has the properties

$$\delta \boldsymbol{r} \neq \boldsymbol{0} \,, \quad \delta t \equiv 0 \,. \tag{2.46}$$

The symbol δ follows the rules of calculus, i.e., it yields

$$\delta(c\boldsymbol{r}) = c\delta\boldsymbol{r} \,, \quad \delta(\boldsymbol{r}_1 + \boldsymbol{r}_2) = \delta\boldsymbol{r}_1 + \delta\boldsymbol{r}_2 \,, \quad \delta\boldsymbol{r}(\boldsymbol{y}) = \frac{\partial \boldsymbol{r}}{\partial \boldsymbol{y}^{\mathrm{T}}}\delta\boldsymbol{y} \,. \tag{2.47}$$

Thus, the virtual motion of a multibody system reads as

$$\delta \boldsymbol{r}_i = \boldsymbol{J}_{Ti}\delta\boldsymbol{y} \,, \quad \delta \boldsymbol{s}_i = \boldsymbol{J}_{Ri}\delta\boldsymbol{y} \,, \quad i = 1(1)p \,. \tag{2.48}$$

This completes the kinematics for rigid body vehicle systems.

■ Problem 2.6 Kinematic rolling of a cylinder

A rigid cylinder with radius R is rolling on a rigid plane without sliding. Evaluate the constraints, choose generalized coordinates and specify the Jacobian matrices $\boldsymbol{J}_T, \boldsymbol{J}_R$ of the translation and rotation, respectively.

Solution

Using the inertial frame $\{O^I, \boldsymbol{e}_\nu^I\}$ and the body-fixed frame $\{C, \boldsymbol{e}_\nu^K\}$, see Fig. 2.15, the position of the cylinder is uniquely specified by the position vector $\boldsymbol{r}_C^I = [r_1\, r_2\, r_3]^{\mathrm{T}}$ and the rotation matrix $\boldsymbol{S}^{IK}(\alpha, \beta, \gamma)$. The assumed

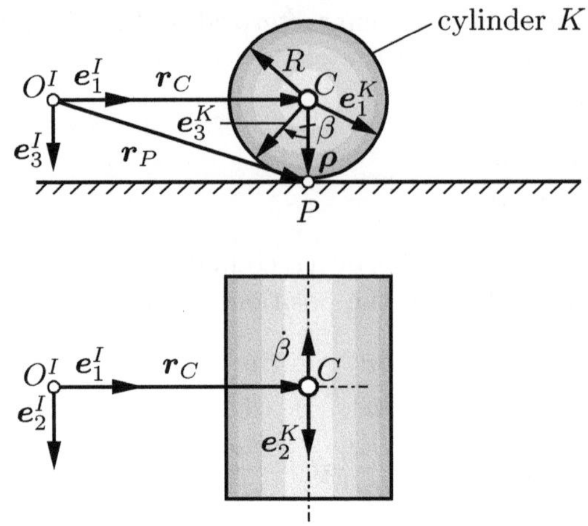

Fig. 2.15. Rolling cylinder

rolling motion without sliding on the plane results in contact lines between cylinder and plane which are parallel all the time. Thus, the center of mass C travels on a straight line, and the corresponding geometric constraint conditions read according to Fig. 2.15 as

$$r_1 \equiv 0\,, \quad r_2 \equiv 0\,, \quad \alpha \equiv 0,\,, \quad \gamma \equiv 0\,. \tag{1}$$

The original position and velocity variables $\{\boldsymbol{r}_C, \boldsymbol{S}^{IK}\}$, $\{\boldsymbol{v}_C, \boldsymbol{\omega}_{IK}\}$ are rewritten as (s$\hat{=}$sin, c$\hat{=}$cos)

$$\boldsymbol{r}_C^I = \begin{bmatrix} r_1 \\ 0 \\ 0 \end{bmatrix}, \quad \boldsymbol{S}^{IK} = -\boldsymbol{\beta}_2 = \begin{bmatrix} \mathrm{c}\beta & 0 & -\mathrm{s}\beta \\ 0 & 1 & 0 \\ \mathrm{s}\beta & 0 & \mathrm{c}\beta \end{bmatrix}, \tag{2}$$

$$\boldsymbol{v}_C^I = \begin{bmatrix} \dot{r}_1 \\ 0 \\ 0 \end{bmatrix}, \quad \boldsymbol{\omega}_{IK}^I = \begin{bmatrix} 0 \\ -\dot{\beta} \\ 0 \end{bmatrix}, \quad \partial \boldsymbol{s}_{IK}^I = \begin{bmatrix} 0 \\ -\partial\beta \\ 0 \end{bmatrix}. \tag{3}$$

The angular velocity vector $\boldsymbol{\omega}_{IK}^I$ or $\partial \boldsymbol{s}_{IK}^I$, respectively, follows immediately from Fig. 2.15 or formally by (2.17) as

$$\tilde{\boldsymbol{\omega}}_{IK}^I = \dot{\boldsymbol{S}}^{IK} (\boldsymbol{S}^{IK})^{\mathrm{T}} = \dot{\beta} \begin{bmatrix} -\mathrm{s}\beta & 0 & -\mathrm{c}\beta \\ 0 & 0 & 0 \\ \mathrm{c}\beta & 0 & -\mathrm{s}\beta \end{bmatrix} \begin{bmatrix} \mathrm{c}\beta & 0 & \mathrm{s}\beta \\ 0 & 1 & 0 \\ -\mathrm{s}\beta & 0 & \mathrm{c}\beta \end{bmatrix} = \begin{bmatrix} 0 & 0 & -\dot{\beta} \\ 0 & 0 & 0 \\ \dot{\beta} & 0 & 0 \end{bmatrix}. \tag{4}$$

Pure rolling requires that all particles of the cylinder coinciding with the contact line are instantaneously reposing (instantaneous center of rotation). Thus, to check the pure rolling condition only one particle will be considered.

Choosing the particle P located under the center of mass C, one gets by the specific position vector $\boldsymbol{\rho}_{CP}^I \equiv \boldsymbol{\rho}^I = [0\ 0\ R]^{\mathrm{T}}$, Fig. 2.15, from the vanishing absolute velocity vector $\boldsymbol{v}_P$ according to (2.21) the relation

$$\boldsymbol{v}_P^I = \boldsymbol{v}_C^I + \tilde{\boldsymbol{\omega}}_{IK}^I \,\boldsymbol{\rho}^I = \mathbf{0} = \begin{bmatrix} \dot{r}_1 \\ 0 \\ 0 \end{bmatrix} + \begin{bmatrix} 0 & 0 & -\dot{\beta} \\ 0 & 0 & 0 \\ \dot{\beta} & 0 & 0 \end{bmatrix} \begin{bmatrix} 0 \\ 0 \\ R \end{bmatrix} = \begin{bmatrix} \dot{r}_1 - R\dot{\beta} \\ 0 \\ 0 \end{bmatrix} = \begin{bmatrix} 0 \\ 0 \\ 0 \end{bmatrix} . \tag{5}$$

It follows in addition to (1) one kinematic constraint,

$$\dot{r}_1 - R\dot{\beta} = 0 , \tag{6}$$

which is integrable,

$$r_1 - r_{10} = R(\beta - \beta_0) . \tag{7}$$

Altogether, (1) and (7) represent $q = 5$ holonomic constraints. The number of degrees of freedom is $f = 6 - q = 1$. As generalized coordinates may be chosen a) $y = r_1$ or b) $\overline{y} = \beta$ depending on the problem to be considered. According to the choice of the generalized coordinate, different Jacobian matrices are found.

a) For $y = r_1$ it follows with (7), (2), (3) from (2.39), (2.40) and (2.43)

$$\beta = \beta(y) = \beta_0 + \frac{1}{R}(y - r_{10}) , \tag{8}$$

$$\boldsymbol{r}_C^I = \boldsymbol{r}_C^I(y) = \begin{bmatrix} y \\ 0 \\ 0 \end{bmatrix} , \quad \partial \boldsymbol{s}_{IK}^I(\beta(y)) = \begin{bmatrix} 0 \\ -\partial\beta(y) \\ 0 \end{bmatrix} , \tag{9}$$

$$\boldsymbol{J}_T = \frac{\partial \boldsymbol{r}_C^I}{\partial y} = \begin{bmatrix} 1 \\ 0 \\ 0 \end{bmatrix} , \quad \boldsymbol{J}_R = \frac{\partial \boldsymbol{s}_{IK}^I}{\partial y} = \begin{bmatrix} 0 \\ -\dfrac{1}{R} \\ 0 \end{bmatrix} . \tag{10}$$

b) For $\overline{y} = \beta$ it follows accordingly

$$r_1 = r_{10} + R(\overline{y} - \beta_0) , \tag{11}$$

$$\boldsymbol{r}_C^I = \boldsymbol{r}_C^I(r_1(\overline{y})) = \begin{bmatrix} r_1(\overline{y}) \\ 0 \\ 0 \end{bmatrix} , \quad \partial \boldsymbol{s}_{IK}^I = \partial \boldsymbol{s}_{IK}^I(\overline{y}) = \begin{bmatrix} 0 \\ -\partial\overline{y} \\ 0 \end{bmatrix} , \tag{12}$$

$$\overline{\boldsymbol{J}}_T = \frac{\partial \boldsymbol{r}_C^I}{\partial r_1} \frac{\partial r_1}{\partial \overline{y}} = \begin{bmatrix} R \\ 0 \\ 0 \end{bmatrix} , \quad \overline{\boldsymbol{J}}_R = \frac{\partial \boldsymbol{s}_{IK}^I}{\partial \overline{y}} = \begin{bmatrix} 0 \\ -1 \\ 0 \end{bmatrix} . \tag{13}$$

■

■ **Problem 2.7 Kinematic hunting of a railway wheelset**

A rigid railway wheelset is rolling without sliding on a rigid track. The wheels are nearly conical (nominal cone angle $2\delta_0$, nominal roll radius r_0) and the track profile is square (gauge $2a$). Evaluate kinematically the lateral motion of the center of mass of the wheelset neglecting all inertia effects.

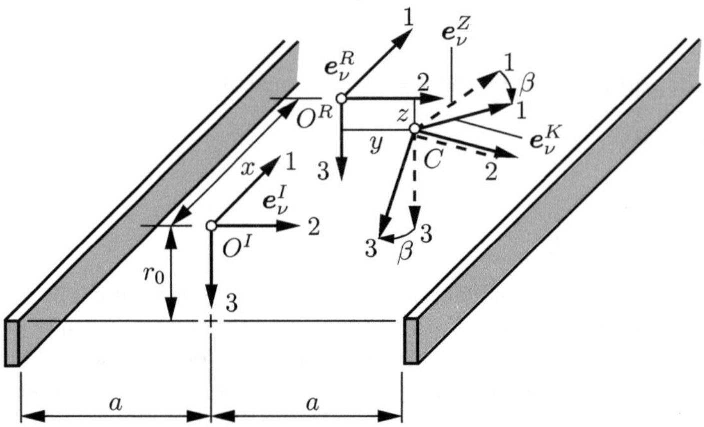

Fig. 2.16. Frame for the description of the wheelset motion

Solution

For the solution there are four frames introduced, see Fig. 2.16: an inertial frame I located in the middle of the track, a reference frame R characterizing the undisturbed wheelset motion, an intermediate frame Z with the contact points, see Problem 2.2, and a body-fixed frame K. The position vector of the center of mass C in frame I, Fig. 2.16, reads as

$$\boldsymbol{r}_C^I = \begin{bmatrix} r_{C1} \\ r_{C2} \\ r_{C3} \end{bmatrix} \equiv \begin{bmatrix} x \\ y \\ z \end{bmatrix} . \tag{1}$$

The wheelset considered has a weakly curvilinear wheel profile only (roll radius r_0 and profile slope δ_0 are related to the nominal position, Fig. 2.17 a)). Due to the assumption of a rigid wheelset and a rigid track the contact condition is punctual. The position of the contact points $P_{l,r}$ ($l \widehat{=}$ left, $r \widehat{=}$ right) follows from a consideration of the plane with the contact points (frame Z, Fig. 2.17 b)) as

$$\boldsymbol{\rho}_l^Z = \begin{bmatrix} 0 \\ -a + \triangle a_l \\ r_l \end{bmatrix} , \quad \boldsymbol{\rho}_r^Z = \begin{bmatrix} 0 \\ a + \triangle a_r \\ r_r \end{bmatrix} . \tag{2}$$

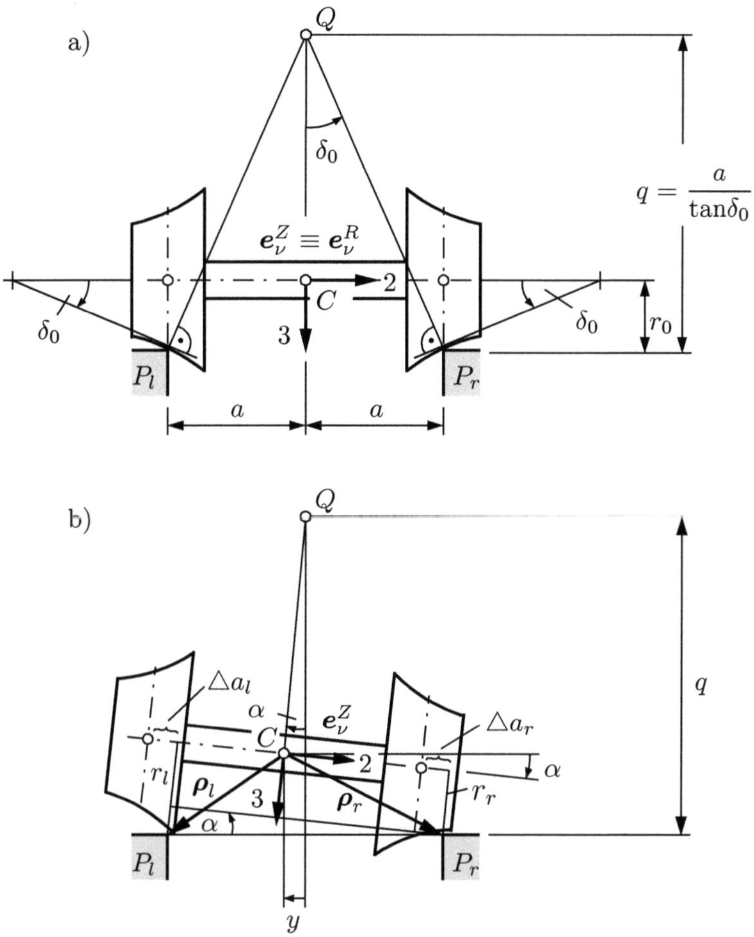

Fig. 2.17. Wheelset geometry: a) nominal position; b) general position

The transformation into frame I according to $\boldsymbol{\rho}^I = \boldsymbol{S}^{IZ}\,\boldsymbol{\rho}^Z$ with $\boldsymbol{S}^{IZ} = \boldsymbol{S}^{IR}\,\boldsymbol{S}^{RZ}$ ($\boldsymbol{S}^{IR} = \boldsymbol{E}$, $\boldsymbol{S}^{RZ}$ see Problem 2.2) results in

$$\boldsymbol{\rho}_l^I = \begin{bmatrix} 1 & -\gamma & 0 \\ \gamma & 1 & -\alpha \\ 0 & \alpha & 1 \end{bmatrix} \begin{bmatrix} 0 \\ -a + \triangle a_l \\ r_l \end{bmatrix} \approx \begin{bmatrix} a\gamma \\ -a \\ r_l \end{bmatrix} , \tag{3}$$

$$\boldsymbol{\rho}_r^I = \begin{bmatrix} 1 & -\gamma & 0 \\ \gamma & 1 & -\alpha \\ 0 & \alpha & 1 \end{bmatrix} \begin{bmatrix} 0 \\ a + \triangle a_r \\ r_r \end{bmatrix} \approx \begin{bmatrix} -a\gamma \\ a \\ r_r \end{bmatrix} . \tag{4}$$

where $\alpha, \gamma \ll 1$ and $\triangle a_{l,r} \ll a$ are taken into account.

The kinematic rolling condition for the wheelset results in a vanishing absolute velocity of the material points of the wheelset which are instantaneously in contact with the track. The application of the rigid body relation (2.21) yields

$$\boldsymbol{v}_{Pl}^{I} = \boldsymbol{v}_{C}^{I} + \tilde{\boldsymbol{\omega}}_{IK}^{I}\, \boldsymbol{\rho}_{l}^{I} = \mathbf{0}\,, \quad \boldsymbol{v}_{Pr}^{I} = \boldsymbol{v}_{C}^{I} + \tilde{\boldsymbol{\omega}}_{IK}^{I}\, \boldsymbol{\rho}_{r}^{I} = \mathbf{0}\,. \tag{5}$$

The velocity of the center of mass $\boldsymbol{v}_C^I$ follows by differentiation from (1). The only missing quantity is therefore the angular velocity $\boldsymbol{\omega}_{IK}^I$ of the wheelset. This quantity follows according to Problem 2.4 considering that the frame K is found relative to frame Z by an elementary rotation around the $\boldsymbol{e}_2^Z$-axis with the rotation angle $-\beta$, Fig. 2.16, as

$$\boldsymbol{\omega}_{IK}^{I} = \begin{bmatrix} \mathrm{c}\gamma \\ \mathrm{s}\gamma \\ 0 \end{bmatrix} \dot{\alpha} + \begin{bmatrix} 0 \\ 0 \\ 1 \end{bmatrix} \dot{\gamma} + \begin{bmatrix} \mathrm{c}\alpha\mathrm{s}\gamma \\ -\mathrm{c}\alpha\mathrm{s}\gamma \\ -\mathrm{s}\alpha \end{bmatrix} \dot{\beta}\,. \tag{6}$$

In the following only linear relations are used valid for small rotations ($\alpha, \gamma \ll 1,\ \dot{\alpha}, \dot{\gamma} \ll \dot{\beta}$) and small displacements ($\triangle a_{l,r} \ll a$). Then, one gets from (6)

$$\boldsymbol{\omega}_{IK}^{I} \approx \begin{bmatrix} \dot{\alpha} + \gamma\dot{\beta} \\ -\dot{\beta} \\ \dot{\gamma} - \alpha\dot{\beta} \end{bmatrix}\,. \tag{7}$$

The evaluation of the rolling condition (5) neglecting all quantities which are small of higher order yields

$$\dot{x} + a\dot{\gamma} - r_l\dot{\beta} = 0\,, \tag{8}$$
$$\dot{y} - r_l(\dot{\alpha} + \gamma\dot{\beta}) = 0\,, \tag{9}$$
$$\dot{z} - a\dot{\alpha} = 0\,, \tag{10}$$
$$\dot{x} - a\dot{\gamma} - r_r\dot{\beta} = 0\,, \tag{11}$$
$$\dot{y} - r_r(\dot{\alpha} + \gamma\dot{\beta}) = 0\,, \tag{12}$$
$$\dot{z} + a\dot{\alpha} = 0\,. \tag{13}$$

Equations (8) to (13) are related to the inertial frame. For a conical wheelset $\delta_0 \neq 0$ the vertical velocities (10) and (13) are depending on the roll angular velocity $\dot{\alpha}$. In the average of (10) and (13) it remains $\dot{z} = 0$ what is confirmed by the geometric relation (19).

By conclusion one gets three essential kinematical relations

$$(8) + (11) \Rightarrow \dot{x} - \frac{1}{2}(r_l + r_r)\dot{\beta} = 0\,, \tag{14}$$

$$(8) - (11) \Rightarrow a\dot{\gamma} - \frac{1}{2}(r_l - r_r)\dot{\beta} = 0\,, \tag{15}$$

$$(9) - (12) \Rightarrow \dot{\gamma} - \frac{1}{2}(r_l + r_r)(\dot{\alpha} + \gamma\dot{\beta}) = 0\,. \tag{16}$$

Further, small rotations around the instantaneous center of rotation Q provide according to Fig. 2.17 b) the geometric relations

$$\alpha \approx \frac{r_l - r_r}{2a} , \tag{17}$$

$$-y \approx \left[q - \frac{1}{2}(r_l + r_r) \right] \alpha , \tag{18}$$

$$z = r_0 - \frac{1}{2}(r_l + r_r) \approx 0 , \tag{19}$$

where it was observed that a positive rotation with angle α results in a negative displacement y. Small rotations γ around the vertical axis do not affect the result (18) since the displacement of y is proportional to γ^2 what means second order. After elimination of the roll radii difference by (17) as well as the roll radii sum according to (19), it remains

$$\dot{x} - r_0 \beta = 0 , \tag{20}$$

$$\dot{\gamma} - \alpha \dot{\beta} = 0 , \tag{21}$$

$$\dot{y} - r_0(\dot{\alpha} + \gamma \dot{\beta}) = 0 , \tag{22}$$

$$y + (q + r_0)\alpha = 0 , \quad q = \frac{a}{\tan \delta_0} , \tag{23}$$

$$z = 0 . \tag{24}$$

This result is valid for conical wheels as well as for weekly curvilinear wheel profiles which may be approximated by cones. Since the kinematical constraints are integrable, there remain five holonomic constraints for the six local position variables. Thus, the wheelset has one degree of freedom. Equation (20) indicates that the velocity of the center of mass of the wheelset in nominal traveling direction corresponds to the velocity of a single wheel, see Problem 2.6. The comparison of (21) with (7) shows that the absolute rotational velocity of the vertical axis of the wheelset is vanishing while (24) states that the height of the center of mass is constant relative to the track for the motion considered. The differential equations (20)-(22) can be solved easily. For this purpose (23) is differentiated and included in (22) resulting in

$$q\dot{\alpha} + r_0 \gamma \dot{\beta} = 0 . \tag{25}$$

Using the rolling condition (20) for the replacement of $\dot{\beta}$, it follows from (21) and (25) a pair of equations,

$$\begin{aligned} \dot{\gamma}(t) - \alpha(t)\frac{\dot{x}(t)}{r_0} &= 0 , \\ \dot{\alpha}(t) + \gamma(t)\frac{\dot{x}(t)}{q} &= 0 . \end{aligned} \tag{26}$$

Now, the independent variable t can be replaced by x as independent variable where the differentiation rule

$$\frac{\mathrm{d}(\bullet)}{\mathrm{d}t} = \frac{\mathrm{d}(\bullet)}{\mathrm{d}x}\frac{\mathrm{d}x}{\mathrm{d}t} \quad \text{or} \quad \dot{(\bullet)} = \dot{x}(\bullet)' \tag{27}$$

has to be considered. Then, from (26) it follows

$$\begin{aligned} \gamma'(x) - \frac{1}{r_0}\alpha(x) = 0 \,, \\ \alpha'(x) + \frac{1}{q}\gamma(x) = 0 \,. \end{aligned} \tag{28}$$

The elimination of $\gamma(x)$ results finally in a differential equation of the second order for the roll angle $\alpha(x)$ as

$$\alpha''(x) + \frac{1}{r_0 q}\alpha(x) = 0 \,. \tag{29}$$

The general solution with the initial conditions $\alpha(0) = -\alpha_0$, $\alpha'(0) = 0$ reads

$$\alpha(x) = -\alpha_0 \cos \Omega x \,, \quad \Omega^2 = \frac{1}{r_0 q} = \frac{\tan \delta_0}{r_0 a} \,, \tag{30}$$

where the spatial angular frequency Ω is introduced. The required lateral motion $y(x)$ follows from (23) as

$$y(x) = y_0 \cos \Omega x \,, \quad y_0 = (q - r_0)\alpha_0 \,. \tag{31}$$

One gets harmonic lateral oscillations with the spatial angular velocity Ω, Fig. 2.18. This motion is also denoted as wave motion, sine motion or hunting motion, respectively. The wavelength λ reads

$$\lambda = \frac{2\pi}{\Omega} = 2\pi\sqrt{\frac{r_0 a}{\tan \delta_0}} \,. \tag{32}$$

This formula was published originally by Klingel (1883) and it is well-known in railway engineering as Klingel formula. In addition to the nominal roll radius r_0 and the half gauge of the track a, the half cone angle δ_0 is an important parameter. With decreasing conicity of the wheels the wave length is increasing. For cylindrical wheels δ_0 the wave length approaches infinity. Then, the wheelset loses the property of self-guidance. Negative cone angles result in an unstable wheelset motion, Fig. 2.19.

The physical reason for a stable wheelset motion, i.e. the self-guidance of the wheelset, can be observed in Fig. 2.18. For conical wheels with increasing roll radii towards the wheelset center, a displacement of the wheelset center of mass to the right results in a larger roll radius at the right wheel than at the left wheel. Therefore, the right wheels overtake the left wheel while traveling on the track, and the displacement of the center of mass is decreasing. If the wheelset center of mass has reached the middle of the track, the axis of the wheelset is inclined and the center of mass moves to the left. Then, the left

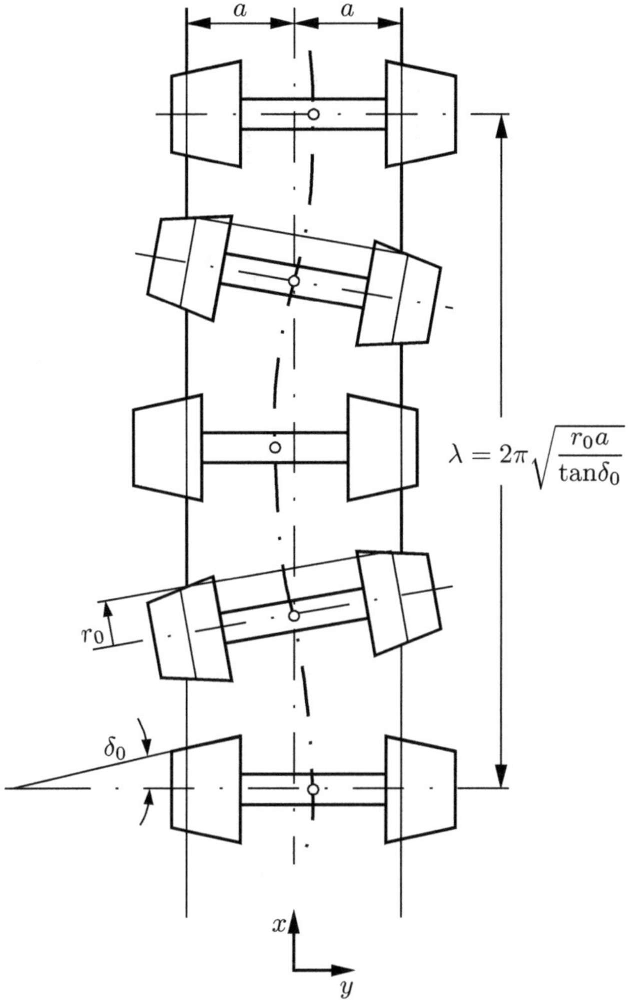

Fig. 2.18. Hunting motion of a railway wheelset

wheels overtakes the right wheel and the conditions are reversed. As a result a stable wheelset motion is obtained. For a wheelset with decreasing roll radii towards the wheelset center, Fig. 2.19, a displacement of the wheelset center of mass from the middle of the track is not decreasing but further increasing with the result of an unstable wheelset motion. The stable motion can be easily verified by an experimental wheelset consisting of two cones, e.g. made by yogurt cups, running on two rods placed an a inclined plane, Fig. 2.20.

With a real wheelset the motion behavior is much more complex. Due to elasticity of wheels and tracks the contact points have to be replaced by

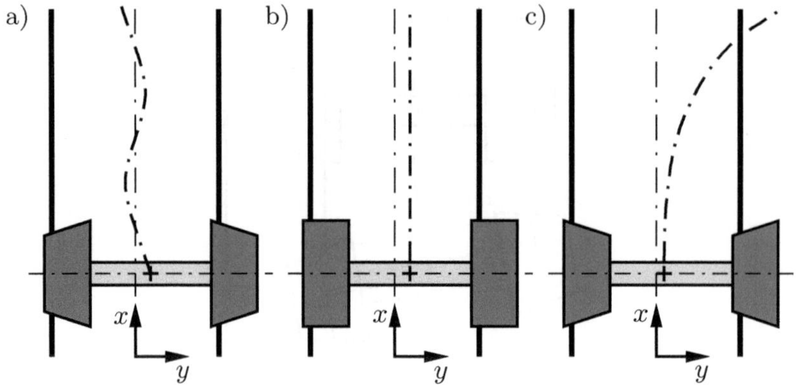

Fig. 2.19. Wheelset motion with different conicities: a) stable motion; b) indifferent motion; c) unstable motion

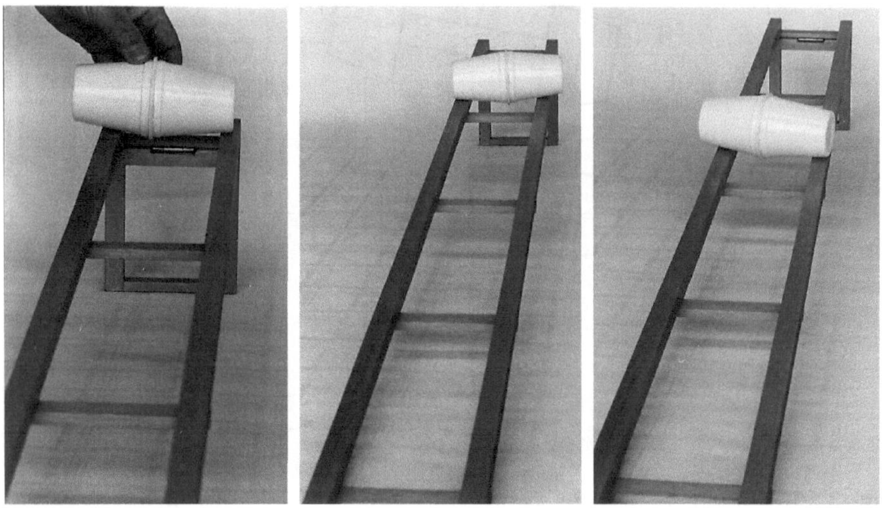

Fig. 2.20. Experimental setup showing the stable wheelset motion

contact patches, and purely kinematic rolling does no longer exist. Furthermore, the wheel and rail profiles are neither conical nor square, and the up to now neglected inertia phenomena are of great influence. ■

2.3 Dynamics

For the generation of the equations of motion of multibody systems, in addition to kinematics, the inertia of the bodies and the acting forces have to be considered. The Newton-Euler approach, also called the synthetic method, uses the free body diagram resulting in full set of local equations which may be reduced by the principles of d'Alembert and Jourdain to the equations of motion. The Lagrangian approach, representing the analytical method, is based on energy considerations and the equations of motions are found directly but without any information on the reaction forces.

2.3.1 Inertia Properties

The inertia of a rigid body K_i is characterized by its mass m_i and its inertia tensor $\boldsymbol{I}_{Ci}$. The coordinates of the inertia tensor read in the body-fixed frame $\{C_i, \boldsymbol{e}_\nu^\mathrm{i}\}$, see Fig. 2.7, as

$$\boldsymbol{I}_{Ci}^{i} = \int\limits_{m_i} (\boldsymbol{\rho}^\mathrm{T}\boldsymbol{\rho}\boldsymbol{E} - \boldsymbol{\rho}\boldsymbol{\rho}^\mathrm{T})\mathrm{d}m = \begin{bmatrix} I_{11} & I_{12} & I_{13} \\ I_{21} & I_{22} & I_{23} \\ I_{31} & I_{32} & I_{33} \end{bmatrix}_{C_i} = \text{const} \,. \tag{2.49}$$

The vector $\boldsymbol{\rho} \equiv \boldsymbol{\rho}^i = [\rho_1 \; \rho_2 \; \rho_3]^\mathrm{T}$ describes a material point P with mass $\mathrm{d}m$ with respect to the center of mass C_i and $\boldsymbol{E}$ means the 3×3-identity matrix. The inertia tensor $\boldsymbol{I}_{Ci}^i$ is symmetric and positive definite, and constant in the body-fixed frame $\boldsymbol{e}_\nu^i$. The diagonal elements $\boldsymbol{I}_{\nu\nu}$ are called moments of inertia, they are nonnegative. The off-diagonal elements are called products of inertia and can be positive, negative or zero.

The coordinates of the inertia tensor depend on the mass distribution and on the choice of the reference frame $\boldsymbol{e}_\nu^i$. For a parallel displacement of the body-fixed frame from the center of mass C_i to an arbitrary body-fixed point O_i characterized by the vector $\boldsymbol{s}$ one gets

$$\boldsymbol{I}_{Oi}^i = \boldsymbol{I}_{Ci}^i + (\boldsymbol{s}^\mathrm{T}\boldsymbol{s}\boldsymbol{E} - \boldsymbol{s}\boldsymbol{s}^\mathrm{T})m_i \,. \tag{2.50}$$

This relation is also known as Huygens-Steiner principle. Thus, the diagonal elements of an inertia tensor are minimal for the center of mass.

For a homogeneous, purely rotational displacement by the rotation matrix $\boldsymbol{S}^{ii'}$ from frame $\boldsymbol{e}_\nu^i$ to $\boldsymbol{e}_\nu^{i'}$ around the center of mass the transformation law for tensors reads as

$$\boldsymbol{I}_{Ci}^i = \boldsymbol{S}^{ii'}\boldsymbol{I}_{Ci}^{i'}\boldsymbol{S}^{i'i} \quad \text{or} \quad \boldsymbol{I}_{Ci}^{i'} = \boldsymbol{S}^{i'i}\boldsymbol{I}_{Ci}^{i}\boldsymbol{S}^{ii'} \,. \tag{2.51}$$

Please note that the inertia tensor may be time-variant if the frame $\{C_i, \boldsymbol{e}_\nu^{i'}\}$ is not body-fixed. This is especially true if the inertial frame is chosen, $i' \equiv I$, due to $\boldsymbol{S}^{iI} = \boldsymbol{S}^{iI}(t)$.

For all reference points there exists a special body-fixed frame in which the off-diagonal elements of the inertia tensor are vanishing, e.g.,

$$\boldsymbol{I_{Ci}} = \mathbf{diag}\,[I_1\ I_2\ I_3] = \mathbf{const}\ . \tag{2.52}$$

The remaining diagonal elements I_ν are called principal moments of inertia with reference to C_i and the corresponding axes are the principal inertia axes. Both quantities follow from the eigenvalue problem

$$(I_\nu \boldsymbol{E} - \boldsymbol{I}^i_{Ci})\boldsymbol{x}_\nu = \boldsymbol{0}\ . \tag{2.53}$$

Thus, the principal moments of inertia are the eigenvalues of the matrix $\boldsymbol{I}^i_{Ci}$ and the eigenvectors $\boldsymbol{x}_\nu = \boldsymbol{e}^{i'}_\nu$ define the principal inertia axes which have to be unit vectors $\boldsymbol{x}^{\mathrm{T}}_\nu \boldsymbol{x}_\nu = 1$. For more details see Schiehlen and Eberhard (2004).

■ **Problem 2.8 Inertia tensor of a railway wheelset**

Evaluate the inertia tensor of a railway wheelset Fig. 2.21 a) by using the strongly simplified model Fig. 2.21 b) with respect to the body-fixed principal axes frame $\{C, \boldsymbol{e}^K_\nu\}$ located in the center of mass C.

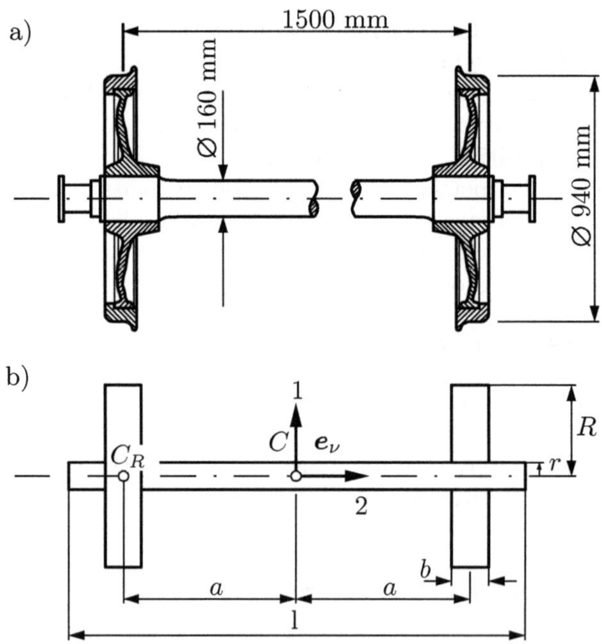

Fig. 2.21. Railway wheelset: a) real structure; b) simplified model

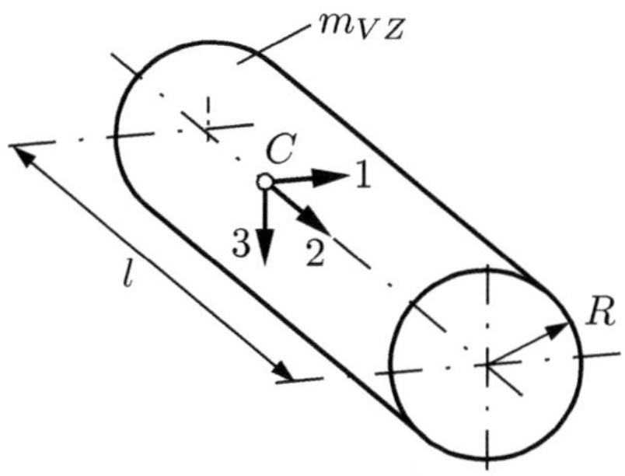

Fig. 2.22. Dimensions of a full cylinder

Solution

The moments of inertia of a full cylinder related to the center of mass C, Fig. 2.22, read as

$$I_{VZ2} = \frac{1}{2} R^2 m_{VZ} , \tag{1}$$

$$I_{VZ1} = I_{VZ3} = \left(\frac{1}{4} R^2 + \frac{1}{12} l^2 \right) m_{VZ} , \tag{2}$$

characterized by mass m_{VZ}, radius R and length l. The moments of inertia of a hollow cylinder related to the center of mass C with mass m_{HZ}, outside radius R, inside radius r and length l are obtained by subtracting the part of the bore with mass m_B,

$$\begin{aligned} I_{HZ2} &= \frac{1}{2} R^2 m_{VZ} - \frac{1}{2} r^2 m_B \\ &= \frac{1}{2} (R^2 R^2 - r^2 r^2) \pi l \rho \\ &= \frac{1}{2} (R^2 + r^2)(R^2 - r^2) \pi l \rho \\ &= \frac{1}{2} (R^2 + r^2) m_{HZ} , \end{aligned} \tag{3}$$

$$\begin{aligned} I_{HZ1} = I_{HZ3} &= \left(\frac{1}{4} R^2 + \frac{1}{12} l^2 \right) m_{VZ} - \left(\frac{1}{4} r^2 + \frac{1}{12} l^2 \right) m_B \\ &= \left[\frac{1}{4} (R^2 + r^2) + \frac{1}{12} l^2 \right] m_{HZ} , \end{aligned} \tag{4}$$

where the density ρ and mass relation

$$m_{HZ} = m_{VZ} - m_B = (R^2 - r^2) \pi l \rho \tag{5}$$

are used.

The principal axes of the wheelset follow from its rotational symmetry as indicated in Fig. 2.21. For the wheelset it yields according to the cylinder $I_1 = I_3$. For the wheelset axle with mass m_W and center of mass $C_W = C$ it remains, see (1) and (2),

$$I_{W2} = \frac{1}{2} r^2 m_W \, , \tag{6}$$

$$I_{W1} = I_{W3} = \left(\frac{1}{4} r^2 + \frac{1}{12} l^2 \right) m_W \, . \tag{7}$$

For one wheel with mass m_R and center of mass C_R it follows from (3) and (4)

$$I'_{R2} = \frac{1}{2}(R^2 + r^2) m_R \, , \tag{8}$$

$$I'_{R1} = I'_{R3} = \left[\frac{1}{4}(R^2 + r^2) + \frac{1}{12} b^2 \right] m_R \, . \tag{9}$$

The transformation from reference point C_R of the wheel to the reference point C of the wheelset by the principle of Huygens-Steiner (2.50) results in

$$I_{R2} = I'_{R2} = \frac{1}{2}(R^2 + r^2) m_R \, , \tag{10}$$

$$I_{R1} = I'_{R1} + a^2 m_R = \left[\frac{1}{4}(R^2 + r^2) \frac{1}{12} b^2 + a^2 \right] m_R = I_{R3} \, . \tag{11}$$

The addition of the shares of the axle and both wheels yields the inertia tensor of the whole wheelset,

$$I_C^K = \begin{bmatrix} I_1 & 0 & 0 \\ 0 & I_2 & 0 \\ 0 & 0 & I_3 \end{bmatrix}_C \, , \tag{12}$$

$$\begin{aligned} I_1 &= I_{W1} + 2 I_{R1} \\ &= \left(\frac{1}{4} r^2 + \frac{1}{12} l^2 \right) m_W + 2 \left[\frac{1}{4}(R^2 + r^2) + \frac{1}{12} b^2 + a^2 \right] m_R \, , \end{aligned} \tag{13}$$

$$I_2 = I_{W2} + 2 I_{R2} = \frac{1}{2} r^2 m_W + (R^2 + r^2) m_R \, , \tag{14}$$

$$I_3 = I_1 \, . \tag{15}$$

For the realistic parameters

$$\begin{aligned} a &= 7.5 \text{ dm}, & \rho &= 7.85 \text{ kg/dm}^3, \\ b &= 1.2 \text{ dm}, & & \\ l &= 20.0 \text{ dm}, & m_W &= 315.7 \text{ kg}, \\ R &= 4.7 \text{ dm}, & & \\ r &= 0.8 \text{ dm}, & m_R &= 634.8 \text{ kg}, \end{aligned}$$

Table 2.4. Moments of inertia of a railway wheelset

Component	Contribution to the principal moment of inertia $I_1 = I_3$ [kgm^2]	%	I_2 [kgm^2]	%
Axle W	105.7	11.8	1.0	0.7
2 wheels (Reference point C_R)	73.7	8.2	144.3	99.3
Steiner-share of both wheels	714.2	80.0	0	0
Total	893.6	100 %	145.3	100 %

the numbers listed in Table 2.4 are found.

It turns out that the largest amount (80 %) of the moment of inertia $I_1 = I_3$ is due to the Huygens-Steiner transformation while the moment of inertia I_2 of the rotation axis is dominated (99.3 %) by the moment of inertia of both wheels. ■

2.3.2 Newton-Euler Equations

After these preliminaries the fundamental the laws of Newton (1687) and Euler (1758), are introduced relating the translational motion represented by the momentum $\boldsymbol{p}$ of a body K to the sum of the external forces $\boldsymbol{f}$ and the rotational motion represented by the moment of momentum $\boldsymbol{h}_O$ to the sum of the external torques $\boldsymbol{l}_O$,

$$\frac{\mathrm{d}^I}{\mathrm{d}t}\boldsymbol{p} = \boldsymbol{f}, \quad \frac{\mathrm{d}^I}{\mathrm{d}t}\boldsymbol{h}_O = \boldsymbol{l}_O \,. \tag{2.54}$$

The time derivatives of the momentum $\boldsymbol{p}$ and the moment of momentum $\boldsymbol{h}_O$ have to be evaluated in the inertial frame I. The common reference point O of the moment of momentum and the resulting external torque may be an inertially fixed point like the origin of the inertial frame, $O \equiv O_I$, or the moving center of mass of the body, $O \equiv C$.

The fundamental laws (2.54) will be now applied to the rigid body K_i, $i = 1(1)p$, of a multibody system and appropriate frames are chosen. First of all, the bodies K_i are dismantled and the constraints are replaced by reaction forces acting then externally on the bodies involved in the same amount but with opposite sign according to the counteraction principle (action = reaction). Further, the center of mass is used as reference point for all bodies, $O \equiv C_i$.

In the inertial frame I momentum and moment of momentum for a rigid body K_i using the inertia properties $m_i\,, \boldsymbol{I}_{Ci}$ read as

$$\boldsymbol{p}_i^I = m_i \boldsymbol{v}_{Ci}^I \,, \quad m_i = \text{const} \,, \tag{2.55}$$
$$\boldsymbol{h}_{Ci}^I = \boldsymbol{I}_{Ci}^I \boldsymbol{\omega}_i^I \,, \quad \boldsymbol{I}_{Ci}^I = \boldsymbol{I}_{Ci}^I(t) \,, \tag{2.56}$$

where $\boldsymbol{v}_{Ci}^I$ and $\boldsymbol{\omega}_i^I$ mean absolute velocities. Introducing (2.55) and (2.56) in (2.54) and omitting the index C one finally gets Newton's and Euler's equations

$$m_i \overset{*}{\boldsymbol{v}}{}_i^i = \boldsymbol{f}_i^I, \qquad m_i = \text{const} \,, \tag{2.57}$$
$$\boldsymbol{I}_i^I \dot{\boldsymbol{\omega}}_i^I + \tilde{\boldsymbol{\omega}}_i^I \boldsymbol{I}_i^I \boldsymbol{\omega}_i^I = \boldsymbol{l}_i^I \,, \qquad \boldsymbol{I}_i^I = \boldsymbol{I}_i^I(t) \,. \tag{2.58}$$

In a second step these equations are transformed in a body-fixed frame resulting in

$$m_i \dot{\boldsymbol{v}}_i^i = \boldsymbol{f}_i^i \,, \qquad m_i = \text{const} \,, \tag{2.59}$$
$$\boldsymbol{I}_i^i \dot{\boldsymbol{\omega}}_i^i + \tilde{\boldsymbol{\omega}}_i^i \boldsymbol{I}_i^i \boldsymbol{\omega}_i^i = \boldsymbol{l}_i^i \,, \qquad \boldsymbol{I}_i^i = \textbf{const} \,. \tag{2.60}$$

Equations (2.57) and (2.58), and (2.59) and (2.60) look completely identical. If there is only one body like in gyrodynamics, then (2.60) is preferable due to the time-invariance of the inertia tensor. In multibody dynamics, however, this advantage is fading.

Equations (2.60) are also known as Euler's equations of gyrodynamics. They can be found with the moment of momentum given in the body-fixed frame also directly from (2.54) using the law of differentiation in a rotating frame (2.28),

$$\frac{\mathrm{d}^I}{\mathrm{d}t} \boldsymbol{h}_i^i = \overset{*}{\boldsymbol{h}}{}_i^i + \tilde{\boldsymbol{\omega}}_i^i \boldsymbol{h}_i^i = \boldsymbol{l}_i^i \,, \quad \boldsymbol{h}_i^i = \boldsymbol{I}_i^i \boldsymbol{\omega}_i^i \,. \tag{2.61}$$

For single rigid bodies it may be of some advantage to represent the moment of momentum in a moving intermediate frame Z with the angular velocity $\omega_Z \equiv \omega_Z$ with respect to the inertial frame. The application of the law of differentiation in a rotating frame (2.28) results in

$$\frac{\mathrm{d}^I}{\mathrm{d}t} \boldsymbol{h}_i^Z = \overset{*}{\boldsymbol{h}}{}_i^Z + \tilde{\boldsymbol{\omega}}_i^Z \boldsymbol{h}_i^Z = \boldsymbol{l}_i^Z \,. \tag{2.62}$$

Finally, in an arbitrarily moving reference frame R Newton's and Euler's equations are also available, see e.g. Schiehlen and Eberhard (2004),

$$m_i \ddot{\boldsymbol{r}}_{Ri} + m_i \left[\overset{**}{\boldsymbol{r}}_R + (\dot{\tilde{\boldsymbol{\omega}}}_R + \tilde{\boldsymbol{\omega}}_R \tilde{\boldsymbol{\omega}}_R) \boldsymbol{r}_{Ri} + 2\tilde{\boldsymbol{\omega}}_R \dot{\boldsymbol{r}}_{Ri} \right] = \boldsymbol{f}_i \,, \tag{2.63}$$

$$\begin{aligned} &\boldsymbol{I}_i \dot{\boldsymbol{\omega}}_{Ri} + \tilde{\boldsymbol{\omega}}_{Ri} \boldsymbol{I}_i \boldsymbol{\omega}_{Ri} \\ &+ [\boldsymbol{I}_i \boldsymbol{\omega}_R + \tilde{\boldsymbol{\omega}}_R \boldsymbol{I}_i \boldsymbol{\omega}_R + \tilde{\boldsymbol{\omega}}_R \boldsymbol{\omega}_{Ri} sp \boldsymbol{I}_i + 2\tilde{\boldsymbol{\omega}}_{Ri} \boldsymbol{I}_i \boldsymbol{\omega}_R] = \boldsymbol{l}_i \,. \end{aligned} \tag{2.64}$$

Now, the coordinates of all vectors and tensors are related to the reference frame R where $\overset{**}{\boldsymbol{r}}_R$ means that the second time derivation has to be performed in the inertial frame, see also Schiehlen and Eberhard (2004). As a matter of fact, a large number of additional inertia forces and torques appear due to the relative motion.

The Newton-Euler equations represent a set of $6p$ scalar equations for $6p$ unknowns which are composed of unknown velocity and position variables and unknown reaction forces and torques. In an unconstrained system reactions do not exist, i.e., there are $6p$ ordinary differential equations (ODEs) to be solved. In a completely constrained system motion does not occur at all, i.e., altogether $6p$ algebraical equations have to be solved. In vehicle dynamics, due to a certain number of constraints between the bodies, motions and reactions appear concurrently featuring a set of differential-algebraical equations (DAEs). However, by the principles of dynamics, a minimal set of f ODEs can be found facilitating the solution and simulation of the problem as shown in Sect. 2.4.

A special case is given if a multibody system performs a plane motion around a common principle axis. Then, the vectors of moment of momentum and angular velocities are parallel, see (2.56). As an example a motion in the $\boldsymbol{e}_1^I$, $\boldsymbol{e}_3^I$-plane is considered where the rotations take place around the principle axis parallel to the $\boldsymbol{e}_2^I$-axis. The six scalar equations of motion (2.57), (2.58) are reduced by

$$\dot{\boldsymbol{v}}_i^I = [\ddot{x}_1 \; 0 \; \ddot{x}_3]^{\mathrm{T}} \,, \quad \dot{\boldsymbol{\omega}}_i^I = [0 \; \dot{\omega}_2 \; 0]^{\mathrm{T}} \,, \tag{2.65}$$

$$\boldsymbol{f}_i^I = [f_1 \; 0 \; f_3]^{\mathrm{T}} \,, \quad \boldsymbol{l}_i^I = [0 \; l_2 \; 0]^{\mathrm{T}} \,, \tag{2.66}$$

to three equations

$$m_i \ddot{x}_1 = f_1 \,, \tag{2.67}$$

$$m_i \ddot{x}_3 = f_3 \,, \tag{2.68}$$

$$I_{i2}^I \dot{\omega}_2 = l_2 \,, \quad I_{i2}^I = I_{i2}^i = const. \tag{2.69}$$

It turns out that moments of inertia are identical and constant what is advantageous. However, it has to be pointed out that the motion remains only planar if the forces and torques satisfy (2.66) all the time.

■ Problem 2.9 Equations of motion of a railway wheelset

The wheelset is an essential component of railway vehicles. Generate the equations of motion of a wheelset using its free body diagram and the inertia tensor found in Problem 2.8. The wheels with standard profile shall be continuously in contact with the standard rails which are assumed to be horizontal and straight.

Solution

The free body diagram with the contact forces acting on the left and right wheel is shown in Fig. 2.23. According to Problem 2.7 the frames I, R, Z and K are used. In the contact points P_l, P_r two additional frames $\{P_l, \boldsymbol{e}_v^l\}$

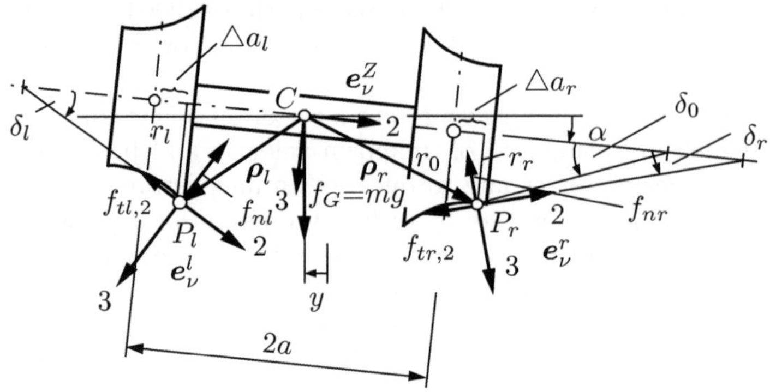

Fig. 2.23. Free body diagram of a railway wheelset

and $\{P_r, \boldsymbol{e}_v^r\}$ are attached where the 2-axes are tangential and the 3-axes are normal to the wheel profile. The $\boldsymbol{e}_1^l$-axis and the $\boldsymbol{e}_1^r$-axis are parallel to each other and parallel to the $\boldsymbol{e}_1^Z$-axis which is normal to the plane of the contact points characterized by the $\boldsymbol{e}_2^Z, \boldsymbol{e}_3^Z$-plane. Thus, the rotation matrices between the frames Z, l and r, respectively, read according to Fig. 2.23 as

$$\boldsymbol{S}^{Zl} \equiv \boldsymbol{\delta}_{l,1} = \begin{bmatrix} 1 & 0 & 0 \\ 0 & \mathrm{c}\delta_l & -\mathrm{s}\delta_l \\ 0 & \mathrm{s}\delta_l & \mathrm{c}\delta_l \end{bmatrix} , \quad \boldsymbol{S}^{Zr} \equiv -\boldsymbol{\delta}_{r,1} = \begin{bmatrix} 1 & 0 & 0 \\ 0 & \mathrm{c}\delta_r & \mathrm{s}\delta_r \\ 0 & -\mathrm{s}\delta_r & \mathrm{c}\delta_r \end{bmatrix} . \tag{1}$$

The rotation matrix $\boldsymbol{S}^{IZ}$ is already known from Problem 2.2

$$\boldsymbol{S}^{IZ} = \boldsymbol{\gamma}_3\boldsymbol{\alpha}_1 = \begin{bmatrix} \mathrm{c}\gamma & -\mathrm{s}\gamma\mathrm{c}\alpha & \mathrm{s}\gamma\mathrm{s}\alpha \\ \mathrm{s}\gamma & \mathrm{c}\gamma\mathrm{c}\alpha & -\mathrm{c}\gamma\mathrm{s}\alpha \\ 0 & \mathrm{s}\alpha & \mathrm{c}\alpha \end{bmatrix} . \tag{2}$$

Then the transformation matrices $\boldsymbol{S}^{Il}$ and $\boldsymbol{S}^{Ir}$ are found as

$$\begin{aligned} \boldsymbol{S}^{Il} &= \boldsymbol{S}^{IZ}\boldsymbol{S}^{Zl} = \boldsymbol{\gamma}_3\boldsymbol{\alpha}_1\boldsymbol{\delta}_{l,1} \\ &= \boldsymbol{\gamma}_3(\boldsymbol{\alpha}+\boldsymbol{\delta}_l)_1 = \begin{bmatrix} \mathrm{c}\gamma & -\mathrm{s}\gamma\mathrm{c}(\alpha+\delta_l) & \mathrm{s}\gamma\mathrm{s}(\alpha+\delta_l) \\ \mathrm{s}\gamma & \mathrm{c}\gamma\mathrm{c}(\alpha+\delta_l) & -\mathrm{c}\gamma\mathrm{s}(\alpha+\delta_l) \\ 0 & \mathrm{s}(\alpha+\delta_l) & \mathrm{c}(\alpha+\delta_l) \end{bmatrix} , \end{aligned} \tag{3}$$

$$\begin{aligned} \boldsymbol{S}^{Ir} &= \boldsymbol{S}^{IZ}\boldsymbol{S}^{Zr} = \boldsymbol{\gamma}_3\boldsymbol{\alpha}_1(-\boldsymbol{\delta}_r)_1 \\ &= \boldsymbol{\gamma}_3(\boldsymbol{\alpha}-\boldsymbol{\delta}_r)_1 = \begin{bmatrix} \mathrm{c}\gamma & -\mathrm{s}\gamma\mathrm{c}(\alpha-\delta_r) & \mathrm{s}\gamma\mathrm{s}(\alpha-\delta_r) \\ \mathrm{s}\gamma & \mathrm{c}\gamma\mathrm{c}(\alpha-\delta_r) & -\mathrm{c}\gamma\mathrm{s}(\alpha-\delta_r) \\ 0 & \mathrm{s}(\alpha-\delta_r) & \mathrm{c}(\alpha-\delta_r) \end{bmatrix} \end{aligned} \tag{4}$$

where the approach of elementary rotation matrices proves to be very efficient due to pooling of the consecutively performed rotations around the 1-axis with the angles $\alpha + \delta_l$ and $\alpha - \delta_r$, respectively.

The forces and torques acting in the contact points on the wheels are assumed to be negative with respect to the contact point coordinates, see Fig. 2.23,

$$\boldsymbol{f}_l^l = -\begin{bmatrix} f_{tl,1} \\ f_{tl,2} \\ f_{nl} \end{bmatrix} , \quad \boldsymbol{f}_r^r = -\begin{bmatrix} f_{tr,1} \\ f_{tr,2} \\ f_{nr} \end{bmatrix} , \tag{5}$$

$$\boldsymbol{l}_l^l = -\begin{bmatrix} 0 \\ 0 \\ l_l \end{bmatrix} , \quad \boldsymbol{l}_r^r = -\begin{bmatrix} 0 \\ 0 \\ l_r \end{bmatrix} . \tag{6}$$

On the rails these forces and torques are acting in positive direction. Now the forces are transformed into the inertial frame

$$\boldsymbol{f}_l^I = \boldsymbol{S}^{Il}\boldsymbol{f}_l^l = \begin{bmatrix} f_{l1} \\ f_{l2} \\ f_{l3} \end{bmatrix} , \quad \boldsymbol{f}_r^I = \boldsymbol{S}^{Ir}\boldsymbol{f}_r^r = \begin{bmatrix} f_{r1} \\ f_{r2} \\ f_{r3} \end{bmatrix} , \tag{7}$$

$$\begin{aligned} f_{l1} &= -\mathrm{c}\gamma f_{tl,1} + \mathrm{s}\gamma \mathrm{c}(\alpha+\delta_l) f_{t1,2} - \mathrm{s}\gamma \mathrm{s}(\alpha+\delta_l) f_{nl} , \\ f_{l2} &= -\mathrm{s}\gamma f_{tl,1} - \mathrm{c}\gamma \mathrm{c}(\alpha+\delta_l) f_{t1,2} + \mathrm{c}\gamma \mathrm{s}(\alpha+\delta_l) f_{nl} , \\ f_{l3} &= \qquad - \mathrm{s}(\alpha+\delta_l) f_{t1,2} - \mathrm{c}(\alpha+\delta_l) f_{nl} , \end{aligned} \tag{8}$$

$$\begin{aligned} f_{r1} &= -\mathrm{c}\gamma f_{tr,1} + \mathrm{s}\gamma \mathrm{c}(\alpha-\delta_r) f_{tr,2} - \mathrm{s}\gamma \mathrm{s}(\alpha-\delta_r) f_{nr} , \\ f_{r2} &= -\mathrm{s}\gamma f_{tr,1} - \mathrm{c}\gamma \mathrm{c}(\alpha-\delta_r) f_{tr,2} + \mathrm{c}\gamma \mathrm{s}(\alpha-\delta_r) f_{nr} , \\ f_{r3} &= \qquad - \mathrm{s}(\alpha-\delta_r) f_{tr,2} - \mathrm{c}(\alpha-\delta_r) f_{nr} . \end{aligned} \tag{9}$$

The translational equations of motion follow from the forces (8) and (9), and the position vector $\boldsymbol{r}_C^I \equiv [x\ y\ z]^{\mathrm{T}}$, see Problem 2.7, simply as

$$m\ddot{x} = f_{l1} + f_{r1} , \tag{10}$$

$$m\ddot{y} = f_{l2} + f_{r2} , \tag{11}$$

$$m\ddot{z} = f_{l3} + f_{r3} + f_G , f_G = mg , \tag{12}$$

The rotational equations of motions are generated more efficiently using the moving reference frame Z as mentioned in (2.62). The evaluation of the torques is more simple in frame Z, and the moments of inertia remain constant due to the rotational symmetry of the wheelset. The torques with reference to the center of mass C follow from the contact torques $\boldsymbol{l}_l, \boldsymbol{l}_r$ and the torques of the contact forces $\boldsymbol{f}_l, \boldsymbol{f}_r$ as

$$\boldsymbol{l}_{Cl}^Z = \boldsymbol{S}^{Zl}\boldsymbol{l}_l^l + \tilde{\boldsymbol{\rho}}_{Cl}^Z \boldsymbol{S}^{Zl}\boldsymbol{f}_l^l = \begin{bmatrix} l_{l1} \\ l_{l2} \\ l_{l3} \end{bmatrix} , \tag{13}$$

$$\boldsymbol{l}_{Cr}^Z = \boldsymbol{S}^{Zr}\boldsymbol{l}_r^r + \tilde{\boldsymbol{\rho}}_{Cr}^Z \boldsymbol{S}^{Zr}\boldsymbol{f}_r^r = \begin{bmatrix} l_{r1} \\ l_{r2} \\ l_{r3} \end{bmatrix} . \tag{14}$$

The position vectors $\boldsymbol{\rho}_{Cl}^Z, \boldsymbol{\rho}_{Cr}^Z$ from the center of mass C to the contact points P_l and P_r, respectively, read according to Fig. 2.23 as

$$\boldsymbol{\rho}_{Cl}^Z = \begin{bmatrix} 0 \\ -a + \Delta a_l \\ r_l \end{bmatrix} , \quad \boldsymbol{\rho}_{Cr}^Z = \begin{bmatrix} 0 \\ a + \Delta a_r \\ r_r \end{bmatrix} . \tag{15}$$

The evaluation of (13) and (14) results in

$$\begin{aligned} l_{l1} &= [(a - \Delta a_l)\mathrm{s}\delta_l + r_l\mathrm{c}\delta_1] f_{tl,2} + [(a - \Delta a_1)\mathrm{c}\delta_l - r_l\mathrm{s}\delta_l] f_{nl} , \\ l_{l2} &= \mathrm{s}\delta_l l_l - r_l f_{tl,1} , \\ l_{l3} &= -\mathrm{c}\delta_l l_l - (a - \Delta a_l) f_{tl,1} , \end{aligned} \tag{16}$$

$$\begin{aligned} l_{r1} &= [(a + \Delta a_r)\mathrm{s}\delta_r + r_r\mathrm{c}\delta_r] f_{tr,2} - [(a + \Delta a_r)\mathrm{c}\delta_r - r_r\mathrm{s}\delta_r] f_{nr} , \\ l_{r2} &= -\mathrm{s}\delta_r l_r - r_r f_{tr,1} , \\ l_{r3} &= -\mathrm{c}\delta_r l_r + (a + \Delta a_r) f_{tr,1} . \end{aligned} \tag{17}$$

For the evaluation of the moment of momentum and its time derivative, see (2.56) and (2.62), the rotational velocities $\boldsymbol{\omega}_{IK}^Z$ and $\boldsymbol{\omega}_{IZ}^Z$ of the wheelset K and the reference system Z with respect to the inertial frame are required. From Problem 2.7, (6) the angular velocity $\boldsymbol{\omega}_{IK}^I$ is known. The transformation in the reference frame Z using (2) results in

$$\begin{aligned} \boldsymbol{\omega}_{IK}^Z = \boldsymbol{S}^{ZI}\boldsymbol{\omega}_{IK}^I &= \begin{bmatrix} \mathrm{c}\gamma & \mathrm{s}\gamma & 0 \\ -\mathrm{c}\alpha\mathrm{s}\gamma & \mathrm{c}\alpha\mathrm{c}\gamma & \mathrm{s}\alpha \\ \mathrm{s}\alpha\mathrm{s}\gamma & -\mathrm{s}\alpha\mathrm{c}\gamma & \mathrm{c}\alpha \end{bmatrix} \begin{bmatrix} \mathrm{c}\gamma & \mathrm{c}\alpha\mathrm{s}\gamma & 0 \\ \mathrm{s}\gamma & -\mathrm{c}\alpha\mathrm{c}\gamma & 0 \\ 0 & -\mathrm{s}\alpha & 1 \end{bmatrix} \begin{bmatrix} \dot\alpha \\ \dot\beta \\ \dot\gamma \end{bmatrix} \\ &= \begin{bmatrix} \dot\alpha \\ \dot\gamma\mathrm{s}\alpha - \dot\beta \\ \dot\gamma\mathrm{c}\alpha \end{bmatrix} . \end{aligned} \tag{18}$$

The rotational velocity $\boldsymbol{\omega}_{IZ}^Z$ for the moving reference frame Z follows from $\boldsymbol{\omega}_{IK}^Z$ by the condition $\dot\beta = 0$,

$$\boldsymbol{\omega}_{IZ}^Z = \begin{bmatrix} \dot\alpha \\ \dot\gamma\mathrm{s}\alpha \\ \dot\gamma\mathrm{c}\alpha \end{bmatrix} . \tag{19}$$

Equation (19) can also be immediately read from Fig. 2.11 of Problem 2.2. With the inertia tensor $\boldsymbol{I}_K^Z = \mathbf{diag}\left[I_1\ I_2\ I_3\right] = \mathbf{const}$ related to the center of mass C and the moment of momentum $\boldsymbol{h}_K^Z = \boldsymbol{I}_K^Z\boldsymbol{\omega}_{IK}^Z$ one gets from (2.62) the result

$$\boldsymbol{I}_K^Z\dot{\boldsymbol{\omega}}_{IK}^Z + \tilde{\boldsymbol{\omega}}_{IZ}^Z\boldsymbol{I}_K^Z\boldsymbol{\omega}_{IK}^Z = \boldsymbol{l}_{Cl}^Z + \boldsymbol{l}_{Cr}^Z \tag{20}$$

or explicitly

$$I_1\ddot{\alpha} + I_1\dot{\gamma}^2 \mathrm{s}\alpha\mathrm{c}\alpha - I_2\dot{\gamma}\mathrm{c}\alpha(-\dot{\beta} + \dot{\gamma}\mathrm{s}\alpha) = l_{l1} + l_{r1} \,, \tag{21}$$

$$I_2(\ddot{\gamma}\mathrm{s}\alpha + \dot{\alpha}\dot{\gamma}\mathrm{c}\alpha - \ddot{\beta}) = l_{l2} + l_{r2} \,, \tag{22}$$

$$I_1(\ddot{\gamma}\mathrm{c}\alpha - 2\dot{\alpha}\dot{\gamma}\mathrm{s}\alpha) + I_2\dot{\alpha}(-\dot{\beta} + \dot{\gamma}\mathrm{s}\alpha) = l_{l3} + l_{r3} \,. \tag{23}$$

In the following two cases are distinguished: a) the ideal case of a purely kinematical rolling with point contact and b) the real case of rolling with elastic contact.

a) In the case of purely kinematical rolling there exist two geometrical constraints between the chosen coordinates $x, y, z, \alpha, \beta, \gamma$ due to the continuous contact between wheels and rails in the contact points P_l, P_r. In addition there exist three kinematical constraints depending on the velocities of the motion. For conical or nearly conical wheels and rectangular rails one gets for small motions the five holonomic constraints (20)-(24) presented in Problem 2.7. Thus, the wheelset has one degree of freedom. The six equations of motion (10)-(12) and (21)-(23) determine the motion on one hand and the unknown force variables on the other hand. In this case the contact torques l_l, l_r do not exit due to the point-shaped contact. For the computation of the remaining six unknown constraint forces $f_{tl,1}$, $f_{tl,2}$, f_{nl}, $f_{tr,1}$, $f_{tr,2}$, f_{nr} there are only five equations at hand. The problem is kinematically underdetermined. The sum $(f_{tl,2} + f_{tr,2})$ can be separated only by additional modeling assumptions like a laterally elastic axis.
b) In the case of elastic contact there are typically chosen two degrees of freedom for both, the lateral and the longitudinal motion characterized by the lateral displacement y and the yaw angle γ as well as the longitudinal displacement x and the angle β, see e.g. Law and Cooperrider (1974). The two geometrical constraints representing the continuous contact in the contact points P_l, P_r are maintained as constraints for the vertical displacement z and the roll angle α. It yields generally

$$z = z(y, \gamma) \,, \quad \alpha = \alpha(y, \gamma) \tag{24}$$

and according to (18), (19) from Problem 2.7 for small displacements

$$z \approx 0 \,, \quad \alpha = \alpha(y) \quad \text{or} \quad \dot{\alpha} = \alpha' \dot{y} \,, \tag{25}$$

where $(\cdot)' \equiv \mathrm{d}(\cdot)/\mathrm{d}y$ means the derivative with respect to the lateral displacement y. For the longitudinal motion of the wheelset additional constraints may be available. They follow from the engineering task and are usually related to the route of the wheelset, e.g., a constant forward velocity of the center of mass, $\dot{x} = \dot{x}_0 = v_0$, and/or a constant angular velocity of the lateral axis, $\dot{\beta} = \dot{\beta}_0 = \Omega = v_0/r_0$.

The elastic contact results in a small contact patch due to local deformations. The force variables $f_{tl,1}, f_{tl,2}, l_l, f_{tr,1}, f_{tr,2}, l_r$ are now known quantities depending on the small slip motions between wheel and rail. The details of the elastic contact law will be presented later in Sect. 3.4.

The normal forces f_{nl}, f_{nr} follow from (12) and (21) considering the geometrical constraints (24) and (25). Neglecting the inertia terms ($\dot{z} \approx 0$, $\dot{\alpha} \approx 0$, $\dot{\gamma} \approx 0$) one obtains in approximation

$$f_{l3} + f_{r3} + f_G = 0 \,, \quad l_{l1} + l_{r1} = 0 \,. \tag{26}$$

From (26) with (8), (9), (16) and (17) or immediately from Fig. 2.23 it follows for small displacements and rotations of the wheelset and small profile inclinations ($\delta_l, \delta_r \ll 1$) in a first approximation

$$\begin{aligned} f_{nl} &\approx \frac{1}{2a}[mga - f_{tl,2}r_l - f_{tr,2}r_r] \,, \\ f_{nr} &\approx \frac{1}{2a}[mga + f_{tl,2}r_l + f_{tr,2}r_r] \,. \end{aligned} \tag{27}$$

Finally, the static normal forces in the contact points P_l, P_r are evaluated in the nominal position with equal wheel profiles (inclination angles $\delta_l = \delta_r = \delta_0$, roll radii $r_l = r_r = r_0$) as well as vanishing tangential contact forces. Then, the force equilibrium in vertical direction yields according to Fig. 2.23

$$f_{nl,0} = f_{nr,0} = f_{n0} = \frac{mg}{2\cos\delta_0} \,. \tag{28}$$

The static loads show the same amount for the left and right wheel with horizontal components in opposite direction. The result (28) agrees with (27) for small inclination angles $\delta \ll 1$. ■

■ Problem 2.10 Inertia forces at a magnetically levitated (maglev) vehicle

A maglev vehicle (mass $m = 120000\,\text{kg}$) travels at a geographical latitude φ with a course angle ψ ($\varphi = 55°, \psi = 0°$) on a straight horizontal track. Then, the vehicle rides on a vertical transition track (radius $r = 20\,\text{km}$) to a straight inclined track with a constant slope of 35 ‰ (slope angle $\beta \approx \tan\beta \approx 0.035$). The vehicle speed $v = 400\,\text{km/h} = 111\,\text{m/s}$ is constant all the time. Evaluate the inertia forces acting on the vehicle

a) on the horizontal track due to the Coriolis force considering the earth rotation,

b) in vertical transition neglecting the earth rotation.

Solution

The solution is based on the momentum principle (2.63)

$$m_i \ddot{\boldsymbol{r}}_{Ri} + m_i[\overset{**}{\boldsymbol{r}}_R + (\dot{\tilde{\boldsymbol{\omega}}}_R + \tilde{\boldsymbol{\omega}}_R\tilde{\boldsymbol{\omega}}_R)\boldsymbol{r}_{Ri} + 2\tilde{\boldsymbol{\omega}}_R\dot{\boldsymbol{r}}_{Ri}] = \boldsymbol{f}_i \,, \tag{1}$$

where all coordinates are related to a corresponding moving frame. The vehicle as a whole is considered as a rigid body ($i = 1$) with mass m.

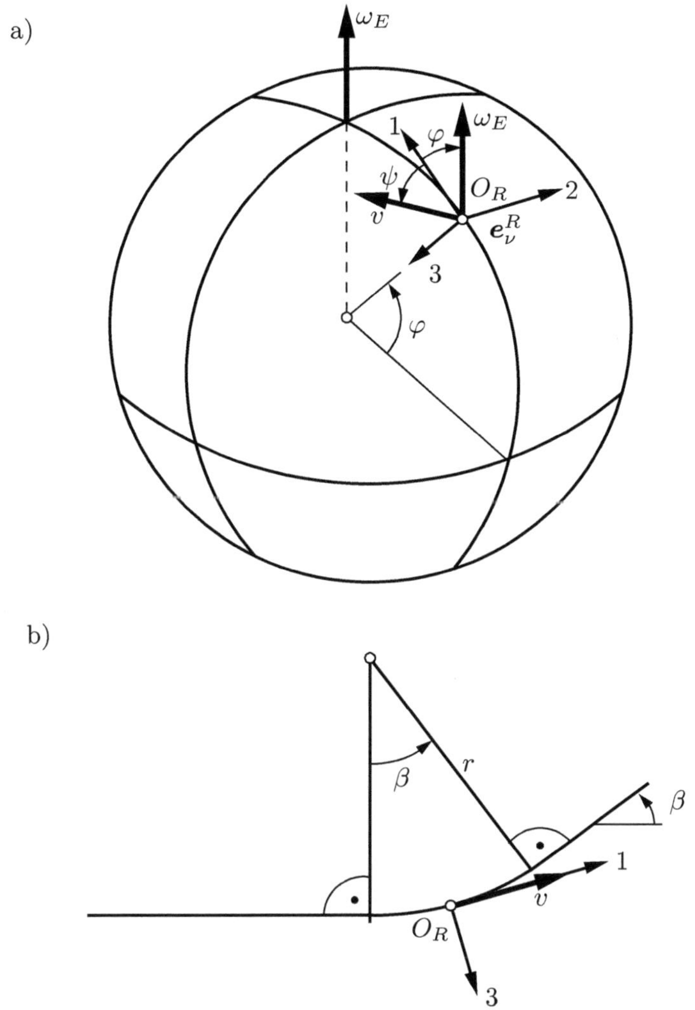

Fig. 2.24. Inertia force acting on a maglev vehicle: a) earth-fixed reference frame; b) vehicle-fixed reference frame.

a) For the computation of the Coriolis forces due to earth rotation an earth-fixed reference system R is used. Its origin O_R is the instantaneous position of the vehicle's center of mass C. The rotational velocity $\boldsymbol{\omega}_R^R$ is given at a geographical latitude φ, Fig. 2.24 a), as

$$\boldsymbol{\omega}_R^R = \begin{bmatrix}\omega_E \cos\varphi & 0 & -\omega_E \sin\varphi\end{bmatrix}^{\mathrm{T}} , \tag{2}$$

where $\boldsymbol{\omega}_E = 7.27 \cdot 10^{-5}\,\mathrm{s}^{-1}$ is the earth's rotational velocity.

The relative vehicle speed $\dot{\boldsymbol{r}}_{RI}^{R}$ reads with a course angle ψ as

$$\dot{\boldsymbol{r}}_{Ri}^{R} = \begin{bmatrix} v\cos\psi & -v\sin\psi & 0 \end{bmatrix}^{\mathrm{T}} . \tag{3}$$

Then, one gets the Coriolis force $\boldsymbol{f}_{Cor}$ acting on the vehicle as

$$\begin{aligned} \boldsymbol{f}_{Cor}^{R} &= -2m\tilde{\boldsymbol{\omega}}_{R}^{R}\dot{\boldsymbol{r}}_{Ri}^{R} = -2mv\omega_E \begin{bmatrix} 0 & \mathrm{s}\varphi & 0 \\ -\mathrm{s}\varphi & 0 & -\mathrm{c}\varphi \\ 0 & \mathrm{c}\varphi & 0 \end{bmatrix} \begin{bmatrix} \mathrm{c}\psi \\ -\mathrm{s}\psi \\ 0 \end{bmatrix} , \\ &= 2mv\omega_E \begin{bmatrix} \mathrm{s}\varphi\mathrm{s}\psi \\ \mathrm{s}\varphi\mathrm{c}\psi \\ \mathrm{c}\varphi\mathrm{s}\psi \end{bmatrix} . \end{aligned} \tag{4}$$

For $\psi = 0$, i.e. a ride to the north, it remains just one non-vanishing force coordinate

$$f_{Cor,2} = 2mv\omega_E \sin\varphi = 1.586\,\mathrm{kN} . \tag{5}$$

The amount of this force results in only 1.35 ‰ of the vehicles weight $f_G = mg$. It has to be mentioned that the Coriolis force acting horizontally on the vehicle is pointing at the northern hemisphere to the right relative to the driving direction.
b) For consideration of the transition track a vehicle-fixed reference frame R is used whose origin O_R coincides with vehicle's center of mass C, Fig. 2.24 b). The rotational velocity $\boldsymbol{\omega}_R^R$ reads as

$$\boldsymbol{\omega}_R^R = \begin{bmatrix} 0 & \omega & 0 \end{bmatrix}^{\mathrm{T}} , \quad \omega = v/r = \mathrm{const} . \tag{6}$$

The inertia force $\boldsymbol{f}_T$ subject to the guidance motion follows due to $\boldsymbol{r}_{Ri}^R \equiv 0$ as

$$\boldsymbol{f}_T^R = -m\overset{**}{\boldsymbol{r}}{}_R^R , \quad \overset{**}{\boldsymbol{r}}{}_R^R = \begin{bmatrix} 0 \; 0 \; -v^2/r \end{bmatrix}^{\mathrm{T}} , \tag{7}$$

where the guidance acceleration $\overset{**}{\boldsymbol{r}}_R$ matches the normal acceleration of the trajectory. The only non-vanishing force coordinate reads as

$$f_{T3} = mv^2/r = 73.9\,\mathrm{kN} . \tag{8}$$

The inertia force $\boldsymbol{f}_{T3}$ amounts to 6.28 % of the vehicle.

For a multibody vehicle, $i = 1(1)p$, with small dimensions $|\boldsymbol{r}_{Ri}| \ll r$ and small relative velocities $|\dot{\boldsymbol{r}}| \ll v$, it turns out that the dominant part of the inertia forces is due to the guidance acceleration $\overset{**}{\boldsymbol{r}}_R$,

$$\boldsymbol{f}_{Ti}^R = -m_i\overset{**}{\boldsymbol{r}}{}_R^R . \tag{9}$$

This inertia force is acting on all bodies of the vehicle as an additional weight component during driving on the transition track. In the presented case the additional weight component amounts to 6.28 % of the corresponding dead weight. However, the time of this action is small, $t = \beta/\omega = \beta r/v = 6.31\,\mathrm{s}$. ■

2.3.3 Principles of d'Alembert and Jourdain

From the Newton-Euler equations the equations of motion can be found by elimination of the reaction forces and torques resulting in a minimal set of ordinary differential equations (ODEs). This is achieved computationally efficient by the principles of dynamics considering the virtual work of a constrained multibody system. For this purpose the external forces acting on the dismantled bodies of the system are subdivided into applied forces $\boldsymbol{f}_i^{(e)}$ and torques $\boldsymbol{l}_i^{(e)}$ as well as constraint or reaction forces $\boldsymbol{f}_i^{(r)}$ and torques $\boldsymbol{l}_i^{(r)}$. The latter ones do not contribute to the virtual work of the system,

$$\delta W^r = \sum_{i=1}^{p} (\boldsymbol{f}_i^{(r)\mathrm{T}} \delta \boldsymbol{r}_i + \boldsymbol{l}_i^{(r)\mathrm{T}} \delta \boldsymbol{s}_i) = 0 , \tag{2.70}$$

where the virtual motions $\delta \boldsymbol{r}_i , \delta \boldsymbol{s}_i$ are known from (2.48). Equation (2.70) can be interpreted as a generalized orthogonality condition. For this purpose, in addition to the generalized coordinates y_k , $k = 1\,(1)\,f$, generalized reaction forces g_j , $j = 1\,(1)\,q$, are introduced and summarized in a $q \times 1$-vector as

$$\boldsymbol{g}(t) = [g_1 \,\dots\dots\dots\, g_q]^{\mathrm{T}} . \tag{2.71}$$

The number of generalized constraint forces is determined by the number q of constraints. The local constraint forces and torques follow from the implicit constraint equations (2.34) as

$$\boldsymbol{f}_i^{(r)\mathrm{T}} = \sum_{j=1}^{q} g_j \frac{\partial \varphi_j}{\partial \boldsymbol{r}_i^{\mathrm{T}}} = \sum_{j=1}^{q} g_j \frac{\partial \varphi_j}{\partial \boldsymbol{x}^{\mathrm{T}}} \frac{\partial \boldsymbol{x}}{\partial \boldsymbol{r}_i^{\mathrm{T}}} = \boldsymbol{g}^{\mathrm{T}} \boldsymbol{F}_{Ti}^{\mathrm{T}} , \tag{2.72}$$

$$\boldsymbol{l}_i^{(r)\mathrm{T}} = \sum_{j=1}^{q} g_j \frac{\partial \varphi_j}{\partial \boldsymbol{s}_i^{\mathrm{T}}} = \sum_{j=1}^{q} g_j \frac{\partial \varphi_j}{\partial \boldsymbol{x}^{\mathrm{T}}} \frac{\partial \boldsymbol{x}}{\partial \boldsymbol{s}_i^{\mathrm{T}}} = \boldsymbol{g}^{\mathrm{T}} \boldsymbol{F}_{Ri}^{\mathrm{T}} , \quad i = 1(1)p . \tag{2.73}$$

The $3 \times q$- Jacobian matrices $\boldsymbol{F}_{Ti} , \boldsymbol{F}_{Ri}$ are defined by (2.72) and (2.73), and now the condition (2.70) can be rewritten as

$$\begin{aligned} \delta W^r &= \boldsymbol{g}^{\mathrm{T}} \sum_{i=1}^{p} (\boldsymbol{F}_{Ti}^{\mathrm{T}} \delta \boldsymbol{r}_i + \boldsymbol{F}_{Ri}^{\mathrm{T}} \delta \boldsymbol{s}_i) \\ &= \boldsymbol{g}^{\mathrm{T}} \sum_{i=1}^{p} (\boldsymbol{F}_{Ti}^{\mathrm{T}} \boldsymbol{J}_{Ti} + \boldsymbol{F}_{Ri}^{\mathrm{T}} \boldsymbol{J}_{Ri}) \, \delta \boldsymbol{y} = 0 . \end{aligned} \tag{2.74}$$

Finally, the global $6p \times q$- distribution matrix $\overline{\boldsymbol{Q}}$, and the global $6p \times f$- Jacobian matrix $\overline{\boldsymbol{J}}$ are introduced

$$\begin{aligned} \overline{\boldsymbol{Q}} &= [\boldsymbol{F}_{T1}^{\mathrm{T}} \,\dots\dots\dots\, \boldsymbol{F}_{Tp}^{\mathrm{T}} \; \boldsymbol{F}_{R1}^{\mathrm{T}} \,\dots\dots\dots\, \boldsymbol{F}_{Rp}^{\mathrm{T}}]^{\mathrm{T}} , \\ \overline{\boldsymbol{J}} &= [\boldsymbol{J}_{T1}^{\mathrm{T}} \,\dots\dots\dots\, \boldsymbol{J}_{Tp}^{\mathrm{T}} \; \boldsymbol{J}_{R1}^{\mathrm{T}} \,\dots\dots\dots\, \boldsymbol{J}_{Rp}^{\mathrm{T}}]^{\mathrm{T}} . \end{aligned} \tag{2.75}$$

Then, one gets from (2.74) simply

$$\overline{\boldsymbol{Q}}^{\mathrm{T}}\overline{\boldsymbol{J}} = \boldsymbol{0}\,, \quad \overline{\boldsymbol{J}}^{\mathrm{T}}\overline{\boldsymbol{Q}} = \boldsymbol{0}\,, \tag{2.76}$$

what clearly shows the generalized orthogonality between motion and constraint. The orthogonality condition or the vanishing virtual work, respectively, is independent from the coordinates chosen, and it is valid for all constrained mechanical systems.

D'Alembert's principle (1743) follows now from the Newton-Euler equations (2.57) and (2.58) after subdividing the external forces

$$\boldsymbol{f}_i = \boldsymbol{f}_i^{(e)} + \boldsymbol{f}_i^{(r)}, \quad \boldsymbol{l}_i = \boldsymbol{l}_i^{(e)} + \boldsymbol{l}_i^{(r)}, \tag{2.77}$$

and considering the orthogonality (2.70) as

$$\sum_{i=1}^{p} [(m_i\dot{\boldsymbol{v}}_i - \boldsymbol{f}_i^{(e)})^{\mathrm{T}}\delta\boldsymbol{r}_i + (\boldsymbol{I}_i\dot{\boldsymbol{\omega}}_i + \tilde{\boldsymbol{\omega}}_i\boldsymbol{I}_i\boldsymbol{\omega}_i - \boldsymbol{l}_i^{(e)})^{\mathrm{T}}\delta\boldsymbol{s}_i] = 0\,. \tag{2.78}$$

Obviously, the reaction forces are eliminated in (2.78).

Analogously Jourdain's principle (1908) can be stated which is based on the fact that the virtual power of the reaction forces is vanishing, too,

$$\delta P^r = \sum_{i=1}^{p} [\boldsymbol{f}_i^{(r)\mathrm{T}}\delta'\boldsymbol{v}_i + \boldsymbol{l}_i^{(r)\mathrm{T}}\delta'\boldsymbol{\omega}_i] = 0\,. \tag{2.79}$$

The virtual velocities $\delta'\boldsymbol{v}_i\,, \delta'\boldsymbol{\omega}_i$ are arbitrary, infinitesimal small variations of the velocities completely compatible with the constraints at any time and at any position. Thus, it yields

$$\delta'\boldsymbol{v}_i \neq 0\,, \quad \delta'\boldsymbol{\omega}_i \neq 0\,, \quad \delta'\boldsymbol{r}_i \equiv 0\,, \quad \delta'\boldsymbol{s}_i \equiv 0\,, \quad \delta' t \equiv 0\,. \tag{2.80}$$

Moreover, the symbol δ' follows the rules of calculus. Then, it remains Jourdain's principle as

$$\sum_{i=1}^{p} [(m_i\dot{\boldsymbol{v}}_i - \boldsymbol{f}_i^{(e)})^{\mathrm{T}}\delta'\boldsymbol{v}_i + (\boldsymbol{I}_i\dot{\boldsymbol{\omega}}_i + \tilde{\boldsymbol{\omega}}_i\boldsymbol{I}_i\boldsymbol{\omega}_i - \boldsymbol{l}_i^{(e)})^{\mathrm{T}}\delta'\boldsymbol{\omega}_i] = 0\,. \tag{2.81}$$

Similar to d'Alembert's principle all the reactions disappeared. However, the virtual displacements are replaced by the virtual velocities and the sometimes cumbersome evaluation of the virtual rotations is dropped. Further, Jourdain's principle handles nonlinear and nonholonomic constraints, too, which may appear in controlled vehicle systems.

In the American literature Jourdain's principle is referred to as Kane's equations and the virtual velocities are denoted as partial velocities, see Kane and Levinson (1985).

Applying D'Alembert's or Jourdain's principles for the generation of the equations of motion, the reactions have not to be considered at all.

2.3.4 Energy Considerations and Lagrange's Equations

An alternative for the generation of the equations of motion is the analytical method by Lagrange (1788) based on energy considerations. The kinetic energy T of a rigid body reads as

$$T = \frac{1}{2} m \boldsymbol{v}_C^2 + \frac{1}{2} \boldsymbol{\omega}^T \boldsymbol{I}_C \boldsymbol{\omega} , \tag{2.82}$$

where the inertia properties $\{ m, \boldsymbol{I}_C \}$ and the velocity properties $\{ \boldsymbol{v}_C, \boldsymbol{\omega} \}$ are related to the center of mass. The kinetic energy is composed by the translational and rotational energy of the body, it is a scalar quantity which may be computed in different frames, too.

The kinetic energy of a multibody system consisting of the bodies K_i, $i = 1(1)p$, comprises the kinetic energy of all bodies as

$$T = \frac{1}{2} \sum_{i=1}^{p} [(\boldsymbol{v}_i^I)^{\mathrm{T}} m_i \boldsymbol{v}_i^I + (\boldsymbol{\omega}_i^I)^{\mathrm{T}} \boldsymbol{I}_i^I \boldsymbol{\omega}_i^I] , \tag{2.83}$$

written consistently in the inertial frame I and related to the center of mass C_i of each body K_i. If the work of the applied forces is independent of the path, then the forces have a potential U and it yields

$$\boldsymbol{f}^{(e)} = -\mathrm{grad}\, U , \tag{2.84}$$

where U is a scalar function of the position. Forces satisfying (2.84) are called conservative, they do not change the total energy of the system. In contrary, non-conservative forces change the total energy, they are called dissipative if the total energy is decreasing. Conservative forces may be due to gravity, $f_G = mg$, or elasticity, $f_F = -ks$. The corresponding potentials read as

$$U_G = -m\, g\, z , \quad U_F = \frac{1}{2} k s^2 , \tag{2.85}$$

where z represents the vertical displacement of the center of mass of a body with mass m in the direction of gravity with acceleration g, and s means the displacement of an elastic spring with coefficient k. A constant may be added to the potentials, i.e., the origin of a potential can be arbitrarily chosen. The potential energy U of a multibody system is given by the sum of the body potentials $U = \sum U_j$. Multibody systems subject to conservative forces only are called conservative systems. For such systems it yields the energy conservation law

$$T + U = T_0 + U_0 = \mathrm{const} . \tag{2.86}$$

The energy conservation law may be derived from Newton's and Euler's law, too, i.e. it does not contain any new information. Its application is advantageous for conservative systems with one degree of freedom to evaluate a relation between the position and velocity variable. If there are two different positions known, the unknown velocity can be found from (2.86).

Based on energy expressions, the equations of motion of multibody systems may also be found. This will be shown for multibody systems with holonomic constraints. In contrary to the Newton-Euler equations, the bodies of the system have not to be dismantled, the system is considered as a whole. For this purpose the generalized coordinates $y_k\,(t)\,,\; k = 1(1)f$, are defined, and the position and the velocity variables (2.38), (2.39) and (2.40) are evaluated. As a result the kinetic energy is available as a function of $y_k\,(t)$ and $\dot{y}_k\,(t)\,,\; k = 1(1)f$,

$$T = T(y_k, \dot{y}_k)\;. \tag{2.87}$$

The applied forces and torques are projected in the direction of the generalized coordinates and composed to the generalized forces

$$q_k = \sum_{i=1}^{p} \left[\left(\frac{\partial \boldsymbol{r}_i^I}{\partial y_k} \right)^{\mathrm{T}} \boldsymbol{f}_i^{(e)I} + \left(\frac{\partial \boldsymbol{s}_i^I}{\partial y_k} \right)^{\mathrm{T}} \boldsymbol{l}_i^{(e)I} \right]\,, \quad k = 1(1)f\;, \tag{2.88}$$

where the Jacobian matrices (2.42) are used. The generalized forces may be also found by a decomposition of the total work of the applied forces and torques

$$\delta W^e = \sum_{i=1}^{p} [\boldsymbol{f}_i^{(e)\mathrm{T}} \delta \boldsymbol{r}_i + \boldsymbol{l}_i^{(e)\mathrm{T}} \delta \boldsymbol{s}_i] = \sum_{k=1}^{f} q_k \delta y_k\;. \tag{2.89}$$

In any case the reaction forces and torques do not appear.

Now the Lagrangian equations of the second kind read as, see e.g. Magnus and Mueller-Slany (2005),

$$\frac{\mathrm{d}}{\mathrm{d}t} \left(\frac{\partial T}{\partial \dot{y}_k} \right) - \frac{\partial T}{\partial y_k} = q_k\;, \quad k = 1(1)f\;. \tag{2.90}$$

For the evaluation of the equations of motion two partial and one total differentiations have to be performed with respect to one scalar function $T\,(y_k, \dot{y}_k)$. As a result the minimal number f of equations of motion is found. However, the reaction forces are completely lost and cannot be regained.

For conservative systems the generalized forces follow immediately from the potential energy

$$q_k = -\frac{\partial U}{\partial y_k}\;. \tag{2.91}$$

From (2.90) and (2.91) it remains

$$\frac{\mathrm{d}}{\mathrm{d}t} \left(\frac{\partial T}{\partial \dot{y}_k} \right) - \frac{\partial T}{\partial y_k} + \frac{\partial U}{\partial y_k} = 0\,, \quad k = 1(1)f\;. \tag{2.92}$$

Introducing the Lagrange function $L = T - U$ also called the kinetic potential, equations (2.92) are even more simplified and reads

$$\frac{\mathrm{d}}{\mathrm{d}t}\left(\frac{\partial L}{\partial \dot{y}_k}\right) - \frac{\partial L}{\partial y_k} = 0 \, , \quad L = T - U \, , \quad k = 1(1)f \, . \tag{2.93}$$

For some engineering applications it is advantageous to use surplus coordinates $\overline{y}_j \, , \, j = 1(1)f + r$, in addition to the f generalized coordinates $y_k \, , \, k = 1(1)f$. Then, there exist r geometric constraints between the surplus coordinates

$$\varphi_n = \varphi_n(\overline{y}_j) = 0 \, , \quad n = 1(1)r \, . \tag{2.94}$$

The equations of motion are now extended by r Lagrangian multipliers $\lambda_n \, , n = 1(1)r$, representing generalized constraint forces

$$\frac{\mathrm{d}}{\mathrm{d}t}\left(\frac{\partial T}{\partial \dot{\overline{y}}_j}\right) - \frac{\partial T}{\partial \overline{y}_j} = \overline{q}_j + \sum_{n=1}^{r} \lambda_n \frac{\partial \varphi_n}{\partial \overline{y}_j} \, , \quad j = 1(1)f + r \, . \tag{2.95}$$

If (2.94) is replaced by (2.34) the Newton-Euler equations are obtained again.

■ Problem 2.11 Lagrangian equations of motion for a differential gear

The differential gear shown in Fig. 2.25 is used for road vehicles to compensate for the differences of the rotation angles of the wheels during cornering. It consists of the driving wheel 1, the crown wheel 2, the driven wheels 3 and 4 as well as of the compensation wheels 5 and 6. The moments of inertia with respect to the corresponding rotation axes including the connected rotating parts are I_i, $i = 1(1)6$. Wheel 1 is driven by the torque M_1, the wheels 3 and 4 are loaded by the torques M_3 and M_4, respectively. Generate Lagrange's equations for the given gear and discuss

a) the case of uniform motion and
b) the case of a suddenly changing load.

Solution

The generation of the equations of motion using Lagrange's equation of the second kind is carried out in five steps.

Step 1: Definition of the number of degrees of freedom and choice of the generalized coordinates. The given differential gear has $f = 2$ degrees of freedom. This is obvious from locking of one degree of freedom. If for example the driven wheel 4 is locked, then there remains a gear with one degree of freedom only. For two non-locked wheels the gear has two degrees of freedom. As generalized coordinates the required rotation angles φ_3 and φ_4 of the driven wheels are chosen. The remaining rotation angles are subject to the following constraints where $a_j, j = 1, 2$ are the corresponding transmission ratios,

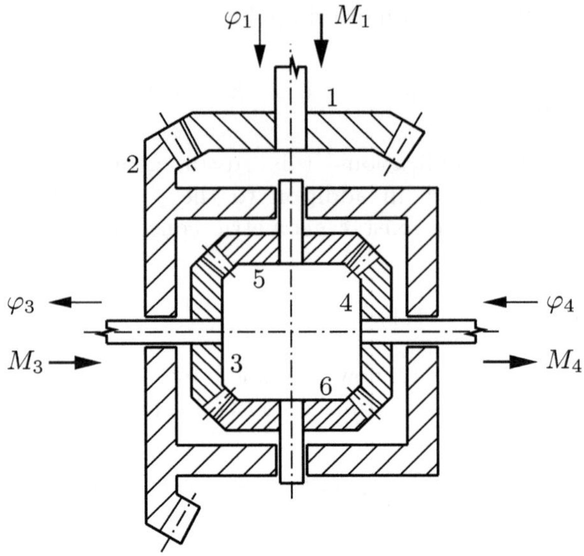

Fig. 2.25. Differential gear

$$
\begin{aligned}
\varphi_1 &= \frac{a_1}{2}(\varphi_3 + \varphi_4)\,, \\
\varphi_2 &= \frac{1}{2}(\varphi_3 + \varphi_4)\,, \\
\varphi_5 &= a_2(\varphi_3 - \varphi_4)\,, \\
\varphi_6 &= a_2(\varphi_4 - \varphi_3)\,.
\end{aligned} \tag{1}
$$

In the case of driving straight ahead it yields $\varphi_2 = \varphi_3 = \varphi_4 = \varphi_1/a_1$. Then the compensation wheels do not rotate around their own axes, $\varphi_5 = \varphi_6 = 0$ even if they are moving together with the crown wheel 2.

Step 2: Computation of the kinetic energy T. In this problem, the kinetic energy is only due to the rotation of the wheels,

$$
T = \frac{1}{2}\sum_{i=1}^{6} I_i \dot{\varphi}_i^2\,. \tag{2}
$$

Using the constraints (1) the kinetic energy is found depending only on the generalized angular velocities $\dot{\varphi}_3$ and $\dot{\varphi}_4$ as

$$
\begin{aligned}
T = \frac{1}{2}\Big[& I_1 \frac{a_1^2}{4}(\dot{\varphi}_3 + \dot{\varphi}_4)^2 + I_2 \frac{1}{4}(\dot{\varphi}_3 + \dot{\varphi}_4)^2 + I_3\dot{\varphi}_3^2 + I_4\dot{\varphi}_4^2 \\
& + I_5 a_2^2(\dot{\varphi}_3 - \dot{\varphi}_4)^2 + I_6 a_2^2(\dot{\varphi}_4 - \dot{\varphi}_3)^2\Big]\,.
\end{aligned} \tag{3}
$$

Due to symmetry of the design $I_3 = I_4$ and $I_5 = I_6$, it remains from (3)

$$T = \frac{1}{2}A(\dot{\varphi}_3^2 + \dot{\varphi}_4^2) + B\dot{\varphi}_3\dot{\varphi}_4 , \tag{4}$$

where the abbreviations

$$A = \frac{I_1 a_1^2}{4} + \frac{I_2}{4} + I_3 + 2I_5 a_2^2 , \tag{5}$$

$$B = \frac{I_1 a_1^2}{4} + \frac{I_2}{4} - 2I_5 a_2^2 \tag{6}$$

are introduced. It holds obviously $A > B$.

Step 3: Computation of the potential energy U or the generalized forces q_k, respectively. The generalized forces are easily obtained considering the virtual work performed by the applied torques acting on the differential gear

$$\delta W^e = M_1\delta\varphi_1 - M_3\delta\varphi_3 - M_4\delta\varphi_4 , \tag{7}$$

where the driving work is positive and the work of the driven wheels is negative. With the constraint $\varphi_1 = (\varphi_3 + \varphi_4)a_1/2$ from (1) it follows

$$\begin{aligned}\delta W^e &= M_1\frac{a_1}{2}(\delta\varphi_3 + \delta\varphi_4) - M_3\delta\varphi_3 - M_4\delta\varphi_4 \\ &= \underbrace{\left(M_1\frac{a_1}{2} - M_3\right)}_{q_3}\delta\varphi_3 + \underbrace{\left(M_1\frac{a_1}{2} - M_4\right)}_{q_4}\delta\varphi_4 .\end{aligned} \tag{8}$$

Thus, the generalized torques read as

$$q_3 = M_1\frac{a_1}{2} - M_3 , \quad q_4 = M_1\frac{a_1}{2} - M_4 . \tag{9}$$

Step 4: Evaluation of the Lagrangian equations of motion. From the fundamental equation (2.90) one gets

$$\frac{\mathrm{d}}{\mathrm{d}t}\left(\frac{\partial T}{\partial\dot{\varphi}_k}\right) - \frac{\partial T}{\partial\varphi_k} = q_k , \quad k = 3,4 . \tag{10}$$

Then, it follows considering (4) and (9) immediately for the equations of motion

$$A\ddot{\varphi}_3 + B\ddot{\varphi}_4 = \frac{1}{2}M_1 a_1 - M_3 , \tag{11}$$

$$B\ddot{\varphi}_3 + A\ddot{\varphi}_4 = \frac{1}{2}M_1 a_1 - M_4 , \tag{12}$$

where the applied torques, $M_j = M_j(\dot{\varphi}_j)$, $j = 1,3,4$, usually depend on the angular velocity.

Step 5: Solution of the equations of motion and discussion. Even if the solution of the equations of motion (11), (12) for given torques $M_j = M_j(\dot{\varphi}_j)$

requires in general numerical time integration, the question stated above can be answered.

a) A uniform motion (stationary driving straight ahead) results in constant angular velocities $\dot{\varphi}_3 = \dot{\varphi}_{3,0} = \text{const}$, $\dot{\varphi}_4 = \dot{\varphi}_{4,0} = \text{const}$. Then, the angular accelerations are vanishing and from (11) and (12) the torque relations

$$M_{3,0} = M_{4,0} = \frac{1}{2} M_{1,0} a_1 \tag{13}$$

are obtained.

b) The consequences of a sudden load change can be estimated if it occurs during uniform motion. Then, (11) and (12) provide for the angular accelerations the following relations

$$\ddot{\varphi}_3 = \frac{\frac{1}{2} M_1 a_1 (A - B) - A M_3 + B M_4}{A^2 - B^2}, \tag{14}$$

$$\ddot{\varphi}_4 = \frac{\frac{1}{2} M_1 a_1 (A - B) - A M_4 + B M_3}{A^2 - B^2}. \tag{15}$$

For a uniform motion ($M_j = M_{j0}$, $j = 1, 3, 4$) the left hand side and the right hand side of (14) and (15) are vanishing. If there is a sudden load change

$$M_3 = M_{3,0} - \Delta M_3 \tag{16}$$

occurring, e.g. by a sudden lowering of the friction at the tire connected to the drive wheel 3, then it follows from (14), (15)

$$\Delta\ddot{\varphi}_3 = \frac{A \Delta M_3}{A^2 - B^2}, \quad \Delta\ddot{\varphi}_4 = \frac{-B \Delta M_3}{A^2 - B^2}. \tag{17}$$

Thus, due to the fact $A^2 > B^2$, the driven wheel 3 is accelerated what may result in a wheelspin of the tire at wheel 3, and the driven wheel 4 is decelerated what may result in a locking of the tire at wheel 4. For very strong load changes of this kind a vehicle may start skidding. ■

2.4 Equations of Motion for Multibody Systems

In Sect. 2.3 there has been presented two methods for the generation of the equations of motion, the synthetical method by Newton-Euler, and the analytical method by Lagrange, respectively. The principal steps in the generation process by the methods of Newton-Euler and Lagrange are shown in Fig. 2.26. Common starting point is a mechanical model of the vehicle composed of the elements of multibody systems. Common result are the equations of motion, they are identical with both methods if the same generalized coordinates are chosen. However, the effort is different. During the generation of the equations of motion using Lagrange's equations there appear terms in $\frac{\mathrm{d}}{\mathrm{d}t}\left(\frac{\partial T}{\partial \dot{y}_i}\right)$

which are afterwards eliminated by $\dfrac{\partial T}{\partial y_i}$ according to (2.90). This means a useless computational effort which is not required with the Newton-Euler approach, see e.g. Schiehlen and Eberhard (2004). On the other hand in the Newton-Euler equations the reactions have to be eliminated. Thus, both of the two approaches have disadvantages which are avoided by a combination of the Newton-Euler equations with the principles presented in Sect. 2.3.3. In any case, the resulting equations of motion are always ordinary differential equations (ODEs). However, their form depends on the type of the multibody system. There are ideal and non-ideal systems, the first ones are characterized by applied forces and torques independent from any reaction while the second ones show a such dependency. E.g., gravitational forces, spring and damper forces are independent from any reaction while sliding friction forces and slip dependent contact forces, regularly found with tires in vehicle dynamics, are a function of the normal or reaction forces, respectively.

Within the class of ideal systems, ordinary and general multibody systems are distinguished. Ordinary multibody systems are due to holonomic constraints and applied forces depending only on position and velocity quantities, they can be always represented by a system of differential equations of the second order. For nonholonomic constraints and/or general force laws one gets general multibody systems.

The equations of ordinary multibody systems read as

$$\boldsymbol{M}(\boldsymbol{y},t)\,\ddot{\boldsymbol{y}}(t) + \boldsymbol{k}(\boldsymbol{y},\dot{\boldsymbol{y}},t) = \boldsymbol{q}(\boldsymbol{y},\dot{\boldsymbol{y}},t)\ , \tag{2.96}$$

where $\boldsymbol{y}$ is the $f \times 1$-position vector of the generalized coordinates, $\boldsymbol{M}$ is the $f \times f$-symmetric inertia matrix, $\boldsymbol{k}$ is a $f \times 1$-vector of generalized gyroscopic forces including the Coriolis and centrifugal forces as well as the gyroscopic torques, and the $f \times 1$-vector $\boldsymbol{q}$ represents generalized applied forces. The equations of motion resulting from Lagrange's method have always the form (2.96) while the Newton-Euler method may require some calculations to get a symmetric inertia matrix.

In vehicle dynamics the deviations $\tilde{\boldsymbol{y}}(t)$ from a reference motion $\boldsymbol{y} = \boldsymbol{y}_R(t)$ are often small,

$$\boldsymbol{y}(t) = \boldsymbol{y}_R(t) + \tilde{\boldsymbol{y}}(t)\ . \tag{2.97}$$

Then, one obtains by a Taylor series expansion under the assumption of differentiable vector functions and skipping of the second and higher order terms from (2.96) the linearized equations of motion, see also Mueller and Schiehlen (1985),

$$\boldsymbol{M}(t)\ddot{\tilde{\boldsymbol{y}}}(t) + \boldsymbol{P}(t)\dot{\tilde{\boldsymbol{y}}}(t) + \boldsymbol{Q}(t)\tilde{\boldsymbol{y}}(t) = \boldsymbol{h}(t)\ , \tag{2.98}$$

where $\boldsymbol{M}(t)$ is the symmetric, positive definite inertia matrix while $\boldsymbol{P}(t)$ and $\boldsymbol{Q}(t)$ characterize the velocity and position dependent forces and the vector $\boldsymbol{h}(t)$ represents the external excitations. If all these matrices are time-invariant

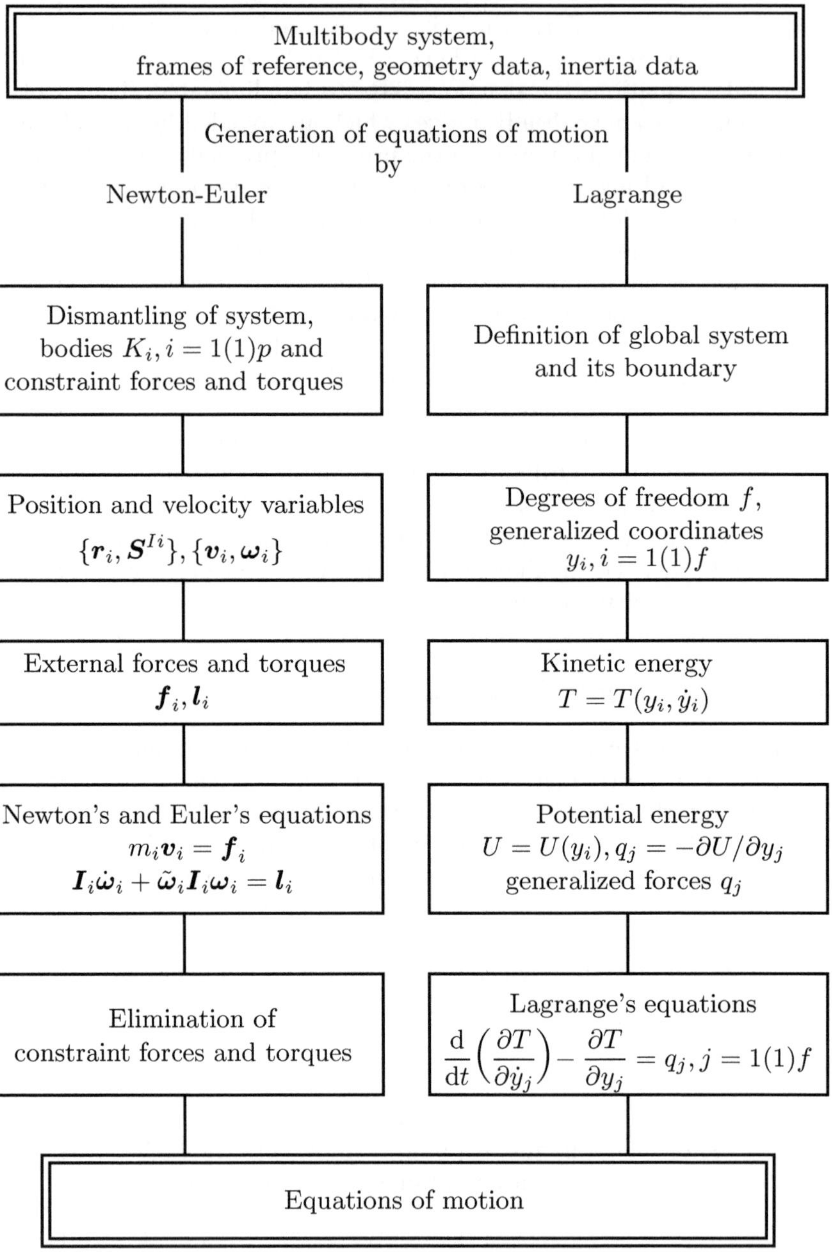

Fig. 2.26. Generation of equations of motion by the methods of Newton-Euler and Lagrange

and subdivided in a symmetrical and skew-symmetrical part, then the equations of motion of a linear ordinary and time-invariant multibody system are found reading as

$$\boldsymbol{M}\ddot{\boldsymbol{y}}(t) + (\boldsymbol{D} + \boldsymbol{G})\dot{\boldsymbol{y}}(t) + (\boldsymbol{K} + \boldsymbol{N})\boldsymbol{y}(t) = \boldsymbol{h}(t) \; , \tag{2.99}$$

where $\tilde{\boldsymbol{y}}$ was simply replaced by $\boldsymbol{y}$ and the $f \times f$-matrices have the properties

$$\boldsymbol{M} = \boldsymbol{M}^{\mathrm{T}} > 0, \; \boldsymbol{D} = \boldsymbol{D}^{\mathrm{T}}, \; \boldsymbol{G} = -\boldsymbol{G}^{\mathrm{T}}, \; \boldsymbol{K} = \boldsymbol{K}^{\mathrm{T}}, \; \boldsymbol{N} = -\boldsymbol{N}^{\mathrm{T}}. \tag{2.100}$$

These matrices have a physical meaning which can be identified after premultiplication of (2.99) from the left by $\dot{\boldsymbol{y}}^{\mathrm{T}}$ resulting in the total time derivative of an energy expression

$$\underbrace{\dot{\boldsymbol{y}}^{\mathrm{T}}\boldsymbol{M}\ddot{\boldsymbol{y}}}_{\frac{\mathrm{d}}{\mathrm{d}t}T} + \underbrace{\dot{\boldsymbol{y}}^{\mathrm{T}}\boldsymbol{D}\dot{\boldsymbol{y}}}_{2R} + \underbrace{\dot{\boldsymbol{y}}^{\mathrm{T}}\boldsymbol{G}\dot{\boldsymbol{y}}}_{0} + \underbrace{\dot{\boldsymbol{y}}^{\mathrm{T}}\boldsymbol{K}\boldsymbol{y}}_{\frac{\mathrm{d}}{\mathrm{d}t}U} + \underbrace{\dot{\boldsymbol{y}}^{\mathrm{T}}\boldsymbol{N}\boldsymbol{y}}_{2S} = \underbrace{\dot{\boldsymbol{y}}^{\mathrm{T}}\boldsymbol{h}}_{P} \; , \tag{2.101}$$

The inertia matrix $\boldsymbol{M}$ determines the kinetic energy $T = \frac{1}{2}\dot{\boldsymbol{y}}^{\mathrm{T}}\boldsymbol{M}\,\dot{\boldsymbol{y}}$ and therefore the inertia forces, from $T > 0$ it follows again the positive definiteness of the inertia matrix. The damping matrix $\boldsymbol{D}$ defines via Rayleigh's dissipation function $R = \frac{1}{2}\dot{\boldsymbol{y}}^{\mathrm{T}}\boldsymbol{D}\,\dot{\boldsymbol{y}}$ the damping forces while the gyro matrix $\boldsymbol{G}$ describes the gyroscopic forces which do not change the total energy of the system. The stiffness matrix determines the potential energy $U = \frac{1}{2}\boldsymbol{y}^{\mathrm{T}}\boldsymbol{K}\,\boldsymbol{y}$ and, therefore, the conservative position forces while the matrix $\boldsymbol{N}$ identifies the circulatory forces also known as nonconservative position forces. Furthermore, S represents the power of the circulatory forces and P describes the power of the external excitation forces. For $\boldsymbol{D} = \boldsymbol{0}, \;\; \boldsymbol{N} = \boldsymbol{0}$ and $\boldsymbol{h} = \boldsymbol{0}$ the multibody system is conservative, i.e., the total energy is constant for all motions,

$$T + U = \text{const} \; . \tag{2.102}$$

The matrix properties (2.100) allow often to check the equations of motion with respect to the physical phenomena involved.

The principal structure of the equation of motion of general MBS is different from (2.96) and (2.99), respectively. Systems with nonholonomic constraints are treated by Schiehlen and Eberhard (2004) and will be in Sect. 9.1.2 exemplified. Examples for general force laws are considered in Chap. 3.

■ Problem 2.12 Equations of motion for the bounce and pitch vibrations of an automobile

The three-body system shown in Fig. 2.27 is qualified for the analysis of the low frequency bounce and pitch vibrations of an automobile in the vertical plane. The vehicle body is modeled as a rigid body while the axles together with the wheels are considered as particles, they are connected with each other by struts consisting of spring and damper. The support on the guideway is handled by the tires which are modeled as spring-damper-system, too. It is assumed that only small pitch motions appear, and the springs and dampers have a linear characteristic. In the equilibrium position the vehicle body shall be completely horizontal. The stiffness of the tires is much larger than the stiffness of the suspension springs. Therefore, the vehicle body is called the sprung mass while the wheels are denoted as unsprung mass. The horizontal guideway is characterized by an uneven profile ζ, the vehicle is travelling with constant speed. For this vehicle model the equations of motion have to be specified.

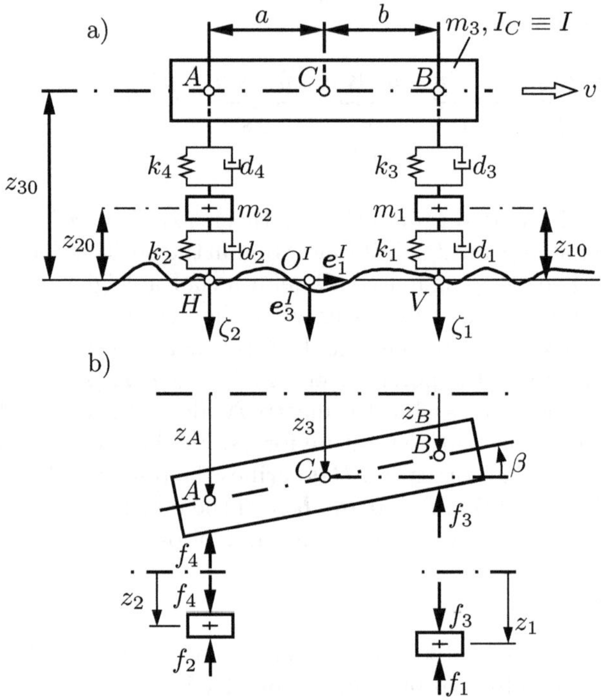

Fig. 2.27. Three body system of an autombile: a) equilibrium position ; b) free body diagram

Solution

All three masses show vertical bouncing motions while the vehicle body in addition shows a pitching motion. All motions are restricted to the vertical $\mathbf{e}_1^I, \mathbf{e}_3^I$-plane, see Fig. 2.27 a). The system has $f = 4$ degrees of freedom characterized by the four generalized coordinates z_1, z_2, z_3 and β. They describe the deviations of the bodies from their reference position which coincides with the equilibrium position on an even road. In the equilibrium position the gravity forces acting on the bodies are compensated by preloading forces of the springs. Therefore, in the free body diagram Fig. 2.27 b) neither gravity forces nor preloading forces of the springs are shown and the coordinates are measured from the equilibrium position indicated by the dot and dash line. Newton's and Euler's equations read for the bodies K_i, $i = 1(1)3$, using the notation of the system parameters in Fig. 2.27 a) as

$$i = 1: \quad m_1 \ddot{z}_1 = -f_1 + f_3 \,, \tag{1}$$
$$i = 2: \quad m_2 \ddot{z}_2 = -f_2 + f_4 \,, \tag{2}$$
$$i = 3: \quad m_3 \ddot{z}_3 = -f_3 - f_4 \,, \tag{3}$$
$$I \ddot{\beta} = b f_3 - a f_4 \,. \tag{4}$$

The corresponding spring and damper forces of the suspensions and the wheels are given by

$$f_j = k_j(z_j - \zeta_j) + d_j(\dot{z}_j - \dot{\zeta}_j) \,, \quad j = 1, 2 \,, \tag{5}$$
$$f_3 = k_3(z_B - z_1) + d_3(\dot{z}_B - \dot{z}_1) \,, \tag{6}$$
$$f_4 = k_4(z_A - z_2) + d_4(\dot{z}_A - \dot{z}_2) \,. \tag{7}$$

The kinematical relation between the displacements z_A, z_B and the coordinates z_3, β follow for small pitch angles $\beta \le 1$, Fig. 2.27 b) as

$$z_A = z_3 + a\beta \,, \quad z_B = z_3 - b\beta \,: \tag{8}$$

Adopting (5)-(8) to (1)-(4) one gets the four equations of motion

$$m_1 \ddot{z}_1 + k_1 z_1 + d_1 \dot{z}_1 - k_3(z_3 - b\beta - z_1) - d_3(\dot{z}_3 - b\dot{\beta} - \dot{z}_1) = k_1 \zeta_1 + d_1 \dot{\zeta}_1 \,, \tag{9}$$

$$m_2 \ddot{z}_2 + k_2 z_2 + d_2 \dot{z}_2 - k_4(z_3 + a\beta - z_2) - d_4(\dot{z}_3 + a\dot{\beta} - \dot{z}_2) = k_2 \zeta_2 + d_2 \dot{\zeta}_2 \,, \tag{10}$$

$$\begin{aligned} &m_3 \ddot{z}_3 + k_3(z_3 - b\beta - z_1) + d_3(\dot{z}_3 - b\dot{\beta} - \dot{z}_1) \\ &+ k_4(z_3 + a\beta - z_2) + d_4(\dot{z}_3 + a\dot{\beta} - \dot{z}_2) = 0 \,, \end{aligned} \tag{11}$$

$$\begin{aligned} &I\ddot{\beta} - bk_3(z_3 - b\beta - z_1) - bd_3(\dot{z}_3 - b\dot{\beta} - \dot{z}_1) \\ &+ ak_4(z_3 + a\beta - z_2) + ad_4(\dot{z}_3 + a\dot{\beta} - \dot{z}_2) = 0 \,. \end{aligned} \tag{12}$$

The equations of motion can be well arranged in matrix form by introduction of the column vector $\boldsymbol{y}$ of the generalized coordinates

$$\boldsymbol{y} = \begin{bmatrix} z_1 & z_2 & z_3 & \beta \end{bmatrix}^{\mathrm{T}} \,. \tag{13}$$

Then, it remains

$$\boldsymbol{M}\ddot{\boldsymbol{y}} + \boldsymbol{D}\dot{\boldsymbol{y}} + \boldsymbol{K}\boldsymbol{y} = \boldsymbol{h} \,, \tag{14}$$

with matrices and vectors

$$\begin{aligned}
&\boldsymbol{M} = \textbf{diag}\left[m_1\ m_2\ m_3\ I\right] \,, \quad \boldsymbol{h} = \left[k_1\zeta_1 + d_1\dot{\zeta}_1 \ \ k_2\zeta_2 + d_2\dot{\zeta}_2 \ \ 0 \ \ 0\right]^{\mathrm{T}} \,, \\
&\boldsymbol{D} = \begin{bmatrix} d_1+d_3 & 0 & -d_3 & bd_3 \\ 0 & d_2+d_4 & -d_4 & -ad_4 \\ -d_3 & -d_4 & d_3+d_4 & -bd_3+ad_4 \\ bd_3 & -ad_4 & -bd_3+ad_4 & b^2d_3+a^2d_4 \end{bmatrix} \,, \\
&\boldsymbol{K} = \begin{bmatrix} k_1+k_3 & 0 & -k_3 & bk_3 \\ 0 & k_2+k_4 & -k_4 & -ak_4 \\ -k_3 & -k_4 & k_3+k_4 & -bk_3+ak_4 \\ bk_3 & -ak_4 & -bk_3+ak_4 & b^2k_3+a^2k_4 \end{bmatrix} \,,
\end{aligned} \tag{15}$$

where the 4×4-matrices $\boldsymbol{M}$, $\boldsymbol{D}$, $\boldsymbol{K}$ describe the inertia, damping and stiffness properties of the automobile while the 4×1-column vector $\boldsymbol{h}$ characterizes the guideway unevenness.

For checking the equations of motions (14)-(17) the fact can be used that the matrices $\boldsymbol{M}$, $\boldsymbol{D}$, $\boldsymbol{K}$ have to be symmetric since gyroscopic and nonconservative forces do not appear in the automobile model. Furthermore, in this problem the matrices $\boldsymbol{K}$ and $\boldsymbol{D}$ have the same structure since all springs and dampers are placed in parallel. Using the same indices for the parallel placed springs and dampers it yields

$$\begin{aligned}
&\boldsymbol{K} = \boldsymbol{K}(k_i) \rightarrow \boldsymbol{D} = \boldsymbol{K}(d_i) \,, \\
&\boldsymbol{D} = \boldsymbol{D}(d_i) \rightarrow \boldsymbol{K} = \boldsymbol{D}(k_i) \,.
\end{aligned} \tag{16}$$

Such kind of intuitive checkings are most valuable for large multibody systems too, and they support the understanding of the mechanical background of the model chosen.

The total wheel loads $\boldsymbol{f}_V$, $\boldsymbol{f}_H$ acting at the front wheel (index V) and the rear wheel (index H) on the guideway are composed of the static loads f_{10}, f_{20} and the dynamic loads f_1, f_2, see (5) as

$$f_V(t) = f_{10} + f_1(t) \,, \quad f_H(t) = f_{20} + f_2(t) \,. \tag{17}$$

The static wheel loads follow from a statical analysis as

$$f_{10} = \left(m_1 + \frac{a}{a+b}m_3\right) g \,, \quad f_{20} = \left(m_2 + \frac{b}{a+b}m_3\right) g \,, \tag{18}$$

while the dynamic wheel loads depend on the solution of the equations of motion subject to the road unevenness. ■

2.5 Formalisms for Multibody Systems

The generation of equations of motion for large multibody systems is a nontrivial task requiring numerous steps during the evaluation of the fundamental relations. Beginning with the space age in the middle of the 1960s the generation of equations of motion was more formalized. The resulting formalisms were used for the development of computer codes for multibody systems, they are the basis of computational multibody dynamics. Twenty-five years later, in 1990, there were known 20 formalisms described in the Multibody System Handbook, Schiehlen (1990). Many of them are still used today.

Multibody system formalisms are based on Newton-Euler equations or Lagrange's equations, respectively, as described in Sect. 2.3 and Sect. 2.4. Regarding the computational procedure, numerical and symbolical formalisms are distinguished. Numerical formalisms supply the elements of the matrices as numbers in the case of linear time-invariant multibody systems (2.99). In the case of linear time-variant systems (2.98) and nonlinear systems (2.96) a numerical formalism provides the numbers in the equations of motion for each time step required by the simulation programme. In contrary, symbolical formalisms generate the equations of motion only once with the computer how it is done with paper and pencil. The advantage is that variations of the system parameters and, for time-variant systems, the running time have to be inserted in the symbolical equations of motion only. Symbolical formalisms are especially helpful for optimizations and control design.

Furthermore, non-recursive and recursive formalisms are distinguished. Recursive formalisms make use of special topology properties of multibody systems.

2.5.1 Non-recursive Formalisms

As an introduction the symbolical formalism Neweul is presented. Neweul is a research software based on the Newton-Euler equations and the principles of d'Alembert and Jourdain, it was developed at the University of Stuttgart in a Fortran based version, see e.g. Kreuzer and Leister (1988). More recently there is also the Maple-Matlab based version Neweul-M^2 available, see Kurz and Eberhard (2009). Neweul generates equations of motion in minimal form (2.96) or (2.99), respectively, which may be solved by any integration code for ordinary differential equations. The formalisms comprises five steps which may be evaluated by hand for smaller multibody systems, too.

Step 1: System specification and input data.
At first the multibody system is defined and treated as a whole. The number of degrees of freedom is determined and the generalized coordinates y_k, $k = 1(1)f$, are chosen. The inertial frame I and the body fixed frames i are defined. Each body K_i , $i = 1(1)p$, is dismantled, and the corresponding inertia parameters $\{ m_i, \boldsymbol{I}_i\}$, the position variables $\{ \boldsymbol{r}_i^I, \boldsymbol{S}^{Ii}\}$ as well as the

applied forces and torques $\{\boldsymbol{f}_i^{(e)I}, \boldsymbol{l}_i^{(e)I}\}$ are specified. The quantities $\boldsymbol{I}_i^I$ and $\boldsymbol{l}_i^{(e)I}$ are related to the corresponding center of mass C_i. The input data read as

$$f,\ \boldsymbol{y} = [y_1, \ldots\ldots, y_j, \ldots\ldots, y_f]^{\mathrm{T}}\ , \tag{2.103}$$

$$p,\ \{\ m_i, \boldsymbol{I}_i^i\}, \left\{\boldsymbol{r}_i^I, \boldsymbol{S}^{Ii}\right\}, \left\{\boldsymbol{f}_i^{(e)I}, \boldsymbol{l}_i^{(e)I}\right\}\ , \quad i = 1(1)p\ , \tag{2.104}$$

$$\boldsymbol{r}_i^I = \left[r_{i1}\, r_{i2}\, r_{i3}\right]^{\mathrm{T}}, \quad \boldsymbol{S}^{Ii} = \boldsymbol{S}^{Ii}(\alpha_i, \beta_i, \gamma_i) \equiv \boldsymbol{S}_i\ , \quad i = 1\,(1)\,p\ . \tag{2.105}$$

The rotation matrices are described by three angles, e.g. the Cardano angles $\alpha_i, \beta_i, \gamma_i$.

Step 2: Element consideration, local equations.
Now the elements of the inertia tensors are computed in the inertial frame I by the transformation

$$\boldsymbol{I}_i^I = \boldsymbol{S}^{Ii} \boldsymbol{I}_i^i \boldsymbol{S}^{iI}. \tag{2.106}$$

Then, there are all quantities available in the inertial frame I which is the only frame further used. Thus, the right upper index is no longer required and just skipped. The local equations of motion for each body K_i read as

$$m_i \dot{\boldsymbol{v}}_i = \boldsymbol{f}_i^{(e)} + \boldsymbol{f}_i^{(r)}\ , \quad i = 1(1)p\ , \tag{2.107}$$

$$\boldsymbol{I}_i \dot{\boldsymbol{\omega}}_i + \tilde{\boldsymbol{\omega}}_i \boldsymbol{I}_i \boldsymbol{\omega}_i = \boldsymbol{l}_i^{(e)} + \boldsymbol{l}_i^{(r)}\ , \qquad i = 1(1)p\ , \tag{2.108}$$

where the external forces and torques are subdivided in the known applied forces $f_i^{(e)}$ and torques $l_i^{(e)}$ and the unknown reactions $f_i^{(r)}$ and $l_i^{(r)}$, cp. Sect. 2.3.2 and 2.3.3. The reactions are eliminated later and, therefore, they have not to be specified for the generation of the equations of motion.

Step 3: Relation between local and global quantities.
The relation between the position (2.105) of a single body K_i and the generalized coordinates (2.103) is given by the holonomic and, in general, also rheonomic constraints. These relations are known from the input data as

$$\boldsymbol{r}_i = \boldsymbol{r}_i(\boldsymbol{y}, t)\ , \quad \boldsymbol{S}_i = \boldsymbol{S}_i(\boldsymbol{y}, t)\ , \quad i = 1(1)p\ , \tag{2.109}$$

and the corresponding velocities $\{\boldsymbol{v}_i, \boldsymbol{\omega}_i\}$ and accelerations $\{\boldsymbol{a}_i, \boldsymbol{\alpha}_i\}$ are computed as

$$\boldsymbol{v}_i = \dot{\boldsymbol{r}}_i = \frac{\partial \boldsymbol{r}_i}{\partial \boldsymbol{y}^{\mathrm{T}}} \dot{\boldsymbol{y}} + \frac{\partial \boldsymbol{r}_i}{\partial t} = \boldsymbol{J}_{Ti}(\boldsymbol{y}, t) \dot{\boldsymbol{y}} + \overline{\boldsymbol{v}}_i(\boldsymbol{y}, t)\ , \tag{2.110}$$

$$\boldsymbol{\omega}_i = \dot{\boldsymbol{s}}_i = \frac{\partial \boldsymbol{s}_i}{\partial \boldsymbol{y}^{\mathrm{T}}} \dot{\boldsymbol{y}} + \frac{\partial \boldsymbol{s}_i}{\partial t} = \boldsymbol{J}_{Ri}(\boldsymbol{y}, t) \dot{\boldsymbol{y}} + \overline{\boldsymbol{\omega}}_i(\boldsymbol{y}, t)\ , \tag{2.111}$$

$$\boldsymbol{a}_i = \dot{\boldsymbol{v}}_i = \boldsymbol{J}_{Ti}(\boldsymbol{y}, t) \ddot{\boldsymbol{y}} + \frac{\partial \boldsymbol{v}_i}{\partial \boldsymbol{y}^{\mathrm{T}}} \dot{\boldsymbol{y}} + \frac{\partial \boldsymbol{v}_i}{\partial t}\ , \tag{2.112}$$

$$\boldsymbol{a}_i = \dot{\boldsymbol{\omega}}_i = \boldsymbol{J}_{Ri}(\boldsymbol{y}, t) \ddot{\boldsymbol{y}} + \frac{\partial \boldsymbol{\omega}_i}{\partial \boldsymbol{y}^{\mathrm{T}}} \dot{\boldsymbol{y}} + \frac{\partial \boldsymbol{\omega}_i}{\partial t}\ . \tag{2.113}$$

For scleronomic constraints the partial time-derivatives are disappearing.

The $3 \times f$-Jacobian matrices $\boldsymbol{J}_{Ti}, \boldsymbol{J}_{Ri}$ of translation and rotation, respectively, present the relation between the local and global coordinates as shown in Sect. 2.2.4 by the virtual motion

$$\delta \boldsymbol{r}_i = \boldsymbol{J}_{Ti} \delta \boldsymbol{y} \,, \quad \delta \boldsymbol{s}_i = J_{Ri} \delta \boldsymbol{y} \,, \quad i = 1(1)p \,. \tag{2.114}$$

These matrices are available from kinematics by relations (2.110) and (2.111). If the velocities are known $\{\boldsymbol{v}_i(\boldsymbol{y}, \dot{\boldsymbol{y}}, t), \boldsymbol{\omega}_i(\boldsymbol{y}, \dot{\boldsymbol{y}}, t)\}$, the Jacobians may be also found from the virtual velocities where position and time are treated as frozen $(\delta r \equiv 0, \delta' v \neq 0, \delta t \equiv 0)$. then it yields, cp. (2.110) and (2.111),

$$\delta' \boldsymbol{v}_i = \boldsymbol{J}_{Ti} \delta' \dot{\boldsymbol{y}} \,, \quad \delta' \boldsymbol{\omega}_i = \boldsymbol{J}_{Ri} \delta' \dot{\boldsymbol{y}} \,, \quad i = 1(1)p \,. \tag{2.115}$$

After these preparatory computations the local equations (2.107) and (2.108) of all bodies K_i of the multibody system are evaluated as functions of the generalized coordinates and their derivatives.

Step 4: System consideration, global equations.
At first the local equations, depending on the generalized coordinates, are composed in global matrices and vectors. For this purpose the $6p \times 6p$-diagonal matrix $\overline{\overline{\boldsymbol{M}}}$ of the inertia quantities is introduced

$$\overline{\overline{\boldsymbol{M}}} = \mathbf{diag}[m_1 \boldsymbol{E}, m_2 \boldsymbol{E}, \ldots, m_p \boldsymbol{E}, \boldsymbol{I}_1, \boldsymbol{I}_2, \ldots, \boldsymbol{I}_p] \,, \tag{2.116}$$

where $\boldsymbol{E}$ means the 3×3-unit matrix. Further, the $6p \times 1$-vectors $\overline{\boldsymbol{q}}^{(e)}, \overline{\boldsymbol{q}}^{(r)}, \overline{\boldsymbol{k}}$ are used to summarize all applied, reaction, Coriolis and gyroscopic forces and torques. These three vectors are defined as follows

$$\overline{\boldsymbol{q}} = [\boldsymbol{f}_1^{\mathrm{T}}, \ldots\ldots, \boldsymbol{f}_p^{\mathrm{T}} \,, \boldsymbol{l}_1^{\mathrm{T}}, \ldots\ldots, \boldsymbol{l}_p^{\mathrm{T}}]^{\mathrm{T}} \,. \tag{2.117}$$

And finally the global $6p \times f$-Jacobian matrix is introduced as

$$\overline{\boldsymbol{J}} = [\boldsymbol{J}_{T1}^{\mathrm{T}}, \ldots\ldots, \boldsymbol{J}_{Tp}^{\mathrm{T}} \,, \boldsymbol{J}_{R1}^{\mathrm{T}}, \ldots\ldots, \boldsymbol{J}_{Rp}^{\mathrm{T}}]^{\mathrm{T}}. \tag{2.118}$$

The global Newton-Euler equations are now represented as one $6p \times 1$-vector equation

$$\overline{\overline{\boldsymbol{M}}} \, \overline{\boldsymbol{J}} \ddot{\boldsymbol{y}} + \overline{\boldsymbol{k}}(\boldsymbol{y}, \dot{\boldsymbol{y}}, t) = \overline{\boldsymbol{q}}^{(e)}(\boldsymbol{y}, \dot{\boldsymbol{y}}, t) + \overline{\boldsymbol{q}}^{(r)} \,. \tag{2.119}$$

These $6p$ equations are reduced to the minimal number of f ordinary differential equations by left pre-multiplication with the transposed $f \times 6p$-Jacobian matrix $\overline{\boldsymbol{J}}^{\mathrm{T}}$,

$$\overline{\boldsymbol{J}}^{\mathrm{T}} \overline{\overline{\boldsymbol{M}}} \, \overline{\boldsymbol{J}} \ddot{\boldsymbol{y}} + \overline{\boldsymbol{J}}^{\mathrm{T}} \overline{\boldsymbol{k}}(\boldsymbol{y}, \dot{\boldsymbol{y}}, t) = \overline{\boldsymbol{J}}^{\mathrm{T}} \overline{\boldsymbol{q}}^{(e)}(\boldsymbol{y}, \dot{\boldsymbol{y}}, t) \,, \tag{2.120}$$

where the term $\overline{\boldsymbol{J}}^{\mathrm{T}} \boldsymbol{q}^{(r)}$ is vanishing due to (2.70).

Summarizing the matrix products, one gets the equations of motion with a symmetric inertia matrix,

$$\begin{aligned} &\boldsymbol{M}(\boldsymbol{y},t)\ddot{\boldsymbol{y}}(t)+\boldsymbol{k}(\boldsymbol{y},\dot{\boldsymbol{y}},t)=\boldsymbol{q}(\boldsymbol{y},\dot{\boldsymbol{y}},t)\ , \\ &\boldsymbol{M}=\boldsymbol{M}^{\mathrm{T}}=\overline{\boldsymbol{J}}^{\mathrm{T}}\overline{\overline{\boldsymbol{M}}}\,\overline{\boldsymbol{J}}\ ,\quad \boldsymbol{k}=\overline{\boldsymbol{J}}^{\mathrm{T}}\overline{\boldsymbol{k}}\ ,\quad \boldsymbol{q}=\overline{\boldsymbol{J}}^{\mathrm{T}}\overline{\boldsymbol{q}}^{(e)}\ , \end{aligned} \tag{2.121}$$

in complete agreement with (2.96).

Optional **step 5:** Computation of the reaction forces and torques. In contrary to Lagrange's approach, the reactions can be regained if the global distribution matrix $\overline{\boldsymbol{Q}}$ according to (2.75) and the generalized constraint forces vector $\boldsymbol{g}$ from (2.71) are employed. Then, it yields

$$\overline{\boldsymbol{q}}^{(r)}=\overline{\boldsymbol{Q}}\boldsymbol{g}. \tag{2.122}$$

Due to the orthogonality condition (2.76) the left pre-multiplication of (2.119) with $\overline{\boldsymbol{Q}}^{\mathrm{T}}\overline{\overline{\boldsymbol{M}}}^{-1}$ results in linear, completely algebraic reaction equations

$$\begin{aligned} &\hat{\boldsymbol{N}}(\boldsymbol{y},t)\boldsymbol{g}(t)+\hat{\boldsymbol{q}}(\boldsymbol{y},\dot{\boldsymbol{y}},t)=\hat{\boldsymbol{k}}(\boldsymbol{y},\dot{\boldsymbol{y}},t)\ , \\ &\hat{\boldsymbol{N}}=\hat{\boldsymbol{N}}^{\mathrm{T}}=\overline{\boldsymbol{Q}}^{\mathrm{T}}\overline{\overline{\boldsymbol{M}}}^{-1}\overline{\boldsymbol{Q}}\ ,\quad \hat{\boldsymbol{q}}=\overline{\boldsymbol{Q}}^{\mathrm{T}}\overline{\overline{\boldsymbol{M}}}^{-1}\overline{\boldsymbol{q}}^{(e)}\ ,\quad \hat{\boldsymbol{k}}=\overline{\boldsymbol{Q}}^{\mathrm{T}}\overline{\overline{\boldsymbol{M}}}^{-1}\overline{\boldsymbol{k}}\ , \end{aligned} \tag{2.123}$$

where $\hat{\boldsymbol{N}}$ is the symmetric, generally positive definite $q\times q$-reaction matrix while the $q\times 1$-vectors $\hat{\boldsymbol{q}}$, $\hat{\boldsymbol{k}}$ show the influence of the applied and gyroscopic forces and torques on the reactions. On the other hand, by elimination of $\ddot{\boldsymbol{y}}$ in (2.119) one gets

$$\overline{\boldsymbol{q}}^{(r)}=\overline{\boldsymbol{Q}}\boldsymbol{g}=\left[\boldsymbol{E}-\overline{\overline{\boldsymbol{M}}}\,\overline{\boldsymbol{J}}(\overline{\boldsymbol{J}}^{\mathrm{T}}\overline{\overline{\boldsymbol{M}}}\,\overline{\boldsymbol{J}})^{-1}\overline{\boldsymbol{J}}^{\mathrm{T}}\right](\overline{\boldsymbol{k}}-\overline{\boldsymbol{q}}^{(e)})\ . \tag{2.124}$$

Then, by partitioning of (2.124) only the required constraint forces and torques may be computed.

In vehicle dynamics often contact forces are found resulting in nonideal multibody systems. Then, the f equations of motion (2.121) have to be solved simultaneously with the q equations of reaction (2.123).

The presented five steps show that the Neweul formalism is based on the Newton-Euler equations, however, they are supplemented by typical features of the analytical method like generalized coordinates and generalized reactions. The required computations include summation, multiplication and differentiation of vectors and matrices, simplification of trigonometrical expressions and linearization of expressions. These computations are symbolically performed in Neweul. But the underlying formalism may be also executed by any formula manipulation software like Maple or Matlab.

The most widely used numerical formalism in vehicle engineering is Adams Multibody Dynamics (2009). The software Adams generates numerical solutions of the original non-reduced Newton-Euler equations (2.57) and (2.58) rewritten as

$$\overline{\overline{M}}\ddot{\boldsymbol{x}} + \boldsymbol{k}(\boldsymbol{x}, \dot{\boldsymbol{x}}, t) = \overline{\boldsymbol{q}}^{(e)}(\boldsymbol{x}, \dot{\boldsymbol{x}}, t) + \overline{\boldsymbol{Q}}\boldsymbol{g} \ , \quad \overline{\boldsymbol{Q}}^{\mathrm{T}} = -\frac{\partial \boldsymbol{\varphi}(\boldsymbol{x}, t)}{\partial \boldsymbol{x}^{\mathrm{T}}} = -\boldsymbol{\varphi}_x \tag{2.125}$$

where the $6p \times 1$-vector $\boldsymbol{x}(t)$ is composed by Cartesian coordinates and the $6p \times q$-distribution matrix $\overline{\boldsymbol{Q}}$ of the reactions is found from the implicit constraints (2.34) written in matrix notation or (2.75), respectively. After two total time derivatives of (2.34) a set of $6p + q$ linear equations remains for the unknowns $\ddot{\boldsymbol{x}}$ and $\boldsymbol{g}$,

$$\begin{bmatrix} \overline{\overline{M}} & \boldsymbol{\varphi}_x^{\mathrm{T}} \\ \boldsymbol{\varphi}_x & \boldsymbol{0} \end{bmatrix} \begin{bmatrix} \ddot{\boldsymbol{x}} \\ \boldsymbol{g} \end{bmatrix} = \begin{bmatrix} \boldsymbol{q}^{(e)} - \boldsymbol{k} \\ -\boldsymbol{\varphi}_t - \dot{\boldsymbol{\varphi}}_x \dot{\boldsymbol{x}} \end{bmatrix} . \tag{2.126}$$

Here, $\overline{\overline{M}}$ is a blockdiagonal $6p \times 6p$-matrix which allows the application of sparse matrix techniques. Eqs. (2.126) represent differential-algebraical equations which are solved by Adams with special integration codes. As with all commercial codes a graphical interface is used, and mathematical equations are not to be handled. For more details search for Adams Multibody Dynamics (2009).

2.5.2 Recursive Formalisms

For time integration of holonomic systems the inertia matrix in (2.96) or (2.121), respectively, has to be inverted what is numerically costly for systems with many degrees of freedom,

$$\ddot{\boldsymbol{y}}(t) = \boldsymbol{M}^{-1}(\boldsymbol{y}, t)\left[\boldsymbol{q}(\boldsymbol{y}, \dot{\boldsymbol{y}}, t) - \boldsymbol{k}(\boldsymbol{y}, \dot{\boldsymbol{y}}, t)\right] . \tag{2.127}$$

Recursive algorithms avoid this matrix inversion. A fundamental requirement, however, is a chain or tree topology of the multibody system as shown in Fig. 2.28. Contributions on recursive algorithms without loop topologies are due, e.g., to Hollerbach (1980), Bae and Haug (1987a), Brandl et al. (1988), Schiehlen (1991), (2006).

2.5.2.1 Kinematics

Recursive kinematics uses the relative motion between two neighboring bodies and the related constraints as shown in Fig. 2.29. The absolute translational and rotational velocity vector $\boldsymbol{w_i}$ of body i, also denoted as twist, is related to the absolute velocity vector $\boldsymbol{w_{i-1}}$ of body $i-1$ and the generalized coordinates $\boldsymbol{y_i}$ of the joint i between this two bodies. It yields

$$\underbrace{\begin{bmatrix} \boldsymbol{v}_{Oi} \\ \boldsymbol{\omega}_i \end{bmatrix}}_{\boldsymbol{w}_i} = \underbrace{\boldsymbol{S}^{i,i-1} \begin{bmatrix} \boldsymbol{E} & -\tilde{\boldsymbol{r}}_{Oi-1,Oi} \\ \boldsymbol{0} & \boldsymbol{E} \end{bmatrix}}_{\boldsymbol{C}_i} \underbrace{\begin{bmatrix} \boldsymbol{v}_{Oi-1} \\ \boldsymbol{\omega}_{i-1} \end{bmatrix}}_{\boldsymbol{w}_{i-1}} + \underbrace{\boldsymbol{S}^{i,i-1} \begin{bmatrix} \boldsymbol{J}_{Ti} \\ \boldsymbol{J}_{Ri} \end{bmatrix}}_{\boldsymbol{J}_i} \dot{\boldsymbol{y}}_i \tag{2.128}$$

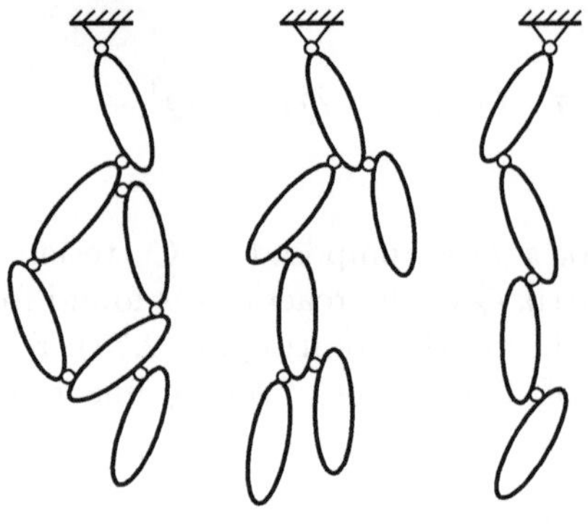

Fig. 2.28. Topologies of multibody systems

Using the fundamentals of relative motion of rigid bodies, it remains for the absolute acceleration

$$\boldsymbol{b}_i = \boldsymbol{C}_i \boldsymbol{b}_{i-1} + \boldsymbol{J}_i \ddot{\boldsymbol{y}}_i + \beta_i(\dot{\boldsymbol{y}}_i, \boldsymbol{w}_{i-1}) , \quad i = 1(1)p . \tag{2.129}$$

where the vector $\boldsymbol{b}_i$ summarizes the translational and rotational accelerations of body i as well.

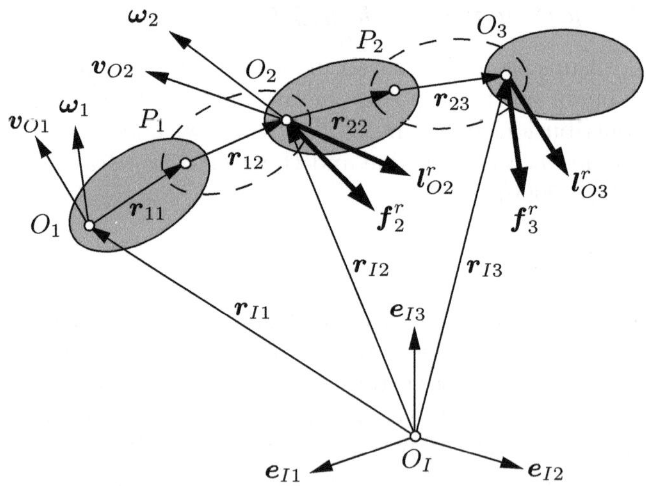

Fig. 2.29. Absolute twists and constraint wrenches for neighboring bodies

For the total system one gets for the absolute acceleration in matrix notation

$$\boldsymbol{b} = \boldsymbol{C}\,\boldsymbol{b} + \boldsymbol{J}\ddot{\boldsymbol{y}} + \boldsymbol{\beta} \tag{2.130}$$

where the geometry matrix $\boldsymbol{C}$ is a lower block-subdiagonal matrix and the Jacobian matrix $\boldsymbol{J}$ is a block-diagonal matrix as follows

$$\boldsymbol{C} = \begin{bmatrix} \boldsymbol{0} & \boldsymbol{0} & \boldsymbol{0} & \cdots & \boldsymbol{0} \\ \boldsymbol{C}_2 & \boldsymbol{0} & \boldsymbol{0} & \cdots & \boldsymbol{0} \\ \boldsymbol{0} & \boldsymbol{C}_3 & \boldsymbol{0} & \cdots & \boldsymbol{0} \\ \vdots & \vdots & \ddots & \ddots & \boldsymbol{0} \\ \boldsymbol{0} & \boldsymbol{0} & \boldsymbol{0} & \boldsymbol{C}_p & \boldsymbol{0} \end{bmatrix}, \quad \boldsymbol{J} = \begin{bmatrix} \boldsymbol{J}_1 & \boldsymbol{0} & \boldsymbol{0} & \cdots & \boldsymbol{0} \\ \boldsymbol{0} & \boldsymbol{J}_2 & \boldsymbol{0} & \cdots & \boldsymbol{0} \\ \boldsymbol{0} & \boldsymbol{0} & \boldsymbol{J}_3 & \cdots & \boldsymbol{0} \\ \vdots & \vdots & \ddots & \ddots & \vdots \\ \boldsymbol{0} & \boldsymbol{0} & \boldsymbol{0} & \boldsymbol{0} & \boldsymbol{J}_p \end{bmatrix}. \tag{2.131}$$

From (2.130) it follows the non-recursive form of the absolute accelerations as

$$\boldsymbol{b} = (\boldsymbol{E} - \boldsymbol{C})^{-1}\,\boldsymbol{J}\,\ddot{\boldsymbol{y}} + \overline{\boldsymbol{\beta}} \tag{2.132}$$

where the global Jabobian matrix $\overline{\boldsymbol{J}}$ is found again, see (2.75) or (2.118),

$$\overline{\boldsymbol{J}} = (\boldsymbol{E} - \boldsymbol{C})^{-1}\,\boldsymbol{J} = \begin{bmatrix} \boldsymbol{J}_1 & \boldsymbol{0} & \boldsymbol{0} & \cdots & \boldsymbol{0} \\ \boldsymbol{C}_2\boldsymbol{J}_1 & \boldsymbol{J}_2 & \boldsymbol{0} & \cdots & \boldsymbol{0} \\ \boldsymbol{C}_3\boldsymbol{C}_2\boldsymbol{J}_1 & \boldsymbol{C}_3\boldsymbol{J}_2 & \boldsymbol{J}_3 & \cdots & \boldsymbol{0} \\ \vdots & \vdots & \vdots & \ddots & \vdots \\ * & * & * & \cdots & \boldsymbol{J}_p \end{bmatrix}. \tag{2.133}$$

Due to the chain topology the global Jacobian matrix is a lower triangular matrix.

2.5.2.2 Newton-Euler Equations

Newton's and Euler's equations are now written for body i in its body-fixed frame at the joint position O_i using the absolute accelerations and the external forces and torques summarized in the vector $\boldsymbol{q}_i$, denoted as wrench, acting on the body with holonomic constraints:

$$\underbrace{\begin{bmatrix} m_i\boldsymbol{E} & m_i\tilde{\boldsymbol{r}}_{OiCi}^{\mathrm{T}} \\ m_i\tilde{\boldsymbol{r}}_{OiCi} & \boldsymbol{I}_{Oi} \end{bmatrix}}_{\boldsymbol{M}_i = \text{const}} \underbrace{\begin{bmatrix} \boldsymbol{a}_{Oi} \\ \boldsymbol{\alpha}_i \end{bmatrix}}_{\boldsymbol{b}_i} + \underbrace{\begin{bmatrix} m_i\tilde{\boldsymbol{\omega}}_i\tilde{\boldsymbol{\omega}}_i\boldsymbol{r}_{OiCi} \\ \tilde{\boldsymbol{\omega}}_i\boldsymbol{I}_{Oi}\boldsymbol{\omega}_i \end{bmatrix}}_{\boldsymbol{k}_i} = \underbrace{\begin{bmatrix} \boldsymbol{f}_i \\ \boldsymbol{l}_{Oi} \end{bmatrix}}_{\boldsymbol{q}_i}. \tag{2.134}$$

Moreover, the external forces are composed of applied forces $\boldsymbol{q}_i^{(e)}$ and constraints forces $\boldsymbol{q}_i^{(r)}$ where the generalized constraint forces $\boldsymbol{g}_i$ of the joint i and $\boldsymbol{g}_{i+1}$ of the joint $i+1$ appear:

$$\boldsymbol{q}_i = \boldsymbol{q}_i^{(e)} + \boldsymbol{q}_i^{(r)}, \quad \boldsymbol{q}_i^{(r)} = \boldsymbol{Q}_i\boldsymbol{g}_i - \boldsymbol{C}_{i+1}^{\mathrm{T}}\boldsymbol{Q}_{i+1}\boldsymbol{g}_{i+1}, \quad i = 1(1)p. \tag{2.135}$$

2.5.2.3 Equations of Motion

For the total system a set of $18p$ scalar equations remains from (2.130), (2.134) and (2.135) as

$$\boldsymbol{b} = \overline{\boldsymbol{J}}\ddot{\boldsymbol{y}} + \overline{\boldsymbol{\beta}}\,, \tag{2.136}$$

$$\overline{\overline{\boldsymbol{M}}}\boldsymbol{b} + \overline{\boldsymbol{k}} = \boldsymbol{q}^{(e)} + \boldsymbol{q}^{(r)}\,, \tag{2.137}$$

$$\boldsymbol{q}^{(r)} = (\boldsymbol{E} - \boldsymbol{C})^{\mathrm{T}}\boldsymbol{Q}\boldsymbol{g} = \overline{\boldsymbol{Q}}\boldsymbol{g} \tag{2.138}$$

with $18p$ unknowns in the vectors $\boldsymbol{b}, \boldsymbol{y}, \boldsymbol{q}^{(r)}, \boldsymbol{g}$.

Now (2.136) and (2.138) are inserted in (2.137) and the global orthogonality $\overline{\boldsymbol{J}}^{\mathrm{T}}\overline{\boldsymbol{Q}} = \boldsymbol{0}$ is used again resulting in the standard form (2.121) of the equations of motion. The mass matrix is completely full, again, and the vector $\boldsymbol{k}$ depends not only on the generalized velocities but also on the absolute velocities,

$$\boldsymbol{M} = \begin{bmatrix} \boldsymbol{J}_1^{\mathrm{T}}(\boldsymbol{M}_1 + \boldsymbol{C}_2^{\mathrm{T}}(\boldsymbol{M}_2 + \boldsymbol{C}_3^{\mathrm{T}}\boldsymbol{M}_3\boldsymbol{C}_3)\boldsymbol{C}_2)\boldsymbol{J}_1 & \boldsymbol{J}_1^{\mathrm{T}}\boldsymbol{C}_2^{\mathrm{T}}(\boldsymbol{M}_2 + \boldsymbol{C}_3^{\mathrm{T}}\boldsymbol{M}_3\boldsymbol{C}_3)\boldsymbol{J}_2 & \boldsymbol{J}_1^{\mathrm{T}}\boldsymbol{C}_2^{\mathrm{T}}\boldsymbol{C}_3^{\mathrm{T}}\boldsymbol{M}_3\boldsymbol{J}_3 \\ \boldsymbol{J}_2^{\mathrm{T}}(\boldsymbol{M}_2 + \boldsymbol{C}_3^{\mathrm{T}}\boldsymbol{M}_3\boldsymbol{C}_3)\boldsymbol{C}_2\boldsymbol{J}_1 & \boldsymbol{J}_1^{\mathrm{T}}(\boldsymbol{M}_2 + \boldsymbol{C}_3^{\mathrm{T}}\boldsymbol{M}_3\boldsymbol{C}_3)\boldsymbol{J}_2 & \boldsymbol{J}_2^{\mathrm{T}}\boldsymbol{C}_3^{\mathrm{T}}\boldsymbol{M}_3\boldsymbol{J}_3 \\ \boldsymbol{J}_3^{\mathrm{T}}\boldsymbol{M}_3\boldsymbol{C}_3\boldsymbol{C}_2\boldsymbol{J}_1 & \boldsymbol{J}_3^{\mathrm{T}}\boldsymbol{M}_3\boldsymbol{C}_3\boldsymbol{J}_2 & \boldsymbol{J}_3^{\mathrm{T}}\boldsymbol{M}_3\boldsymbol{J}_3 \end{bmatrix}, \tag{2.139}$$

$$\boldsymbol{k} = \boldsymbol{k}(\boldsymbol{y}, \dot{\boldsymbol{y}}, \boldsymbol{w})\,. \tag{2.140}$$

However, the mass matrix shows now a characteristic structure which can be used for a Gauss transformation.

2.5.2.4 Recursion

There are three steps required to obtain the generalized accelerations.

1. Forward recursion to get the absolute motion starting with $i = 1$.
2. Backward recursion using a Gauss transformation starting with $i = p$. As a result the system

   $$\hat{\boldsymbol{M}}\,\ddot{\boldsymbol{y}} + \hat{\boldsymbol{k}} = \hat{\boldsymbol{q}} \tag{2.141}$$

   is obtained where $\hat{\boldsymbol{M}}$ is a lower triangular matrix

   $$\hat{\boldsymbol{M}} = \begin{bmatrix} \boldsymbol{J}_1^{\mathrm{T}}\tilde{\boldsymbol{M}}_1\boldsymbol{J}_1 & 0 & 0 \\ \boldsymbol{J}_2^{\mathrm{T}}\tilde{\boldsymbol{M}}_2\boldsymbol{C}_2\boldsymbol{J}_1 & \boldsymbol{J}_2^{\mathrm{T}}\tilde{\boldsymbol{M}}_2\boldsymbol{J}_2 & 0 \\ \boldsymbol{J}_3^{\mathrm{T}}\tilde{\boldsymbol{M}}_3\boldsymbol{C}_3\boldsymbol{C}_2\boldsymbol{J}_1 & \boldsymbol{J}_3^{\mathrm{T}}\tilde{\boldsymbol{M}}_3\boldsymbol{C}_3\boldsymbol{J}_2 & \boldsymbol{J}_3^{\mathrm{T}}\tilde{\boldsymbol{M}}_3\boldsymbol{J}_3 \end{bmatrix}, \tag{2.142}$$

   the block elements of which follow from the recursion formula.

   $$\tilde{\boldsymbol{M}}_{i-1} = \boldsymbol{M}_{i-1} + \boldsymbol{C}_i^{\mathrm{T}}(\tilde{\boldsymbol{M}}_i - \tilde{\boldsymbol{M}}_i\boldsymbol{J}_i(\boldsymbol{J}_i^{\mathrm{T}}\tilde{\boldsymbol{M}}_i\boldsymbol{J}_i)^{-1}\boldsymbol{J}_i^{\mathrm{T}}\tilde{\boldsymbol{M}}_i)\boldsymbol{C}_i\,. \tag{2.143}$$

3. Forward recursion for the generalized accelerations starting with $i = 1$. The recursion requires some computational overhead. Therefore, the recursive algorithms are more efficient than the matrix inversion (2.127) only for more than $p = 8 - 10$ bodies.

There are also some extensions of the recursive approach to loop topologies, see Bae and Haug (1987b) and Saha and Schiehlen (2001). Further, it has to be mentioned that there are two commercial codes based on recursive formalisms: Simpack (2009) and RecurDyn (2009). More details are available on the corresponding websites.

■ Problem 2.13 Equations of motion of a drawbar trailer

For the trailer shown in Fig. 2.30 the nonlinear equations are required. The system consists of a vehicle running on a horizontal plane straight ahead with constant velocity v_0 and the trailer represented by the drawbar 1 and the wheel 2. The drawbar 1 with the inertia properties m_1, $\boldsymbol{I}_{C1}$ is connected to the vehicle by a rotational/translational joint G, which is supported by two springs with the stiffness k_y and k_γ. In the neutral position of the joint the spring forces and torques are vanishing due to the symmetry of the design. At the other end of the drawbar wheel 2 with radius r and inertia parameters $\{m_2, \boldsymbol{I}_{C2}\}$ is mounted. The wheel axis is horizontal, and the joint G and the wheel center C_2 have the same height above ground. For the mathematical modeling as shown in Fig. 2.30 the frames I, R and Z as well as the body-fixed frames 1 and 2 are used. The trailer is defined as a two-body system with drawbar and wheel. The joint G at the traction vehicle has the forward velocity v_0, and the wheel contact point A due to the slip the contact forces and torques f_{t1}, f_{t2}, l_3 are acting. The friction in the joint and the bearing are neglected.

Solution

The solution follows the formalism presented in Sect. 2.5.1 and comprises of five steps.

Step 1: The trailer consists of $p = 2$ bodies and has $f = 3$ degrees of freedom. The lateral joint displacement y, the drawbar angle γ and the rotation angle β of the elastic wheel are chosen as generalized coordinates. First of all the position variables of drawbar 1 and wheel 2 are evaluated using the geometrical dimensions as

$$\boldsymbol{r}_1^I = \begin{bmatrix} v_0 t - bc\gamma \\ y - bs\gamma \\ 0 \end{bmatrix}, \quad \boldsymbol{S}^{I1} = \boldsymbol{S}^{IZ} = \boldsymbol{\gamma}_3 = \begin{bmatrix} c\gamma & -s\gamma & 0 \\ s\gamma & c\gamma & 0 \\ 0 & 0 & 1 \end{bmatrix}, \tag{1}$$

$$\boldsymbol{r}_2^I = \begin{bmatrix} v_0 t - ac\gamma \\ y - as\gamma \\ 0 \end{bmatrix}, \quad \boldsymbol{S}^{I2} = \boldsymbol{S}^{IZ}\boldsymbol{S}^{Z2} = \boldsymbol{\gamma}_3(-\boldsymbol{\beta}_2), \tag{2}$$

where sine and cosine are abbreviated as "s" and "c". The absolute velocities follow by differentiation of the position variables or from the elementary angular velocities, respectively, as

$$\boldsymbol{v}_1^I = \begin{bmatrix} v_0 + b\dot{\gamma}\mathrm{s}\gamma \\ \dot{y} - b\dot{\gamma}\mathrm{c}\gamma \\ 0 \end{bmatrix}, \quad \boldsymbol{\omega}_1^I = \begin{bmatrix} 0 \\ 0 \\ \dot{\gamma} \end{bmatrix}, \tag{3}$$

$$\boldsymbol{v}_2^I = \begin{bmatrix} v_0 + a\dot{\gamma}\mathrm{s}\gamma \\ \dot{y} - a\dot{\gamma}\mathrm{c}\gamma \\ 0 \end{bmatrix}, \quad \boldsymbol{\omega}_2^Z = \begin{bmatrix} 0 \\ -\dot{\beta} \\ \dot{\gamma} \end{bmatrix}, \quad \boldsymbol{\omega}_2^I = \boldsymbol{S}^{IZ}\boldsymbol{\omega}_2^Z = \begin{bmatrix} \dot{\beta}\mathrm{s}\gamma \\ -\dot{\beta}\mathrm{c}\gamma \\ \dot{\gamma} \end{bmatrix}. \tag{4}$$

The force variables are defined in Fig. 2.30 b). The applied forces and torques are due to gravity, spring and contact phenomena while the reaction forces and torques follow from cutting the drawbar and the wheel free. For both bodies related to their centers of mass it remains

$$\boldsymbol{f}_1^{(e)1} = \begin{bmatrix} -k_y y\mathrm{s}\gamma \\ -k_y y\mathrm{c}\gamma \\ m_1 g \end{bmatrix}, \quad \boldsymbol{l}_1^{(e)1} = \begin{bmatrix} 0 \\ 0 \\ -k_\gamma\gamma - k_y yb\mathrm{c}\gamma \end{bmatrix}, \tag{5}$$

$$\boldsymbol{f}_1^{(r)1} = \begin{bmatrix} f\mathrm{c}\gamma - f_{s1} \\ -f\mathrm{s}\gamma - f_{s2} \\ -f_{G0} - f_{G1} \end{bmatrix}, \quad \boldsymbol{l}_1^{(r)1} = \begin{bmatrix} l - l_{s1} \\ f_{G0}b - f_{G1}(a-b) \\ -fb\mathrm{s}\gamma + f_{s2}(a-b) - l_{s3} \end{bmatrix}, \tag{6}$$

$$\boldsymbol{f}_2^{(e)Z} = \begin{bmatrix} -f_{t1} \\ -f_{t2} \\ m_2 g \end{bmatrix}, \quad \boldsymbol{l}_2^{(e)Z} = \begin{bmatrix} f_{t2}r \\ -f_{t1}r \\ -l_3 \end{bmatrix}, \tag{7}$$

$$\boldsymbol{f}_2^{(r)Z} = \begin{bmatrix} f_{s1} \\ f_{s2} \\ f_{G1} - f_n \end{bmatrix}, \quad \boldsymbol{l}_2^{(r)Z} = \begin{bmatrix} l_{s1} \\ 0 \\ l_{s3} \end{bmatrix}. \tag{8}$$

The transformation in the inertial frame I using (1) results in

$$\boldsymbol{f}_1^{(e)I} = \begin{bmatrix} 0 \\ -k_y y \\ m_1 g \end{bmatrix}, \quad \boldsymbol{l}_1^{(e)I} = \begin{bmatrix} 0 \\ 0 \\ -k_\gamma\gamma - k_y yb\mathrm{c}\gamma \end{bmatrix}, \tag{9}$$

$$\boldsymbol{f}_1^{(r)I} = \begin{bmatrix} f - f_{s1}\mathrm{c}\gamma + f_{s2}\mathrm{s}\gamma \\ -f_{s1}\mathrm{s}\gamma - f_{s2}\mathrm{c}\gamma \\ -f_{G0} - f_{G1} \end{bmatrix},$$

$$\boldsymbol{l}_1^{(r)I} = \begin{bmatrix} (l - l_{s1})\mathrm{c}\gamma - f_{G0}b\mathrm{s}\gamma + f_{G1}(a-b)\mathrm{s}\gamma \\ (l - l_{s1})\mathrm{s}\gamma + f_{G0}b\mathrm{c}\gamma - f_{G1}(a-b)\mathrm{c}\gamma \\ -fb\mathrm{s}\gamma + f_{s2}(a-b) - l_{s3} \end{bmatrix}, \tag{10}$$

as well as

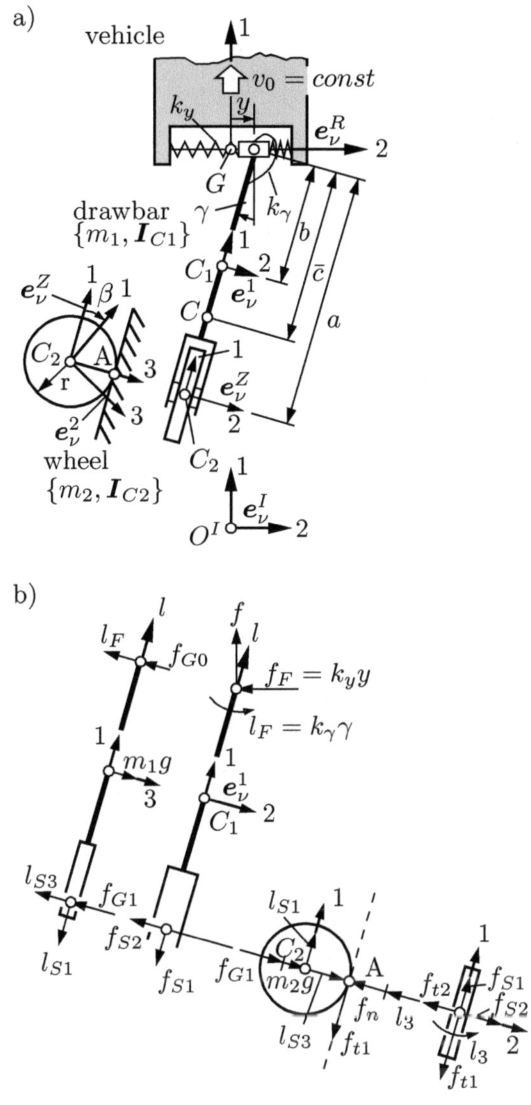

Fig. 2.30. Two body system drawbar and wheel: a) model; b) free body diagram

$$\boldsymbol{f}_2^{(e)I} = \begin{bmatrix} -f_{t1}\mathrm{c}\gamma + f_{t2}\mathrm{s}\gamma \\ -f_{t1}\mathrm{s}\gamma - f_{t2}\mathrm{c}\gamma \\ m_2 g \end{bmatrix}, \quad \boldsymbol{l}_2^{(e)I} = \begin{bmatrix} f_{t2}r\mathrm{c}\gamma + f_{t1}r\mathrm{s}\gamma \\ f_{t2}r\mathrm{s}\gamma - f_{t1}r\mathrm{c}\gamma \\ -l_3 \end{bmatrix}, \tag{11}$$

$$\boldsymbol{f}_2^{(r)I} = \begin{bmatrix} f_{s1}\mathrm{c}\gamma - f_{s2}\mathrm{s}\gamma \\ f_{s1}\mathrm{s}\gamma + f_{s2}\mathrm{c}\gamma \\ f_{G1} - f_n \end{bmatrix}, \quad \boldsymbol{l}_2^{(r)I} = \begin{bmatrix} l_{s1}\mathrm{c}\gamma \\ l_{s1}\mathrm{s}\gamma \\ l_{s3} \end{bmatrix}. \tag{12}$$

Step 2: The inertia tensors of both bodies have diagonal form in the corresponding body-fixed principal axes frame. Due to the rotational symmetry of the wheel this is also true in the reference frame Z,

$$\boldsymbol{I}_{C1}^{I} = \mathbf{diag}[I_{1,1}\ I_{1,2}\ I_{1,3}], \quad \boldsymbol{I}_{C2}^{Z} = \mathbf{diag}[I_{2,1}\ I_{2,2}\ I_{2,3}], \quad I_{2,1} = I_{2,3} = I. \tag{13}$$

The transformation in the inertial frame I results according to (2.106) using the rotational matrices (1) in

$$\boldsymbol{I}_{Ci}^{I} = \begin{bmatrix} I_{i,1}\mathrm{c}^2\gamma + I_{i,2}\mathrm{s}^2\gamma & (I_{i,1} - I_{i,2})\mathrm{s}\gamma\mathrm{c}\gamma & 0 \\ (I_{i,1} - I_{i,2})\mathrm{s}\gamma\mathrm{c}\gamma & I_{i,1}\mathrm{s}^2\gamma + I_{i,2}\mathrm{c}^2\gamma & 0 \\ 0 & 0 & I_{i,3} \end{bmatrix}, \quad i = 1, 2. \tag{14}$$

The Newton-Euler equations (2.107), (2.108) of both bodies read explicitly using (3), (4), (9)-(14) as

$$m_1 \begin{bmatrix} b\ddot{\gamma}\mathrm{s}\gamma + b\dot{\gamma}^2\mathrm{c}\gamma \\ \ddot{y} - b\ddot{\gamma}\mathrm{c}\gamma + b\dot{\gamma}^2\mathrm{s}\gamma \\ 0 \end{bmatrix} = \begin{bmatrix} 0 \\ -k_y y \\ m_1 g \end{bmatrix} + \begin{bmatrix} f - f_{s1}\mathrm{c}\gamma + f_{s2}\mathrm{s}\gamma \\ -f_{s1}\mathrm{s}\gamma - f_{s2}\mathrm{c}\gamma \\ -f_{G0} - f_{G1} \end{bmatrix}, \tag{15}$$

$$m_2 \begin{bmatrix} a\ddot{\gamma}\mathrm{s}\gamma + a\dot{\gamma}^2\mathrm{c}\gamma \\ \ddot{y} - a\ddot{\gamma}\mathrm{c}\gamma + a\dot{\gamma}^2\mathrm{s}\gamma \\ 0 \end{bmatrix} = \begin{bmatrix} -f_{t1}\mathrm{c}\gamma + f_{t2}\mathrm{s}\gamma \\ -f_{t1}\mathrm{s}\gamma - f_{t2}\mathrm{c}\gamma \\ m_2 g \end{bmatrix} + \begin{bmatrix} f_{s1}\mathrm{c}\gamma - f_{s2}\mathrm{s}\gamma \\ f_{s1}\mathrm{s}\gamma + f_{s2}\mathrm{c}\gamma \\ f_{G1} - f_n \end{bmatrix}, \tag{16}$$

$$\begin{bmatrix} 0 \\ 0 \\ I_{1,3}\ddot{\gamma} \end{bmatrix} = \begin{bmatrix} 0 \\ 0 \\ -k_\gamma\gamma - k_y y b\mathrm{c}\gamma \end{bmatrix} + \begin{bmatrix} (l - l_{s1})\mathrm{c}\gamma - f_{G0}b\mathrm{s}\gamma + f_{G1}(a-b)\mathrm{s}\gamma \\ (l - l_{s1})\mathrm{s}\gamma + f_{G0}b\mathrm{c}\gamma - f_{G1}(a-b)\mathrm{c}\gamma \\ -fb\mathrm{s}\gamma + f_{s2}(a-b) - l_{s3} \end{bmatrix}, \tag{17}$$

$$\begin{bmatrix} I_{2,2}\ddot{\beta}\mathrm{s}\gamma + I_{2,2}\dot{\beta}\dot{\gamma}\mathrm{c}\gamma \\ -I_{2,2}\ddot{\beta}\mathrm{c}\gamma + I_{2,2}\dot{\beta}\dot{\gamma}\mathrm{s}\gamma \\ I\ddot{\gamma} \end{bmatrix} = \begin{bmatrix} f_{t2}r\mathrm{c}\gamma + f_{t1}r\mathrm{s}\gamma \\ f_{t2}r\mathrm{s}\gamma - f_{t1}r\mathrm{c}\gamma \\ -l_3 \end{bmatrix} + \begin{bmatrix} l_{s1}\mathrm{c}\gamma \\ l_{s1}\mathrm{s}\gamma \\ l_{s3} \end{bmatrix}. \tag{18}$$

The translational motion of drawbar and wheel are characterized by (15), (16) while (17) and (18) describe the rotations of these bodies. Altogether these are 12 equations for the evaluation of $f = 3$ generalized coordinates as well as $q = 9$ generalized constraint forces.

Step 3: The Jacobian matrices of translation and rotation are found most easily for this problem by rewriting the velocities for a fixed traction vehicle using (3) and (4)

$$\delta' \boldsymbol{v}_1 = \begin{bmatrix} bs\gamma\delta\dot{\gamma} \\ \delta\dot{y} - bc\gamma\delta\dot{\gamma} \\ 0 \end{bmatrix}, \quad \delta' \boldsymbol{\omega}_1 = \begin{bmatrix} 0 \\ 0 \\ \delta\dot{\gamma} \end{bmatrix}, \tag{19}$$

$$\delta' \boldsymbol{v}_2 = \begin{bmatrix} as\gamma\delta\dot{\gamma} \\ \delta\dot{y} - ac\gamma\delta\dot{\gamma} \\ 0 \end{bmatrix}, \quad \delta' \boldsymbol{\omega}_2 = \begin{bmatrix} s\gamma\delta\dot{\beta} \\ -c\gamma\delta\dot{\beta} \\ \delta\dot{\gamma} \end{bmatrix}. \tag{20}$$

By separation of the variation $\delta'\dot{\boldsymbol{y}} = [\delta\dot{y}\ \delta\dot{\beta}\ \delta\dot{\gamma}]^{\mathrm{T}}$ the virtual velocity vectors (19) and (20), see also (2.115) follow as

$$\delta' \boldsymbol{v}_1 = \boldsymbol{J}_{T1}\delta'\dot{\boldsymbol{y}} = \begin{bmatrix} 0 & 0 & bs\gamma \\ 1 & 0 & -bc\gamma \\ 0 & 0 & 0 \end{bmatrix} \begin{bmatrix} \delta\dot{y} \\ \delta\dot{\beta} \\ \delta\dot{\gamma} \end{bmatrix},$$

$$\delta' \boldsymbol{\omega}_1 = \boldsymbol{J}_{R1}\delta'\dot{\boldsymbol{y}} = \begin{bmatrix} 0 & 0 & 0 \\ 0 & 0 & 0 \\ 0 & 0 & 1 \end{bmatrix} \begin{bmatrix} \delta\dot{y} \\ \delta\dot{\beta} \\ \delta\dot{\gamma} \end{bmatrix}, \tag{21}$$

$$\delta' \boldsymbol{v}_2 = \boldsymbol{J}_{T2}\delta'\dot{\boldsymbol{y}} = \begin{bmatrix} 0 & 0 & as\gamma \\ 1 & 0 & -ac\gamma \\ 0 & 0 & 0 \end{bmatrix} \begin{bmatrix} \delta\dot{y} \\ \delta\dot{\beta} \\ \delta\dot{\gamma} \end{bmatrix},$$

$$\delta' \boldsymbol{\omega}_2 = \boldsymbol{J}_{R2}\delta'\dot{\boldsymbol{y}} = \begin{bmatrix} 0 & s\gamma & 0 \\ 0 & -c\gamma & 0 \\ 0 & 0 & 1 \end{bmatrix} \begin{bmatrix} \delta\dot{y} \\ \delta\dot{\beta} \\ \delta\dot{\gamma} \end{bmatrix}. \tag{22}$$

Step 4: The Jacobian matrices $\boldsymbol{J}_{T1}$, $\boldsymbol{J}_{T2}$, $\boldsymbol{J}_{R1}$, $\boldsymbol{J}_{R2}$ are now available from (21), (22), and they are composed to the global Jacobian matrix $\overline{\boldsymbol{J}}$ as

$$\begin{aligned} \overline{\boldsymbol{J}}^{\mathrm{T}} &= \left[\boldsymbol{J}_{T1}^{\mathrm{T}} \vdots \boldsymbol{J}_{T2}^{\mathrm{T}} \vdots \boldsymbol{J}_{R1}^{\mathrm{T}} \vdots \boldsymbol{J}_{R2}^{\mathrm{T}}\right] \\ &= \begin{bmatrix} 0 & 1 & 0 & \vdots & 0 & 1 & 0 & \vdots & 0\,0\,0 & \vdots & 0 & 0 & 0 \\ 0 & 0 & 0 & \vdots & 0 & 0 & 0 & \vdots & 0\,0\,0 & \vdots & s\gamma & -c\gamma & 0 \\ bs\gamma & -bc\gamma & 0 & \vdots & as\gamma & -ac\gamma & 0 & \vdots & 0\,0\,1 & \vdots & 0 & 0 & 1 \end{bmatrix}. \end{aligned} \tag{23}$$

To extract the equations of motion of the trailer depending on the generalized coordinates and their time derivatives only, the reaction force variables have to be eliminated form the twelve Newton-Euler equations (15)-(18). The elimination procedure is a nontrivial task even if the problem treated is a comparatively simple system. This elimination procedure can be executed systematically by means of the Neweul formalism. By premultiplication of (15)-(18) written as column matrices with the matrix $\overline{\boldsymbol{J}}^{\mathrm{T}}$, see (23), one gets immediately the three equations of motion. The column grouping of the Jacobian matrices in (23) has to correspond with respect to translation, rotation and body index with the row grouping of the Newton-Euler equations (15)-(18) what is already true for this problem. Thus, for example, the first equation of motion of the trailer follows by addition of the second equation in (15) with the second equation in (16) resulting in

$$(m_1 + m_2)\ddot{y} + (m_1 b + m_2 a)(-\ddot{\gamma}\mathrm{c}\gamma + \dot{\gamma}^2\mathrm{s}\gamma) = -k_y y - f_{t1}\mathrm{s}\gamma - f_{t2}\mathrm{c}\gamma \,. \quad (24)$$

Accordingly, the two additional equations of motion of the two-body system are found as

$$I_{2,2}\ddot{\beta} = f_{t1} r \,, \quad (25)$$
$$(I_{1,3} + I + m_1 b^2 + m_2 a^2)\ddot{\gamma} - (m_1 b + m_2 a)\ddot{y}\mathrm{c}\gamma = -k_\gamma \gamma - l_3 + a f_{t2} \,. \quad (26)$$

The matrix representation of (24)-(26) is found using the abbreviations

$$m = m_1 + m_2 \,, \quad m\bar{c} = m_1 b + m_2 a \,, \quad I_G = I_{1,3} + I + m_1 b^2 + m_2 a^2 \,, \quad (27)$$

where m is the total mass, I_G is the total moment of inertia with respect to the $\boldsymbol{e}_3^R$-axis in the joint G and $\bar{c}$ means the distance of the total center of mass C from G, see Fig. 2.30 a). Thus, the equations of motion have the form (2.121),

$$\underbrace{\begin{bmatrix} m & 0 & -\bar{c}m\mathrm{c}\gamma \\ 0 & I_{2,2} & 0 \\ -\bar{c}m\mathrm{c}\gamma & 0 & I_G \end{bmatrix}}_{\boldsymbol{M}(y)} \underbrace{\begin{bmatrix} \ddot{y} \\ \ddot{\beta} \\ \ddot{\gamma} \end{bmatrix}}_{\ddot{\boldsymbol{y}}} + \underbrace{\begin{bmatrix} \bar{c}m\dot{\gamma}^2\mathrm{s}\gamma \\ 0 \\ 0 \end{bmatrix}}_{\boldsymbol{k}(\boldsymbol{y},\dot{\boldsymbol{y}})} = \underbrace{\begin{bmatrix} -k_y y - f_{t1}\mathrm{s}\gamma - f_{t2}\mathrm{c}\gamma \\ r f_{t1} \\ -k_\gamma\gamma - l_3 + a f_{t2} \end{bmatrix}}_{\boldsymbol{q}(\boldsymbol{y},\dot{\boldsymbol{y}})} \,. \quad (28)$$

The symmetric inertia matrix $\boldsymbol{M}$ can also be found directly by a congruence transformation as

$$\boldsymbol{M} = \overline{\boldsymbol{J}}^{\mathrm{T}}\overline{\overline{\boldsymbol{M}}}\,\overline{\boldsymbol{J}} \,, \quad \overline{\overline{\boldsymbol{M}}} = \mathbf{diag}[m_1\boldsymbol{E}, m_2\boldsymbol{E}, \boldsymbol{I}_{C1}^I, \boldsymbol{I}_{C2}^I] \,, \quad (29)$$

The equations of motion (24)-(26) may be also evaluated for the assembled two-body system cut only at the joint G and the contact point A. Equation (24) follows from Newton's equation for the $\boldsymbol{e}_2^I$-direction where the horizontal velocity $v_{C2} = (\dot{y} - \bar{c}\dot{\gamma}\mathrm{c}\gamma)$ of the total center of mass appears. Eqs. (25) and (26) follow for the application of Euler's equations for the $\boldsymbol{e}_2^Z$-axis and $\boldsymbol{e}_3^I$-axis, respectively, where the acceleration of the reference point G has to be considered according to (2.134). However, such a consideration of the assembled trailer does not provide the reaction force variables within the system.

Step 5: The nine generalized constraint forces read for this problem as: f_{G0}, f_{G1}, f_L, f_n l, l_{s1}, l_{s2}, f, f_{s1}, f_{s2}. First of all the four static equations from (15)-(18) can be used to evaluate the reaction force variables f_{G0}, f_{G1}, f_n and $l - l_{s1}$ rewritten as

$$\begin{bmatrix} 1 & 1 & 0 & 0 \\ 0 & 1 & 1 & 0 \\ -bs\gamma & (a-b)s\gamma & 0 & c\gamma \\ bc\gamma & -(a-b)c\gamma & 0 & s\gamma \end{bmatrix} \begin{bmatrix} f_{G0} \\ f_{G1} \\ f_n \\ l - l_{s1} \end{bmatrix} = \begin{bmatrix} m_1 g \\ m_2 g \\ 0 \\ 0 \end{bmatrix} . \tag{30}$$

The solutions are

$$f_{G0} = \frac{a-b}{a} m_1 g \, , \quad f_{G1} = \frac{b}{a} m_1 g \, , \quad f_n = \left(\frac{b}{a} m_1 + m_2 \right) g \, , \quad l = l_{s1} \, . \tag{31}$$

This result can easily confirmed by applying the static equilibrium conditions based on Fig. 2.30 b). The remaining equations can be represented in form (2.119). Then, the five missing reaction force variables can be computed formally as presented in Sect. 2.5. In this case the computation is even simpler due to the special structure of the equations. From (18) it follows

$$(18.1)c\gamma + (18.2)s\gamma \Rightarrow l_{s1} = I_{2,2}\dot{\beta}\dot{\gamma} - f_{t2}r \, , \tag{32}$$

$$(18.3) \Rightarrow l_{s3} = I\ddot{\gamma} + l_3 \, . \tag{33}$$

The acceleration vector $\ddot{\boldsymbol{y}}$ may be obtained from (28) using Cramer's rule with the quotient of two determinants. Thus, one gets

$$\ddot{y} = \frac{D_1(\boldsymbol{y}, \dot{\boldsymbol{y}})}{D(\boldsymbol{y})} \, , \quad \ddot{\gamma} = \frac{D_3(\boldsymbol{y}, \dot{\boldsymbol{y}})}{D(\boldsymbol{y})} \, , \quad D = \det[\boldsymbol{M}(\boldsymbol{y})] \, . \tag{34}$$

Comparing (33) and (34) results in

$$l_{s3} = \frac{D_3}{D} I + l_3 \, . \tag{35}$$

The still missing reaction forces f, f_{s1}, f_{s2} are obtained from (15), (16) as

$$\underbrace{\begin{bmatrix} 0 & m_1 bs\gamma \\ m_1 & -m_1 bc\gamma \\ 0 & m_2 as\gamma \\ m_2 & m_2 ac\gamma \end{bmatrix}}_{\overline{\overline{\boldsymbol{M}}}^* \overline{\boldsymbol{J}}^*} \underbrace{\begin{bmatrix} \ddot{y} \\ \ddot{\gamma} \end{bmatrix}}_{\ddot{\boldsymbol{y}}^*} + \underbrace{\begin{bmatrix} m_1 bc\gamma \\ m_1 bs\gamma \\ m_2 ac\gamma \\ m_2 as\gamma \end{bmatrix}}_{\overline{\boldsymbol{k}}^*} \dot{\gamma}^2 =$$

$$\underbrace{\begin{bmatrix} 0 \\ -k_y y \\ -f_{t1}c\gamma + f_{t2}s\gamma \\ -f_{t1}s\gamma - f_{t2}c\gamma \end{bmatrix}}_{\overline{\boldsymbol{q}}^{(e)*}} + \underbrace{\begin{bmatrix} 1 & -c\gamma & s\gamma \\ 0 & -s\gamma & -c\gamma \\ 0 & c\gamma & -s\gamma \\ 0 & s\gamma & c\gamma \end{bmatrix}}_{\overline{\boldsymbol{Q}}^*} \underbrace{\begin{bmatrix} f \\ f_{s1} \\ f_{s2} \end{bmatrix}}_{\boldsymbol{g}^*} . \tag{36}$$

Equations (36) represent a reduced number of the Newton-Euler equations (2.119) which may be solved by left-multiplication with the matrix

$$\boldsymbol{Q}^{+} = \begin{bmatrix} 1\,0 & 1 & 0 \\ 0\,0 & \mathrm{c}\gamma & \mathrm{s}\gamma \\ 0\,0 & -\mathrm{s}\gamma & \mathrm{c}\gamma \end{bmatrix} , \tag{37}$$

where it yields $\boldsymbol{Q}^{+}\overline{\boldsymbol{Q}}^{*} = \boldsymbol{E}$. Then, the required generalized constraint forces read as

$$\boldsymbol{g}^{*} = \boldsymbol{Q}^{+}[\overline{\overline{\boldsymbol{M}}}^{*}\overline{\boldsymbol{J}}^{*}\ddot{\boldsymbol{y}}^{*} + \overline{\boldsymbol{k}}^{*} - \overline{\boldsymbol{q}}^{(e)*}] , \quad \ddot{\boldsymbol{y}}^{*} = [D_1 D_3]^{\mathrm{T}}/D . \tag{38}$$

In more general problems, however, numerical methods from linear algebra have to be applied.

In an additional step the equations of motion and the equations of reaction are linearized with the respect to the reference state

$$y_R = 0 , \quad \gamma_R = 0 , \quad \dot{\beta}_R = \Omega = v_0/r . \tag{39}$$

Then, it yields

$$\begin{aligned} &y = y_R + \tilde{y} = \tilde{y} , \quad \gamma = \gamma_R + \tilde{\gamma} = \tilde{\gamma} , \quad \tilde{\gamma} \ll 1 , \\ &\dot{y} = \dot{\tilde{y}} , \quad \dot{\beta} = \Omega + \dot{\tilde{\beta}} , \quad \dot{\gamma} = \dot{\tilde{\gamma}} , \quad \dot{\tilde{\beta}} \ll \Omega , \end{aligned} \tag{40}$$

where all position and velocity variables characterized by the tilde are small of the first order. During linearization all variables small of second order and higher are neglected. The linearized equations of motion follow immediately from (28) considering $\mathrm{s}\tilde{\gamma} = \tilde{\gamma}$, $\mathrm{c}\tilde{\gamma} = 1$, $\dot{\tilde{\gamma}}^2 = 0$ as

$$\begin{bmatrix} m & 0 & -\bar{c}m \\ 0 & I_{2,2} & 0 \\ -\bar{c}m & 0 & I_G \end{bmatrix} \begin{bmatrix} \ddot{\tilde{y}} \\ \ddot{\tilde{\beta}} \\ \ddot{\tilde{\gamma}} \end{bmatrix} + \begin{bmatrix} k_y & 0 & 0 \\ 0 & 0 & 0 \\ 0 & 0 & k_\gamma \end{bmatrix} \begin{bmatrix} \tilde{y} \\ \tilde{\beta} \\ \tilde{\gamma} \end{bmatrix} = \begin{bmatrix} -f_{t1}\gamma - f_{t2} \\ r f_{t1} \\ a f_{t2} - l_3 \end{bmatrix} . \tag{41}$$

Obviously the inertia and stiffness matrix are time-invariant and symmetric. The static reaction force variables (31) remain unchanged, the other variables read as

$$\begin{aligned} l = l_{s1} &= I_{2,2}\Omega\dot{\tilde{\gamma}} - f_{t2} r , \\ l_{s3} &= I\ddot{\tilde{\gamma}} + l_3 , \\ f &= f_{t1} - f_{t2}\tilde{\gamma} , \\ f_{s1} &= f_{t1} , \\ f_{s2} &= f_{t2} + m_2(\ddot{\tilde{y}} - a\ddot{\tilde{\gamma}}) , \end{aligned} \tag{42}$$

where the acceleration quantities are given by

$$\ddot{\tilde{y}} = \tilde{D}_1/\tilde{D}\ , \tilde{D}_1 = \bar{c}m(af_{t2} - l_3 - k_\gamma\tilde{\gamma}) - I_G(f_{t1}\tilde{\gamma} + f_{t2} + k_y\tilde{y})\ ,$$
$$\ddot{\tilde{\gamma}} = \tilde{D}_3/\tilde{D}\ , \tilde{D}_3 = m(af_{t2} - l_3 - k_\gamma\tilde{\gamma}) - \bar{c}m(f_{t1}\tilde{\gamma} + f_{t2} + k_y\tilde{y})\ ,$$
$$\tilde{D} = m(I_G - m\bar{c}^2)\ . \tag{43}$$

According to (41) the longitudinal motion characterized by $\tilde{\beta}$ and the lateral motion represented by $\tilde{y}$ and $\tilde{\gamma}$ are completely decoupled for small motions.

■

■ Problem 2.14 Application of Neweul formalism on a vehicle model with $f = 10$ degrees of freedom

The research multibody dynamics tool Neweul is applied for the generation of the equations of motion of a three dimensional model of an automobile. The model presented in Fig. 2.31 consists of the vehicle body, a semi-trailing arm rear suspension and two independent front wheel suspensions. The wheels as well as the rear axle suspension are mounted elastically to the vehicle body. The semi-trailing arm rear suspension consists of one axle carrier and two control arms.

The inertia parameters as well as the 10 generalized coordinates are shown in Fig. 2.31. The geometrical parameters as well as the position of the connecting points of the elastic coupling elements are presented in Fig. 2.32. The guideway excitations are characterized by the four variables $WS1$, $WS2$, $WS3$, $WS4$. The numerical parameters are summarized in Table 2.5.

The essential input steps will be shown but not completely listed. A detailed description is available from Kreuzer and Leister (1988).

Position vector of generalized coordinates

The position vector reads as

$$\boldsymbol{y} = [\mathrm{ZK\ \ AK\ \ BK\ \ ZT\ \ AT\ \ BT\ \ BSL\ \ BSR\ \ ZVL\ \ ZVR}]^{\mathrm{T}} \tag{1}$$

Frames

Altogether 18 frames are used. As an example the first frame is shown. The frame No.1 describes the vehicle body by rotation and translation.

KOSART: S	Classification of frame
KOSYNA: AUFBAU	Name of frame
KOSYNA: I	Name of reference frame

Number of elementary rotations: 2

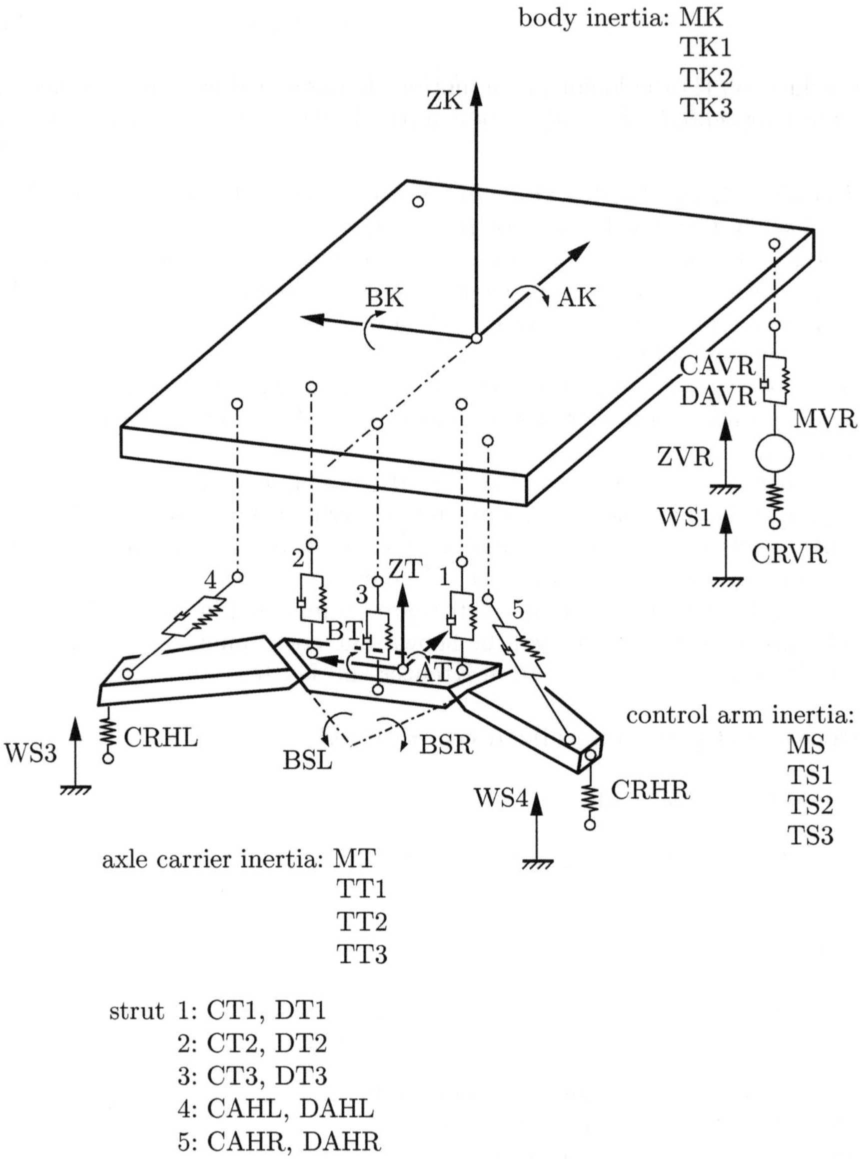

Fig. 2.31. Vehicle model with 10 degrees of freedom

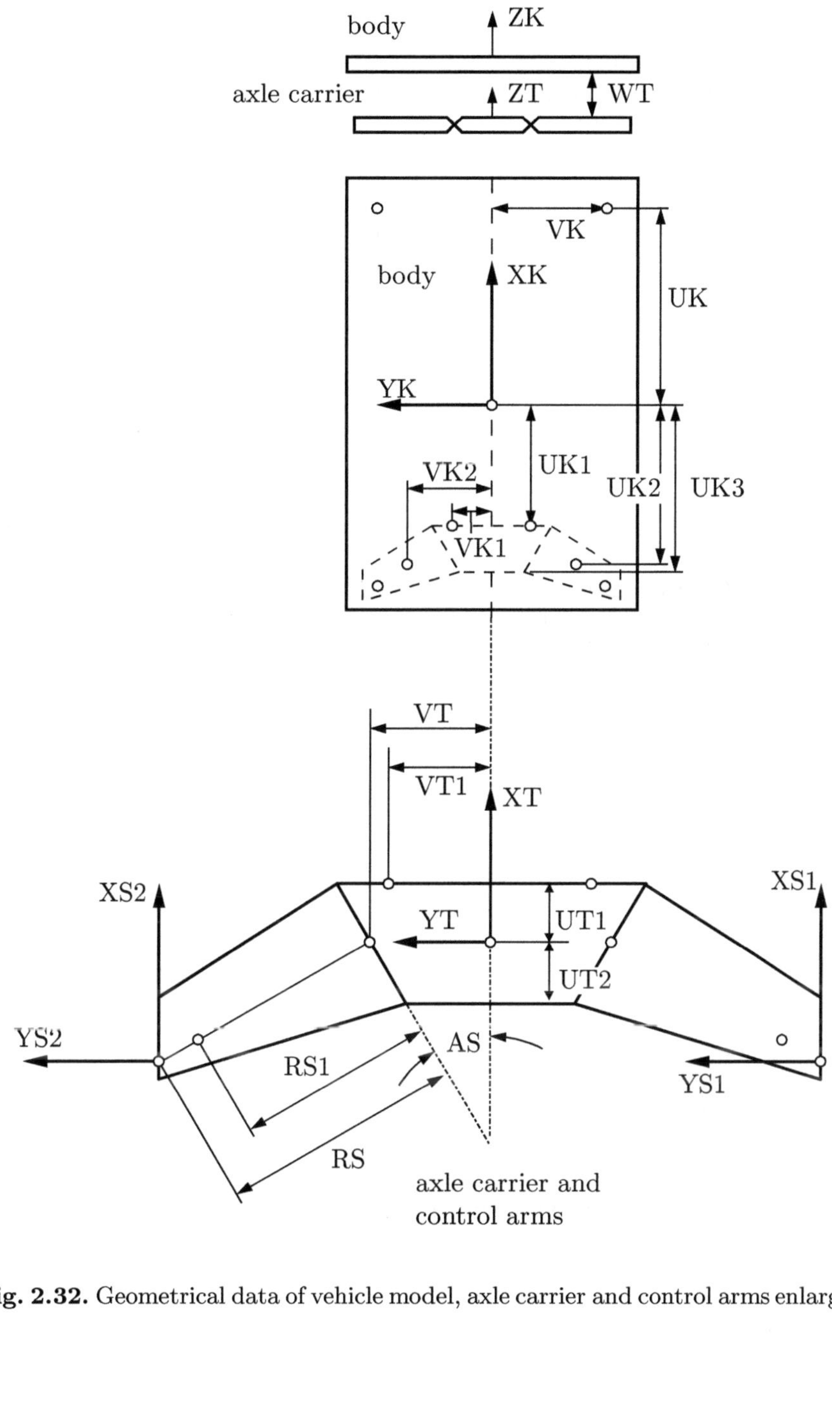

Fig. 2.32. Geometrical data of vehicle model, axle carrier and control arms enlarged

Table 2.5. Numerical parameters

AS	= 30 grd	CT1	= 250000 N/m
RS	= 0.45 m	CT2	= 250000 N/m
RS1	= 0.37 m	CT3	= 250000 N/m
UK	= 1.384 m	DAHL	= 2100 kg/s
UK1	= 1.089 m	DAHR	= 2100 kg/s
UK2	= 1.28 m	DAVL	= 1900 kg/s
UK3	= 1.299 m	DAVR	= 1900 kg/s
UT1	= 0.07 m	DT1	= 320 kg/s
UT2	= 0.14 m	DT2	= 320 kg/s
VK	= 0.744 m	DT3	= 320 kg/s
VK1	= 0.31 m	MK	= 1130 kg
VK2	= 0.564 m	MS	= 42 kg
VT	= 0.356 m	MT	= 25 kg
VT1	= 0.31 m	MVL	= 42.5 kg
WT	= 0.305 m	MVR	= 42.5 kg
CAHL	= 23300 N/m	TK1	= 504 kgm^2
CAHR	= 23300 N/m	TK2	= 1840 kgm^2
CAVL	= 15500 N/m	TK3	= 0 kgm^2
CAVR	= 15500 N/m	TS1	= 0.8 kgm^2
CRHL	= 200000 N/m	TS2	= 0.5 kgm^2
CRHR	= 200000 N/m	TS3	= 0.4 kgm^2
CRVL	= 180000 N/m	TT1	= 1.5 kgm^2
CRVR	= 180000 N/m	TT2	= 2 kgm^2
		TT3	= 0 kgm^2

Elementary rotation No.1
rotation axis 1
WINK = AK
rotation angle

Elementary rotation No.2
rotation axis 2
WINK = BK
rotation angle

Number of subvectors of translation: 1

Subvector No.1
KOSYNA: I frame of subvector
Subvector
R(1) = 0
R(2) = 0
R(3) = ZK

Inertia Data

Inertia data for body-fixed frame
KOSYNA: AUFBAU

Mass
MASS = MK

Inertia tensor
KOSYNA: AUFBAU frame for definition of the inertia tensor
I(1, 1) = TK1
I(2, 1) = 0
I(2, 2) = TK2
I(3, 1) = 0
I(3, 2) = 0
I(3, 3) = TK2

Applied forces and torques

External/internal force or torque
FLEART: IK Kind of force/torque

Frames affected by the force or torque
KOSYNA: NODE1 1. System (Action positive)
KOSYNA: NODE2 2. System (Action negative; for internal forces/ torques only)
Frame used for force/torque definition
KOSYNA: I

Force/torque inputvector
FLE(1) = 0
FLE(2) = 0
FLE(3) = SU5

Equations of linearized motion

There are only the nonvanishing elements printed.

Inertia matrix

M(1,1) = 1130.

M(2,2) = 504.

M(3,3) = 1840.

M(4,4) = 109.

M(5,5) = 49.811185

M(6,4) = 18.9
M(6,6) = 7.2525

M(7,4) = -18.9
M(7,5) = -14.786766
M(7,6) = -4.5025
M(7,7) = -9.23

M(8,4) = -18.9
M(8,5) = 14.786766
M(8,6) = -4.5025

M(7,7) = 9.23

M(9,9) = 42.5

M(10,10) = 42.5

Damping matrix

D(1,1) = 8960.
D(1,3) = 1229.44
D(1,4) = -5160.
D(1,6) = -777.
D(1,7) = 777.
D(1,8) = 777.
D(1,9) = -1900.
D(I,10) = -1900.

D(2,2) = 3500.944
D(2,5) = -1663.83
D(2,7) = 438.228
D(2,8) = -438.228
D(2,9) = -1413.6
D(2,10) = 1413.6

D(3,1) = 1229.44
D(3,3) = 15458.971
D(3,4) = -6488.64
D(3,6) = -1003.968
D(3,7) = 994.56
D(3,8) = 994.56
D(3,9) = 2629.6
D(3,10) = 2629.6

D(4,1) = -5160.
D(4,3) = -6488.64
D(4,4) = 5160.
D(4,6) = 777.
D(4,7) = -777.
D(4,8) = -777.

D(5,2) = -1663.83
D(5,5) = 1983.2423
D(5,7) = -525.58564
D(5,8) = 525.58564

D(6,1) = -777.
D(6,3) = -1003.968
D(6,4) = 777.
D(6,6) = 153.153
D(6,7) = -143.745
D(6,8) = -143.745

D(7,1) = 777.
D(7,2) = 438.228
D(7,3) = 994.56
D(7,4) = -777.
D(7,5) = -525.58564
D(7,6) = -143.745
D(7,7) = 287.49

D(8,1) = 777.
D(8,2) = -438.228
D(8,3) = 994.56
D(8,4) = -777.
D(8,5) = 525.58564
D(8,6) = -143.745
D(8,8) = 287.49

D(9.1) = -1900.
D(9,2) = -1413.6
D(9,3) = 2629.6
D(9.9) = 1900.

D(10,1) = -1900.
D(10,2) = 1413.6
D(10,3) = 2629.6
D(10,10) = 1900.

Stiffness matrix

K(1,1)	= 827600.
K(1,3)	= 885994.
K(1,4)	= -796600.
K(1,6)	= -8621.
K(1,7)	= 8621.
K(1,8)	= 8621.
K(1,9)	= -15500.
K(I,10)	= -15500.
K(2,2)	= 80032.89
K(2,5)	= -65828.188
K(2,7)	= 4862.244
K(2,8)	= -4862.244
K(2,9)	= -11532.
K(2,10)	= 11532.
K(3,1)	= 885994.
K(3,3)	= 1150539.3
K(3,4)	= -928898.
K(3,6)	= -18384.88
K(3,7)	= 11034.88
K(3,8)	= 11034.88
K(3,9)	= 21452.
K(3,10)	= 21452.
K(4,1)	= -796600.
K(4,3)	= -928898.
K(4,4)	= 1196600.
K(4,6)	= 98621.
K(4,7)	= -98621.
K(4,8)	= -98621.
K(5,2)	= -65828.188
K(5,5)	= 291806.36
K(5,7)	= -72945.527
K(5,8)	= 72945.527
K(6,1)	= -8621.
K(6,3)	= -18384.88
K(6,4)	= 98621.
K(6,6)	= 29194.885
K(6,7)	= -21844.885
K(6,8)	= -21844.885
K(7,1)	= 8621.
K(7,2)	= 4862.244
K(7,3)	= 11034.88
K(7,4)	= -98621.
K(7,5)	= -72945.527
K(7,6)	= -21844.885
K(7,7)	= 43689.77
K(8,1)	= 8621.
K(8,2)	= -4862.244
K(8,3)	= 11034.88
K(8,4)	= -98621.
K(8,5)	= 72945.527
K(8,6)	= -21844.885
K(8,8)	= 43689.77
K(9.1)	= -15500.
K(9,2)	= -11532.
K(9,3)	= 21452.
K(9.9)	= 195500.
K(10,1)	= -15500.
K(10,2)	− 11532.
K(10,3)	= 21452.
K(10,10)	= 195500.

Vector of excitations

H(4)	$= 200000. * WS3 + 200000. * WS4$
H(5)	$= 149142.29 * WS3 - 149142.29 * WS4$
H(6)	$= 45000. * WS3 + 45000. * WS4$
H(7)	$= -90000. * WS3$
H(8)	$= -90000. * WS4$
H(9)	$= 180000. * WS2$
H(10)	$= 180000. * WS1$

■

3

Models for Support and Guidance Systems

The tasks of the vehicle support and guidance system are

- support the vehicle on the track and guiding the vehicle along the track,
- isolation of the vehicle body including passengers and goods from disturbances caused by track unevenness and external loads e. g. wind gusts.

The first task affects directly the driving safety, whereas the second task is related to the driving comfort. To ensure a high driving safety, the support and guidance system has to provide a firm connection to the track. This is in contradiction to the demand for a high driving comfort which requests small accelerations of the vehicle body and, therefore, requires a soft suspension. For most vehicles, the resulting conflict of these tasks is solved by dividing the support and guidance system into two subsystems: the primary suspension system consisting of the components in the immediate vicinity of the track and the secondary suspension system connecting the vehicle body and the primary suspension system. This results in the basic vehicle set up shown in Fig. 3.1. In general, the design of the primary and the secondary suspension system cannot be performed independently from each other.

The primary and the secondary suspension system may consist of passive and/or active elements. An element is passive if an external energy source is not needed. On the other hand, an active element requires an external energy source and a system to control the energy transfer as well.

In the following, models of secondary suspension systems are presented, like passive spring and damper combinations, as well as models of the most important passive and active primary suspensions systems. In particular, wheels for road and rail vehicles, and magnetic actuators for magnetically levitated vehicles are presented.

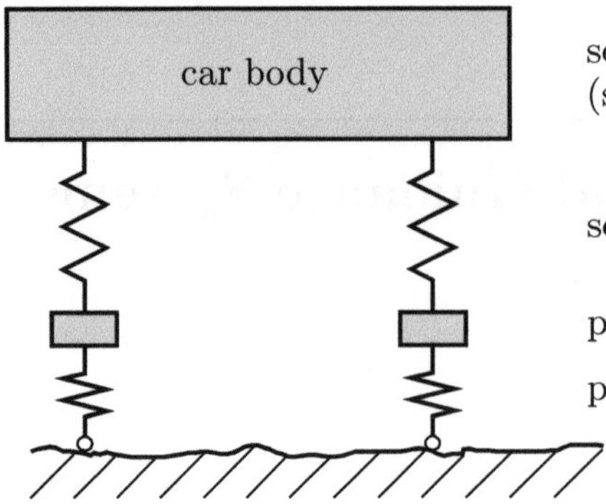

Fig. 3.1. Basic vehicle set-up

3.1 Models for Passive Spring and Damper Systems

Spring and damper elements are classified linear or nonlinear depending on the relation of the corresponding forces and moments to the displacements and velocities which may be linear or nonlinear. These elements can be used in different combinations, thus, a great variety of the resulting force laws exist. Simple linear elements are described first followed later by some nonlinear elements. Only elements for translational motions are considered here, but the transfer to rotational elements is straight forward.

In Table 3.1 the mechanical models for different combinations of linear spring and damper elements are shown. Using the element parameters, the corresponding applied forces can be expressed by the displacement s and their time derivative $\dot{s}$, respectively. The displacement s is measured with respect to the position of the unloaded springs.

The force law for a single displacement-proportional spring and for a single velocity-proportional damper can be found in row 1 and 2 of Table 3.1. This type of damping occurs in the motion of high viscous fluids or in damping by the use of eddy currents. Because of the equilibrium condition, the total force f of parallel combination of springs and/or dampers with the corresponding forces f_i is the sum of these forces,

$$f = \sum_{i=1}^{n} f_i \,. \tag{3.1}$$

Using this relationship the force law in Table 3.1, row 3 - 5, are derived. From the free body diagram it follows that in series combination of elements, every element is loaded with the same force f which is equal to the total force. In a series combination of identical elements, the total displacement s and the total velocity $\dot{s}$, respectively, is given as the sum of the individual values,

Table 3.1. Linear force laws for different combinations of linear spring and damper elements

Type	Mechanical Model	Characteristics	Force Law
1. Linear spring	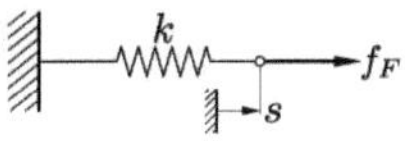	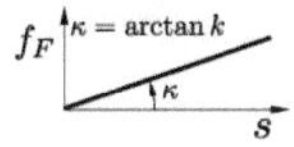	$f_F = ks$
2. Linear damper	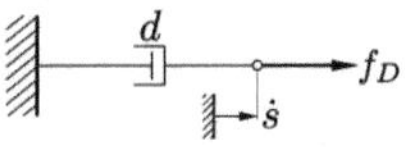	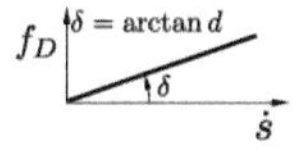	$f_D = d\dot{s}$
3. Parallel set of springs	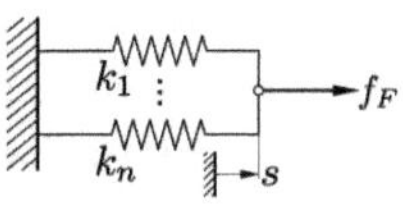	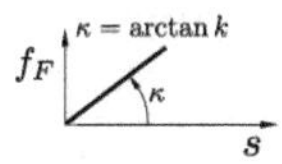	$f_F = \sum_{i=1}^{n} k_i s = ks$ $k = \sum_{i=1}^{n} k_i$
4. Parallel set of dampers	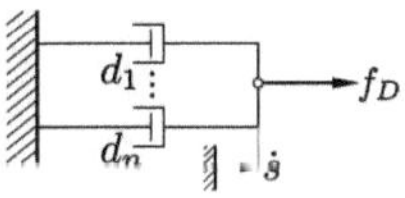	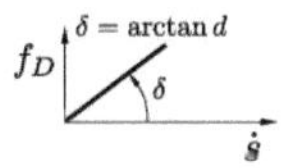	$f_D = \sum_{i=1}^{n} d_i \dot{s} = d\dot{s}$ $d - \sum_{i=1}^{n} d_i$
5. Parallel set of spring and damper	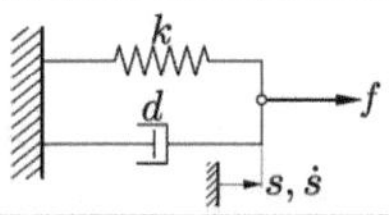		$f = d\dot{s} + ks$
6. Series set of springs	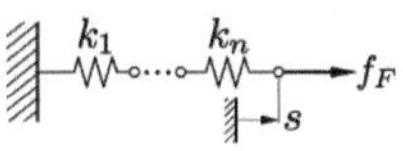	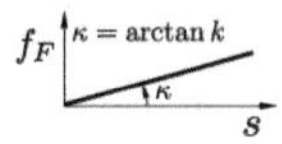	$s = \sum_{i=1}^{n} \frac{f_F}{k_i} = \frac{f_F}{k}$ $f_F = ks\,, \frac{1}{k} = \sum_{i=1}^{n} \frac{1}{k_i}$
7. Series set of dampers	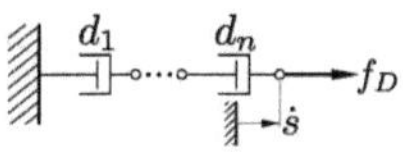	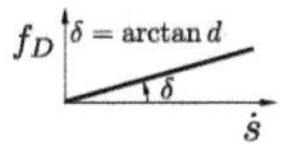	$\dot{s} = \sum_{i=1}^{n} \frac{f_D}{d_i} = \frac{f_D}{d}$ $f_D = d\dot{s}\,, \frac{1}{d} = \sum_{i=1}^{n} \frac{1}{d_i}$
8. Series set of spring and damper	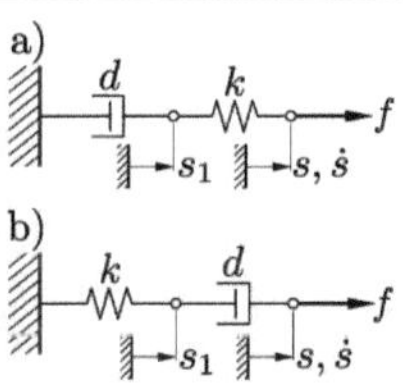	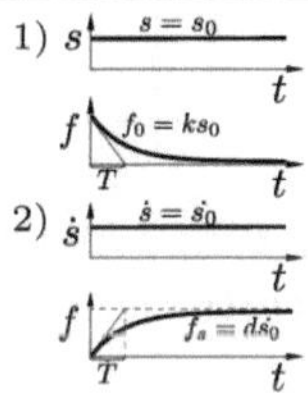	$\dot{f} + \frac{k}{d} f = k\dot{s}$ $f(t=0) = f_0\,, T = \frac{d}{k}$
9. Series and parallel set of springs and dampers	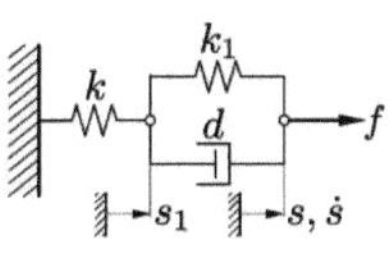	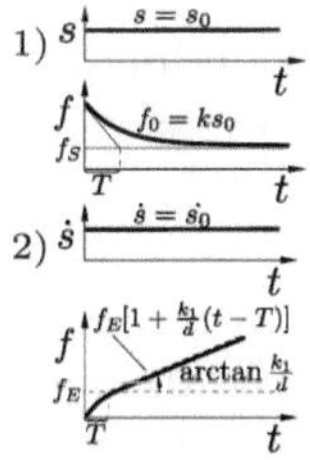	$\dot{f} + \frac{1}{d}(k + k_1) f = \frac{1}{d} k k_1 s + k\dot{s}$ $f(t=0) = f_0\,, T = \frac{d}{k + k_1}$ $f_S = \frac{k k_1}{k + k_1} s_0\,, f_E = k\dot{s}_0 T$

$$s = \sum_{i=1}^{n} s_i \,, \quad \dot{s} = \sum_{i=1}^{n} \dot{s}_i \,. \tag{3.2}$$

Using these relationships, the force law in Table 3.1, row 6 and 7, are found. In the cases considered so far, the forces f can be expressed as functions of s and $\dot{s}$ in the form

$$f = f(s) \,, \quad f = f(\dot{s}) \,, \quad f = f(s, \dot{s}) \,. \tag{3.3}$$

In terms of the control theory, these relations describe a proportional (P), a differential (D) or a proportional plus differential (PD) behavior.

The force law for a series combination of different elements is more complex. For example in case a), Table 3.1, row 8, the damper force is

$$f_D = d\dot{s}_1 \tag{3.4}$$

and the spring force reads as

$$f_F = k(s - s_1) \tag{3.5}$$

with the displacement s_1 of the central node. Because of the series combination, both elements have the same force due to equilibrium, $f_D = f_F = f$. Differentiation of Eq. (3.5) and elimination of $\dot{s}_1$ leads to the force law

$$\dot{f}(t) + \frac{k}{d} f(t) = k\dot{s}(t) \,, \quad f(0) = f_0 \,. \tag{3.6}$$

The force $f(t)$ is now described by a linear differential equation of first order. The general solution

$$f(t) = \mathrm{e}^{-\frac{k}{d}t} \left[f_0 + \int_0^t k\dot{s}(\tau)\mathrm{e}^{\frac{k}{d}\tau} \mathrm{d}\tau \right] \tag{3.7}$$

depends on the initial condition f_0 and on the disturbance function $\dot{s}(t)$. Two examples are presented to illustrate the behavior of the solution

1) If a displacement step is applied at the initial time $t = 0$ ($s = s_0 = \text{const}$, $\dot{s} = 0$ for $t > 0$), the corresponding initial value of the force is $f_0 = ks_0$. Thus, it follows from Eq. (3.7)

$$f(t) = ks_0 \mathrm{e}^{-\frac{k}{d}t} \,. \tag{3.8}$$

Due to the compliance of the damper, the force is decreasing from its initial value $f(t = 0) = f_0$ to the final value $f(t \to \infty) = 0$.

2) If a velocity step is applied at the initial time $t = 0$ ($\dot{s} = \dot{s}_0 = \text{const}$, $s(t) = \dot{s}_0 t$ for $t > 0$), it follows from Eq. (3.7)

$$f(t) = d\dot{s}_0 \left(1 - \mathrm{e}^{-\frac{k}{d}t}\right) . \tag{3.9}$$

The initial value of the force is now $f_0 = 0$. The spring is loaded and the force $f(t)$ increases until its stationary final value $f(t \to \infty) = f_S = d\dot{s}_0$ is reached. The force is then a pure damper force.

In both cases the differential equation (3.6) represents a balancing process of the force, starting from a constant initial value to a constant final value. The formal integration of Eq. (3.6) results in

$$f(t) + \frac{k}{d}\int\limits_0^t f(\bar{t})\mathrm{d}\bar{t} = k[s(t) - s(0)] + f_0 . \tag{3.10}$$

In terms of the control theory, the force law has an integral (I) behaviour. Equation (3.6) is also valid for case b) of Table 3.1, row 8.

A generalization of the force law Eq. (3.6) results in the spring-damper-combination shown in Table 3.1, row 9,

$$\dot{f}(t) + \frac{1}{d}(k + k_1)f(t) = \frac{1}{d}kk_1 s(t) + k\dot{s}(t) , \quad f(0) = f_0 . \tag{3.11}$$

Now the response to a displacement step denoted by 1) results in a stationary final value not equal to zero, whereas the response to a velocity step denoted by 2) is a force growing with time. Such models with a spring in parallel to one or more series of a spring and a damper are frequently used to model the dynamic behavior of rubber elements, which show relaxation and creep.

In vehicle engineering, the force laws of most of the spring and damper elements used deviate from linear behavior. Table 3.2 shows a selection of nonlinear force laws including the corresponding characteristics. The characteristics are frequently odd, $f(s) = -f(-s)$, $f(\dot{s}) = -f(-\dot{s})$. This can be expressed mathematically by using the signum-function

$$\mathrm{sgn}x = \begin{cases} +1 & \text{for} \quad x > 0 \\ -1 & \text{for} \quad x < 0 \end{cases} . \tag{3.12}$$

Nonlinear springs with progressive or degressive characteristics can be found in Table 3.2, row 1. Nonlinear characteristics can also be obtained by certain combinations of linear elements, like a spring with clearance or a spring preloaded in the design position, e. g. Table 3.2, row 2 and 3. Dampers with a quadratic characteristic are models for turbulent damping in fluids with a low viscosity, cp. Table 3.2, row 4. Friction effects between rigid bodies are of particular interest. These effects can be modeled by Coulomb's friction

Table 3.2. Nonlinear force laws for different combinations of spring and damper elements as well as Coulomb sliders

Type	Mechanical Model	Characteristics	Force Law
1. Nonlinear spring	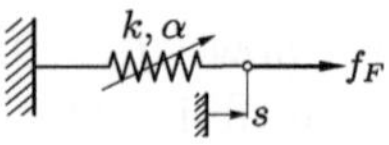	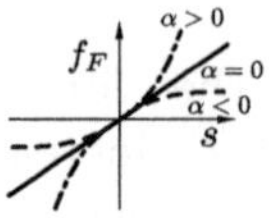	$f_F = ks + \alpha s^3$ $\alpha > 0$ progressive $\alpha = 0$ linear $\alpha < 0$ degressive
2. Linear spring with gap (gap-width $2a$)	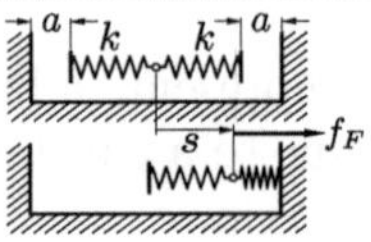	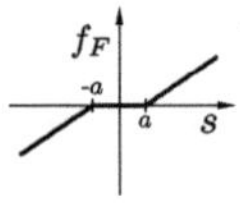	$f_F = \begin{cases} k(s-a), s \geq a \\ 0, -a \leq s \leq a \\ k(s+a), s \leq -a \end{cases}$
3. Preloaded linear springs (preloading displacement a)	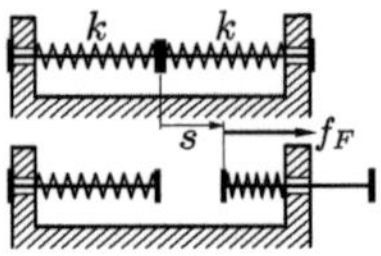	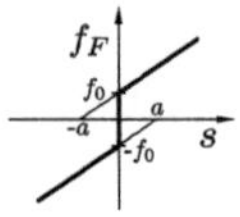	$f_F = \begin{cases} k(s+a), s > 0 \\ k(s-a), s < 0 \end{cases}$ $f_0 = ka$
4. Quadratic damper	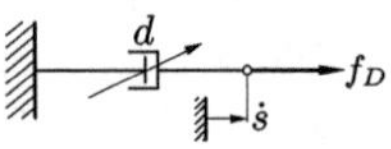	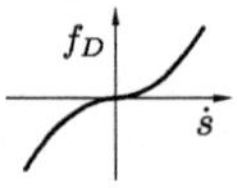	$f_D = d\dot{s}^2 \mathrm{sgn}\dot{s}$
5. Coulomb slider	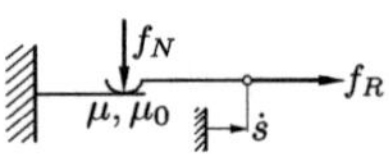	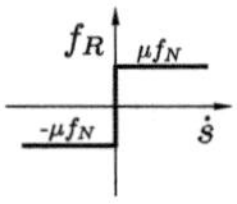	Stick: $\lvert f_R \rvert \leq \mu_0 f_N$, $\dot{s} = 0$ Slip: $f_R = \mu f_N \mathrm{sgn}\dot{s}$, $\dot{s} \neq 0$
6. Coulomb slider in series with a linear spring	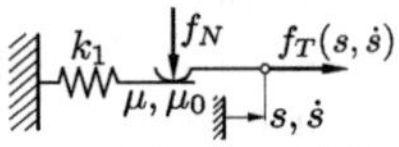 $s_1 = s(t^*), \dot{s_1} = \dot{s}(t^*)$ $t^* \widehat{=}$ Sticking begins	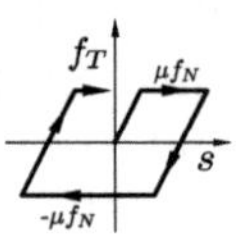	Stick: $(t^* = 0 : s_1 = 0 \,, \mathrm{sgn}\dot{s}_1 = 0)$ $f_T = k_1(s - s_1) - \mu f_N \mathrm{sgn}\dot{s}_1$ $\lvert f_T \rvert \leq \mu f_N$ Slip: $f_T = \mu f_N \mathrm{sgn}\dot{s}$
7. Parallel set of Coulomb sliders in series with linear springs	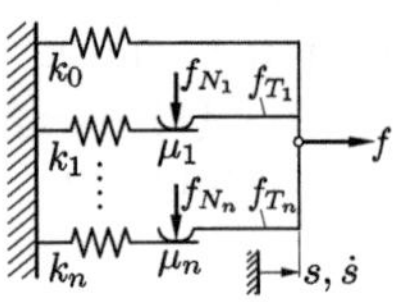	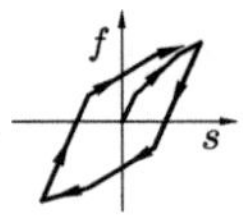	$f = k_0 s + \sum\limits_{i=1}^{n} f_{T_i}(s, \dot{s}_i)$ Stick (cp. 6.): $f_{T_i} =$ $k_i(s - s_i) - \mu_i f_{N_i} \mathrm{sgn}\dot{s}_i$, $\lvert f_{T_i} \rvert \leq \mu_i f_{N_i}$ Slip: $f_{T_i} = \mu_i f_{N_i} \mathrm{sgn}\dot{s}$
8. Spring with a displacement-dependent spring coefficient	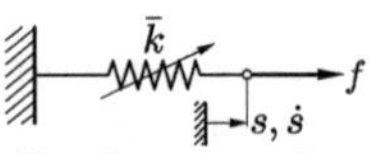 $\bar{k} = \bar{k}(z) = k e^{-\vartheta z}$ $z = z(t, \tau)$ $= \int_\tau^t \lvert \dot{s}(\bar{\tau}) \rvert d\bar{\tau}$	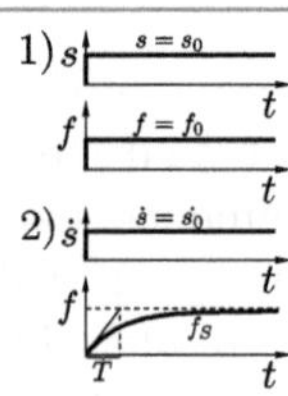	$\dot{f} + \vartheta\dot{s}\ \mathrm{sgn}(\dot{s}) f = k\dot{s}$, $f(t=0) = f_0$, 1) $s = s_0 : f_0 = k s_0 \mathrm{e}^{-\vartheta \lvert s_0 \rvert}$ 2) $\dot{s} = \dot{s}_0$: $T = \dfrac{\mathrm{sgn}\dot{s}_0}{\vartheta \dot{s}_0}$, $f_S = \dfrac{k}{\vartheta} \mathrm{sgn}\dot{s}_0$

law, cp. Table 3.2, row 5, where it must be strictly distinguished between stick (static coefficient of friction μ_0) and slip (coefficient of kinematic friction $\mu < \mu_0$). The static friction force is a reaction force that cannot exceed a certain limit, whereas the kinematic friction force is an applied force depending on the normal force which is a reaction force by itself. A series combination of a Coulomb friction element with a linear spring results in a simple mechanical model for an element showing hysteretic effects. Often $\mu_0 = \mu$ is assumed, than the hysteretic loop characteristic shown in Table 3.2, row 6, can be found. The parallel combination of Coulomb sliders in series with springs allows an adjustment of the computed hysteretic loop to measured data, cp. Table 3.2, row 7. Nonlinear elements can also have an integral action included in the force law. The spring in Table 3.2, row 8, has a displacement-dependent spring constant $\bar{k} = \bar{k}(z)$ with the always positive displacement of the spring $z = z(t,\tau)$ during the time interval $\tau \leq \bar{\tau} \leq t$

$$z = z(t\,,\tau) = \int_{\tau}^{t} |\dot{s}(\bar{\tau})|\,\mathrm{d}\bar{\tau} \quad . \tag{3.13}$$

In general, the corresponding force law follows from

$$f(t) = f_0 + \int_0^t \bar{k}(t\,,\tau)\dot{s}(\tau)\mathrm{d}\tau \; . \tag{3.14}$$

For $\bar{k}(t,\tau) = k = \text{const}$, the linear force law is obtained

$$f(t) = ks(t) \tag{3.15}$$

with $s(0) = f_0/k$. For a spring characteristic exponentially decreasing with the displacement z,

$$\bar{k}(z) = k\mathrm{e}^{-\vartheta z} \tag{3.16}$$

with a constant ϑ, it follows from Eq. (3.14)

$$f(t) = f_0 + \int_0^t k\,\mathrm{e}^{-\vartheta \int_{\tau}^{t} |\dot{s}(\bar{\tau})|d\bar{\tau}}\,\dot{s}(\tau)\mathrm{d}\tau \; . \tag{3.17}$$

It can be recognized that the spring force $f(t)$ at the time t depends on the previous motion. With the increase of the time duration $(t-\tau)$, the time history of the velocities $\dot{s}(\tau)$ dating back become less important for the response.

Differentiation of Eq. (3.14) with respect to time

$$\dot{f}(t) = \frac{\mathrm{d}}{\mathrm{d}t}\int_0^t \bar{k}(t\,,\tau)\dot{s}(\tau)\mathrm{d}\tau = \bar{k}(t\,,t)\dot{s}(t) + \int_0^t \frac{\partial}{\partial t}\bar{k}(t\,,\tau)\dot{s}(\tau)\mathrm{d}\tau \tag{3.18}$$

in combination with Eqs. (3.16) and (3.13) leads to a differential equation for the spring force

$$\dot{f}(t) + \vartheta \dot{s}(t)\mathrm{sgn}(\dot{s}) f(t) = k\dot{s}(t) , \quad f(0) = f_0 . \tag{3.19}$$

In contrast to Eq. (3.6), the spring force $f(t)$ is governed by a piecewise linear differential equation of first order. The response to a displacement step ($s = s_0 =$ const, $\dot{s} = 0$ for $t > 0$) is the constant spring force

$$f = f_0 = ks_0 \mathrm{e}^{-\vartheta |s_0|} , \tag{3.20}$$

whereas the response to a velocity step ($\dot{s} = \dot{s}_0 =$ const for $t > 0$) shows a similar behavior like the linear differential equation Eq. (3.6), but the stationary final value is a constant slip force, $f(t \to \infty) = f_s = k/\vartheta \,\mathrm{sgn}\dot{s}_0$, cp. Table 3.2, row 8.

The element described by Eq. (3.19) was used by Kranz (1983) in a parallel combination with a linear spring as a model for trapezoidal and parabolic springs of a freight wagon. A comparison with measured data showed that the resulting hysteretic behavior is described correctly by this model. The model parameters ϑ and k can be identified easily by the measured static spring characteristics.

For all elements given in Table 3.1 and 3.2 the appropriate parameters like the spring constants, damping constants and coefficients of friction are needed. In many cases, the spring constants can be determined analytically, whereas the damping constants and coefficients of friction have to be found by experiments.

■ Problem 3.1 Mathematical model of a layered leaf spring

The static force-displacement characteristic shown in Fig. 3.2 a) of a preloaded layered truck leaf spring was measured. The measured parameters are, Cebon (1986):

$$\begin{aligned}
k_o &= k_u = k_1 = 60\,\mathrm{N/mm} , \\
\beta_o &= \beta_u = \beta = 1.2\,\mathrm{mm} , \\
f_o &= 460\,\mathrm{N} , \\
f_u &= -410\,\mathrm{N} , \\
s_o &= 7.3\,\mathrm{mm} .
\end{aligned}$$

Determine a mathematical model using a spring with a displacement-dependent spring constant.

Solution

First of all a zero shift of the force coordinate is used to create an odd characteristic, $f(s) = -f(-s)$, with $f = \bar{f} - f_m$ and $f_m = \frac{1}{2}(f_o + f_u) = 25\,\mathrm{N}$. The upper and the lower tangent on the hysteresis loop are parallel (gradient

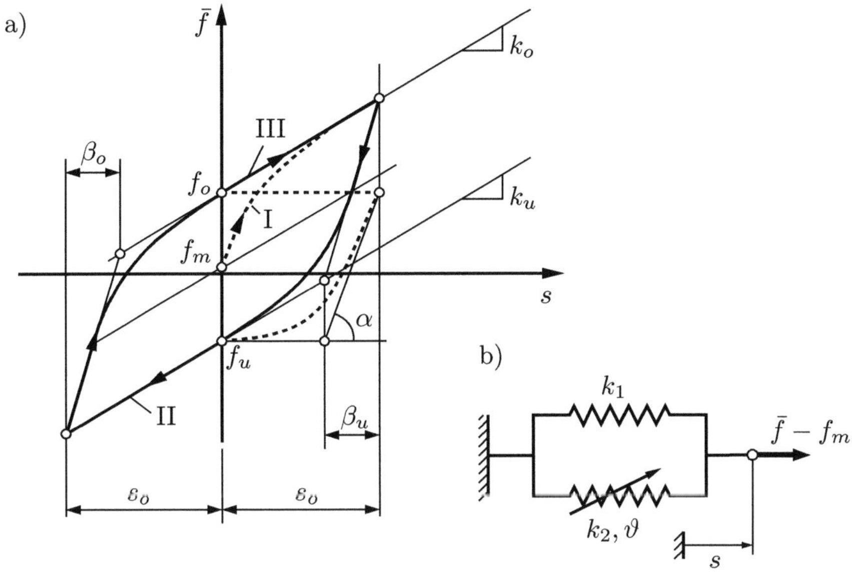

Fig. 3.2. a) Force-displacement characteristic; b) equivalent system for a layered leaf spring

$k_o = k_u = k_1$), thus, a model consisting of a parallel combination of a linear spring 1 (spring constant k_1) and of a displacement-dependent spring 2 (parameters k_2, ϑ) is chosen. This model, see Fig. 3.2 b), represents the leaf spring. The total spring force results from

$$\begin{aligned} f(s) &\equiv \bar{f}(s) - f_m = f_1(s) + f_2(s) \ , \\ f_1(s) &= k_1 s \ . \end{aligned} \tag{1}$$

The spring force $f_2(s)$ is given according to Eq. (3.17) after the transformation of the variables from time domain into space domain,

$$f_2(s) = \int\limits_0^s k_2 \mathrm{e}^{-\vartheta \int\limits_\sigma^s |\mathrm{d}\bar{\sigma}|} \mathrm{d}\sigma \tag{2}$$

with the unknown parameters k_2, ϑ. The solution of this function depends on the displacement path $z = z(s, \sigma) = \int\limits_0^s |\mathrm{d}\bar{\sigma}|$. For path I, see Fig. 3.2 a), with $0 \leq s \leq s_0$, $0 \leq \sigma_\mathrm{I} \leq s (z_\mathrm{I} = s - \sigma_\mathrm{I})$, it follows

$$f_2(s) = \int_0^s k_2 \mathrm{e}^{-\vartheta(s-\sigma_\mathrm{I})} \mathrm{d}\sigma_\mathrm{I} = \frac{k_2}{\vartheta}(1 - \mathrm{e}^{-\vartheta s}) \,, \tag{3}$$

$$f_2(s_0) = \frac{k_2}{\vartheta}(1 - \mathrm{e}^{-\vartheta s_0}) \equiv f^* \,. \tag{4}$$

For path I and path II with $s_0 \geq s \geq -s_0$, it can be obtained after splitting the integral (2) into two separate integrals with the integration variables σ_I and σ_II, respectively,

$$\begin{aligned} &\text{path I:} && 0 \leq \sigma_\mathrm{I} \leq s_0 && (z_\mathrm{I} = s_0 - \sigma_\mathrm{I} + s_0 - s = 2s_0 - s - \sigma_\mathrm{I}) \,, \\ &\text{path II:} && s_0 \geq \sigma_\mathrm{II} \geq s_0 && (z_\mathrm{II} = \sigma_\mathrm{II} - s) \,: \end{aligned}$$

$$\begin{aligned} f_2(s) &= \int_0^{s_0} k_2 \mathrm{e}^{-\vartheta(2s_0 - s - \sigma_\mathrm{I})} \mathrm{d}\sigma_\mathrm{I} + \int_{s_0}^{s} k_2 \mathrm{e}^{-\vartheta(\sigma_\mathrm{II} - s)} \mathrm{d}\sigma_\mathrm{II} \\ &= \frac{k_2}{\vartheta} \left[\left(\mathrm{e}^{-\vartheta(s_0 - s)} - \mathrm{e}^{-\vartheta(2s_0 - s)} \right) - \left(1 - \mathrm{e}^{-\vartheta(s_0 - s)} \right) \right] \,, \end{aligned} \tag{5}$$

$$f_2(-s_0) = \frac{k_2}{\vartheta} \left[\mathrm{e}^{-2\vartheta s_0} \left(1 - \mathrm{e}^{-\vartheta s_0} \right) - \left(1 - \mathrm{e}^{-2\vartheta s_0} \right) \right] \equiv - \left(f^{**} - \mathrm{e}^{-2\vartheta s_0} f^* \right) \,, \tag{6}$$

using the function

$$f^{**} = \frac{k_2}{\vartheta}(1 - \mathrm{e}^{-2\vartheta s_0}) \,. \tag{7}$$

Analogously, for path I, II and III with $-s_0 \leq s \leq s_0$, is calculated after splitting the integral (2) into three separate integral with the integration variables σ_I, σ_II and σ_III, respectively,

$$\begin{aligned} &\text{path I:} && 0 \leq \sigma_\mathrm{I} \leq s_0 && (z_\mathrm{I} = s_0 - \sigma_\mathrm{I} + 2s_0 + s + s_0 = 4s_0 + s - \sigma_\mathrm{I}) \,, \\ &\text{path II:} && s_0 \geq \sigma_\mathrm{II} \geq -s_0 && (z_\mathrm{II} = s_0 + \sigma_\mathrm{II} + s + s_0 = 2s_0 + s + \sigma_\mathrm{II}) \,, \\ &\text{path III:} && -s_0 \leq \sigma_\mathrm{III} \leq s_0 && (z_\mathrm{III} = s - \sigma_\mathrm{III}) \,: \end{aligned}$$

$$\begin{aligned} f_2(s) &= k_2 \left[\int_0^{s_0} \mathrm{e}^{-\vartheta(4s_0 + s - \sigma_\mathrm{I})} \mathrm{d}\sigma_\mathrm{I} + \int_{s_0}^{-s_0} \mathrm{e}^{-\vartheta(2s_0 + s + \sigma_\mathrm{II})} \mathrm{d}\sigma_\mathrm{II} \right. \\ &\quad \left. + \int_{-s_0}^{s} \mathrm{e}^{-\vartheta(s - \sigma_\mathrm{III})} \mathrm{d}\sigma_\mathrm{III} \right] \\ &= \frac{k_2}{\vartheta} \left[\left(\mathrm{e}^{-\vartheta(3s_0 + s)} - \mathrm{e}^{-\vartheta(4s_0 + s)} \right) - \left(\mathrm{e}^{-\vartheta(s_0 + s)} - \mathrm{e}^{-\vartheta(3s_0 + s)} \right) \right. \\ &\quad \left. + (1 - \mathrm{e}^{-\vartheta(s_0 - s)}) \right] \,, \end{aligned} \tag{8}$$

$$f_2(s_0) = \frac{k_2}{\vartheta}\left[\left(e^{-4\vartheta s_0} - e^{-5\vartheta s_0}\right) - \left(e^{-2\vartheta s_0} - e^{-4\vartheta s_0}\right) + \left(1 - e^{-2\vartheta s_0}\right)\right] \equiv f^{**}\left(1 - e^{-2\vartheta s_0}\right) + e^{-4\vartheta s_0} f^{*} \;. \tag{9}$$

Assuming

$$e^{-2\vartheta s_0} \ll 1 \;, \tag{10}$$

the force-displacement-characteristic $f_2(s)$ reaches a steady-state condition after just one load cycle and is bounded by $f_2(-s_0) \approx -k_2/\vartheta \leq f_2(s) \leq f_2(s_0) \approx k_2/\vartheta$. This results in an approximation for the force law of the displacement-dependent spring 2:

$$\text{path II: } (\mathrm{d}s < 0): \quad f_2(s) \approx \frac{k_2}{\vartheta}\left[2e^{-\vartheta(s_0 - s)} - 1\right] \;, \qquad -s_0 \leq s \leq s_0 \;, \tag{11}$$

$$\text{path III: } (\mathrm{d}s > 0): \quad f_2(s) \approx -\frac{k_2}{\vartheta}\left[2e^{-\vartheta(s_0 + s)} - 1\right] \;, \qquad -s_0 \leq s \leq s_0 \;. \tag{12}$$

The gradient $\mathrm{d}f_2(s)/\mathrm{d}s$ in the point $s = +s_0$ of path II and in the point $s = -s_0$ of path III, respectively, is given by

$$\left.\frac{\mathrm{d}f_2(s)}{\mathrm{d}s}\right|_{s=\pm s_0} \equiv \tan\alpha \approx 2k_2 \;. \tag{13}$$

Using the results above the unknown parameters k_2 and ϑ can be identified from measured data, cp. Fig. 3.2 a):

- The vertical distance between the upper and the lower tangent on the hysteresis loop delivers

$$f_o - f_u \approx 2\frac{k_2}{\vartheta} \;. \tag{14}$$

- The projection of the corner points of the hysteresis loop on the horizontal line through the intersections f_o and f_u permits the determination of the gradient given in Eq. (13). Using the projection, the part of the linear spring is eliminated. This results in, cp. Fig. 3.2 a),

$$\tan\alpha = \frac{f_o - f_u}{\beta} \approx 2k_2 \;. \tag{15}$$

Using the given data, it follows from Eqs. (14) and (15)

$$\vartheta \approx \frac{1}{\beta} \approx 0.83\,\mathrm{mm}^{-1} \;, \qquad k_2 \approx \frac{f_0 - f_u}{2\beta} = \frac{1}{2}\tan\alpha = 362.5\,\mathrm{N/mm} \;. \tag{16}$$

Finally, the validity of the assumption of Eq. (10) is checked

$$e^{-2\vartheta s_0} = e^{-2\cdot 0.83\cdot 7.3} = 5.2 \cdot 10^{-6} \ll 1 \ . \tag{17}$$

Thus, it is shown that the used approximations are valid in the considered case. This is due to a sufficiently large displacement s_0 of the measured force-displacement characteristic.

The relatively complicated hysteresis loop of the layered leaf spring can be represented mathematically by the simple 3-parameter-model using Eqs. (1), (11) and (12) with sufficient accuracy. An alternative to the presented model with the displacement-dependent spring 2 is the use of a series combination of a coulomb slider with a linear spring, cp. Table 3.2, row 6. This model has the same number of parameters and is also able to represent a hysteretic behavior. However, the resulting hysteresis loop has four corner points instead of two and therefore the representation of the measured data is not as good as using the representation with the model presented here. Two corner points can be achieved using more elements in series and/or parallel combinations, cp. Table 3.2, row 7. ■

3.2 Models of Force Actuators

Force actuators can be classified into magnetic, electromechanic, hydraulic and pneumatic actuators as well as piezo actuators, depending on the principle of force generation. In automotive systems actuators are used in support systems like servo steering or servo braking, and in active suspensions. Magnetic levitated trains are using the electromagnetic attraction by magnetic actuators for support and guidance. In the following, models of magnetic actuators will be discussed. The structure of the resulting force laws is found for other kinds of actuators, too.

3.2.1 Models of Magnetic Actuators

As usual in modeling it is not possible to find a simple mathematical model for a magnetic actuator that is complete and exact. The models for the magnets are either simple and represent the physical effect insufficiently or they are exact but also complicated and hard to compute. Because of the dependency of the magnet behavior on the actual geometry and the materials used, the axiomatic way of modeling leads to coarse results. Therefore, methods based on measurements are necessary. In the following, nonlinear and linearized models of magnets are presented that can be adapted to measured static and dynamic magnet characteristics. The models are based on a single magnet whose magnetic effect is represented by a single force, see Fig. 3.3. The general equation of the relation between voltage and current of the DC circuit reads as

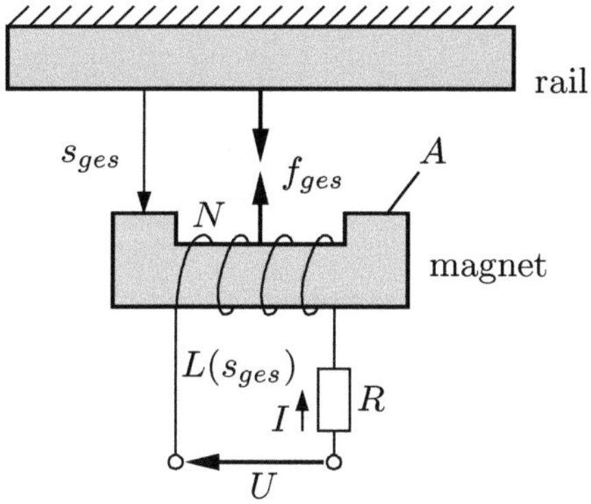

Fig. 3.3. Model of a magnetic actuator

$$U(t) = RI(t) + \frac{\mathrm{d}}{\mathrm{d}t}\Phi_{ges}[I(t), s_{ges}(t)] \,, \tag{3.21}$$

where $U(t)$ represents the acting voltage and $I(t)$ the current in the coil. Furthermore, R is the resistance and Φ_{ges} the magnetic flux which depends on the current I, the gap s_{ges}, the geometry and the material of the magnet. Equation (3.21) balances the voltages of the circuit. The external voltage has to be the sum of potential drops in the circuit. The first potential drop follows from Ohm's law, the second is caused by the electromagnetic induction. The multiplication of Eq. (3.21) with the current $I(t)$ and a time integration results in an energy balance. Because of the equality of magnetic energy and the work of the magnetic forces, the relation between magnetic flux Φ_{ges} and total force f_{ges} reads as

$$f_{ges}(t) = -\frac{\partial}{\partial s_{ges}} \int_0^{I(t)} \Phi_{ges}(i, s_{ges})\mathrm{d}i \,. \tag{3.22}$$

For measurement reasons it is more convenient to introduce the inductance L_{ges} in the magnetic flux

$$\Phi_{ges}(I, s_{ges}) = L_{ges}(I, s_{ges})I \,. \tag{3.23}$$

Considering the change in time of the current $I = I(t)$ and the gap $s_{ges} = s_{ges}(t)$ Eqs. (3.21) and (3.23) yield

$$U(t) = RI(t) + \left(\frac{\partial L_{ges}}{\partial I} I(t) + L_{ges}\right)\dot{I}(t) + \frac{\partial L_{ges}}{\partial s_{ges}} I(t)\dot{s}(t) \,, \tag{3.24}$$

while Eqs. (3.22) and (3.23) result in

$$f_{ges}(t) = -\frac{\partial}{\partial s_{ges}} \int\limits_0^{I(t)} L_{ges}(i, s_{ges}) i \, di \, . \tag{3.25}$$

Equations (3.24) and (3.25) require realistic assumptions for the inductance L_{ges}. The following effects have to be taken into account for the calculation of the inductance L_{ges}:

- gap s_{ges} of the magnet,
- length l_F of the flux lines in iron,
- permeability μ_F of the iron,
- variation of μ_F by the current I (saturation),
- diffusion of the flux lines,
- displacement of flux lines by eddy currents caused by changes of current and movements of the magnet along the rail and
- influences of hysteresis.

A summary of models for the inductance L_{ges} regarding the first five points and the corresponding magnetic forces are given in Table 3.3. Effects of eddy currents and hysteresis are hard to describe analytically. They result not only in a lower inductance but also in higher voltage drops (temperature increase of the magnet) that can be modeled by a raised resistance R. With respect to these effects, the general model (3.24) is not sufficient. The eddy currents and their negative effect can be suppressed by using laminated iron cores and armatures as well as by materials with low electric conductivity. These influences are only listed here, but they will not be discussed in detail.

In the following the nonlinear Eqs. (3.24) and (3.25) will be linearized according to a reference point R. The equations are continuously differentiable. Therefore, an expansion of Taylor series is possible. Small deviations around the reference point are assumed. This assumption is justified since a controller for the stabilization of the system is needed due to the unstable behavior of the magnet. The controller minimizes the deviation from the reference values. With the equations

$$\begin{aligned} U(t) &= U_R + u(t) \, , \qquad I(t) = I_R + i(t) \, , \\ f_{ges}(t) &= f_R + f(t) \, , \quad s_{ges}(t) = s_R + s(t) \end{aligned} \tag{3.26}$$

and the nonlinear Eqs. (3.24), (3.25), one gets the equations for constant reference values as

$$U_R = RI_R \, , \quad f_R = f_{ges}(I_R, s_R) \, , \tag{3.27}$$

marked with the index R. The differences to the constant reference values are described by the linearized equation for voltage and current as

$$\begin{aligned} u(t) &= Ri(t) + \left(\frac{\partial L_{ges}}{\partial I}\bigg|_R I_R + L_{ges}|_R \right) \dot{i}(t) + \frac{\partial L_{ges}}{\partial s_{ges}}\bigg|_R I_R \, \dot{s}(t) \\ &= Ri(t) + K_{\dot{I}} \, \dot{i}(t) - K_{\dot{s}} \, \dot{s}(t) \, . \end{aligned} \tag{3.28}$$

Table 3.3. Four models of the inductance $L_{ges} = L_k$ and the magnetic force $f_{ges} = f_k$ of a magnetic actuator, $k = 1(1)4$

Equivalent network	Inductance $L_{ges} = L_k$	Magnetic force $f_{ges} = f_k$
1) Ideal magnet with the gap inductance L_L		
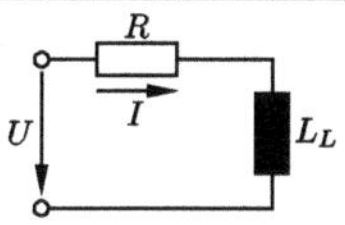	$L_1 = L_L = \dfrac{\mu_0 A N^2}{2 s_{ges}}$	$f_1 = f_L = \dfrac{\mu_0 A N^2}{4}\left(\dfrac{I}{s_{ges}}\right)^2$
2) Magnet with gap and iron inductance, L_L and L_F (equal cross sections $A_L = A_F = A$)		
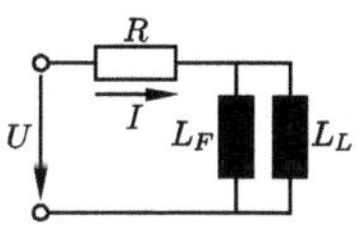	$L_2 = \dfrac{\mu_0 A N^2}{2 s_{ges} + \frac{l_F}{\mu_r}}$ $= L_L \dfrac{1}{1 + \frac{l_F}{2 s_{ges} \mu_r}}$	$f_2 = \mu_0 A N^2 \left(\dfrac{I}{2 s_{ges} + \frac{l_F}{\mu_r}}\right)^2$ $= f_L \dfrac{1}{\left(1 + \frac{l_F}{2 s_{ges} \mu_r}\right)^2}$
3) Magnet with gap, iron and leakage inductance, L_L, L_F and L_s		
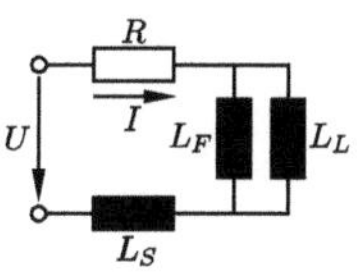	$L_3 = L_N(s_{ges}) + L_s$ $= L_L \dfrac{1}{1 + \frac{l_F}{2 s_{ges} \mu_r}} + L_s$ L_s, L_3 from measurements, $\eta_s = \dfrac{L_s}{L_3}$	$f_3 = \dfrac{1}{2} \dfrac{\partial L_N}{\partial s_{ges}} I^2$ $= f_L \dfrac{1}{\left(1 + \frac{l_F}{2 s_{ges} \mu_r}\right)^2}$
4) Magnet like model 3) regarding the *B*-*H*-curve (partly approximation: $\mu_r \sim 1/I^k$, e. g. $\mu_r = c/I^2$)		
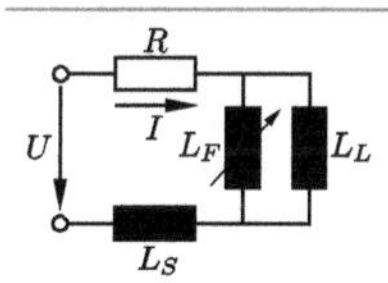	$L_4 = L_N(s_{ges}, I) + L_s$ $= L_L \dfrac{1}{1 + \frac{l_F I^2}{2 s_{ges} c}} + L_s$ c, k from measurements	$f_4 = \dfrac{\partial}{\partial s_{ges}} \int_0^I L(s_{ges}, i) i \, di$ $= f_L \dfrac{1}{1 + \frac{l_F I^2}{2 s_{ges} c}}$

Nomenclature

				Indices	
U	voltage in V	f	magnetic force in N=VAs/m		
I	current in A				
R	resistance in $\Omega = \mathrm{V/A}$	μ_0	inductance constant $\mu_0 = 4\pi \cdot 10^{-7}$ H/m	ges	total
L	inductance in $\mathrm{H} = \Omega\mathrm{s}$			N	efficient
s	gap in m	μ_r	relative permeability	s	leakage
l_F	magn. distance in iron in m	N	number of windings	L	gap
A	surface area of one pole in m^2	η_s	leakage factor	F	iron
		c, k	constants		

For the linearized force to current and gap relation one gets

$$\begin{aligned} f(t) &= \left.\frac{\partial f_{ges}}{\partial I}\right|_R i(t) + \left.\frac{\partial f_{ges}}{\partial s_{ges}}\right|_R s(t) \\ &= K_I \quad i(t) - \quad K_s \quad s(t) \ . \end{aligned} \tag{3.29}$$

The values of $K_{\dot{I}}$, $K_{\dot{s}}$, K_I and K_s are constant and positive, and are related to the reference point R. Due to the nonlinear characteristics the following equation is valid

$$\begin{aligned} K_I = \left.\frac{\partial f_{ges}}{\partial I}\right|_R &= -\frac{\partial}{\partial I}\frac{\partial}{\partial s_{ges}} \left.\int_0^I L_{ges}(i, s_{ges}) i \, di \right|_R \\ &= -\left.\frac{\partial L_{ges}(I, s_{ges})}{\partial s_{ges}}\right|_R I_R = K_{\dot{s}} \ . \end{aligned} \tag{3.30}$$

The constants depend on the model used for the inductance. For example, the idealized model 1) of Table 3.3 yields

$$K_{\dot{I}} = L_{LR} \ , \quad K_{\dot{s}} = K_I = L_{LR}\frac{I_R}{s_R} \ , \quad K_s = L_{LR}\frac{I_R^2}{s_R^2} \ , \quad L_{LR} = \frac{\mu_0 A N^2}{2 s_R} \ . \tag{3.31}$$

From Eq. (3.31) a relationship between the constants can be found

$$\frac{K_I^2}{K_{\dot{I}} K_s} = 1 \ . \tag{3.32}$$

For more realistic magnet models like 3) and 4), the influence of leakage inductance is taken into account. Then Eq. (3.32) is replaced by

$$\frac{K_I^2}{K_{\dot{I}} K_s} = 1 - \eta_s \ , \quad \eta_s = \left.\frac{L_s}{L_{ges}}\right|_R \ , \tag{3.33}$$

with the leakage factor η_s. Equations (3.30)-(3.33) have to be understood as theoretical approximations. In practice, the values of the constants could be found by measured magnet characteristics. The values of K_I and K_s can be easily determined by static force characteristics, Fig. 3.4. By measured voltage characteristics or by approximation using Eqs. (3.30) and (3.33) the remaining parameters are found. Using the linearized Eqs. (3.28) and (3.29) and eliminating of the current i, a linear differential equation of first order follows for the force f,

$$\dot{f}(t) + k_f f(t) = -k_s s(t) - k_{\dot{s}} \dot{s}(t) + k_u u(t) \ , \quad f(0) = f_0 \ . \tag{3.34}$$

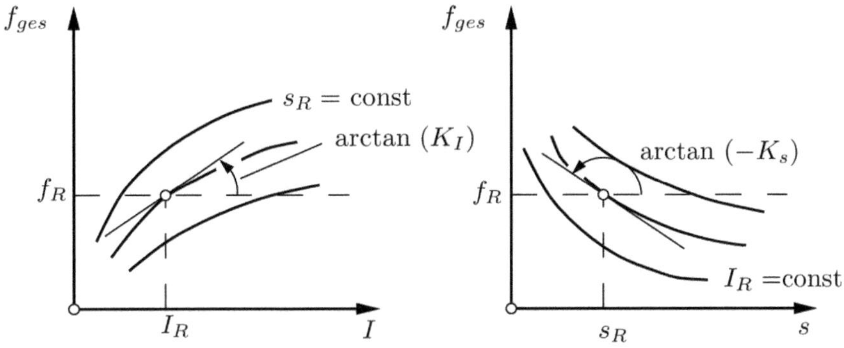

Fig. 3.4. Identification of the constants K_I and K_s out of the static force characteristics

The abbreviations are

$$k_f = \frac{R}{K_{\dot{I}}} , \quad k_s = \frac{R}{K_{\dot{I}}} K_s , \quad k_{\dot{s}} = K_s - \frac{K_I}{K_{\dot{I}}} K_{\dot{s}} , \quad k_u = \frac{K_I}{K_{\dot{I}}} , \tag{3.35}$$

whereas with Eqs. (3.30) and (3.33) follows $k_{\dot{s}} = \eta_s K_s$. Thus, all constants in Eq. (3.35) are positive.

Comparing the force law (3.34) of an active magnetic actuator with the force law (3.11) of a passive spring-damper-system one gets for $u(t) = 0$ the same mathematical structure. But a main difference is the sign of the constant factors with $s(t)$ and $\dot{s}(t)$. The reason is the decreasing absolute value of the magnetic force for $u(t) = 0$ if the values of $s(t)$ and $\dot{s}(t)$ are increasing, cp. Fig. 3.4, while the force in a spring-damper-system is increasing in the same condition. This behavior leads to an unstable reference position of an uncontrolled magnetic actuator. The system can be stabilized by a controller for the voltage $u(t)$. For example a linear controller reads as

$$u(t) = r_1 s(t) + r_2 \dot{s}(t) + r_3 f(t) \tag{3.36}$$

where its input variables $s(t)$, $\dot{s}(t)$ and $f(t)$ are measured continuously. The coefficients r_ν, $\nu = 1(1)3$, represent the gains of the controller. With such a controller the reference position can easily be stabilized. Additionally all characteristics of a passive spring-damper-system can be achieved with a magnetic actuator using Eq. (3.36) and

$$r_1 = \left(k_s + \frac{1}{d} k k_1 \right) \frac{1}{k_u} , \quad r_2 = (k_{\dot{s}} + k) \frac{1}{k_u} ,$$
$$r_3 = \left[k_f - \frac{1}{d}(k + k_1) \right] \frac{1}{k_u} . \tag{3.37}$$

Then, Eq. (3.34) is identical with Eq. (3.11) in the case of small deflections around the reference position. Especially great values of k and k_1 cause a stiff

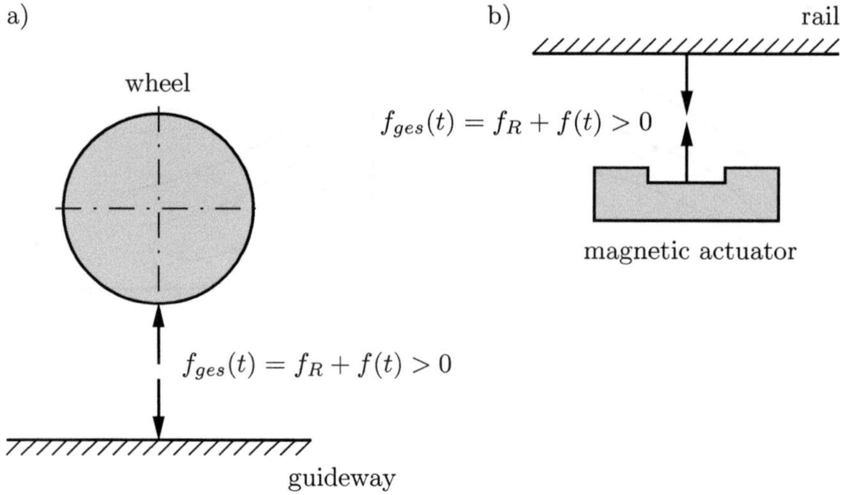

Fig. 3.5. Unilateral contact: a) wheel; b) magnetic actuator

attachment to the reference position resulting in a so-called magnetic wheel because of the similar kinematic behavior, see Gottzein (1984). The comparison of a controlled magnetic actuator with a wheel is also correct regarding to the fact that each system has a one-sided constraint. An electromagnet is generating only attraction forces, a wheel only compression forces to the track, $f_{ges}(t) = f_R + f(t) > 0$, cp. Fig. 3.5. If the dynamic forces $f(t)$ compensate the static reference values f_R, the wheel lifts of the track and accordingly the magnetic actuator sticks to the track.

3.2.2 General Linear Model of Force Actuators

The mathematical structure of the linear magnet model (3.34) can also be found with electromechanic, hydraulic and pneumatic actuators. In the case of small deflections around the reference point without regarding friction and inertia effects of force actuators the following equation is valid generally:

$$c_{\dot{f}}\dot{f}(t) + c_f f(t) = c_s s(t) + c_{\dot{s}}\dot{s}(t) + c_u u(t) , \quad f(0) = f_0 . \tag{3.38}$$

The coefficients $c_{\dot{f}}$, c_f, c_s, $c_{\dot{s}}$, c_u and the control signal $u(t)$ depend on the type of actuator and the design of the system. Because of the mounting of the actuator between two points the dependency on the relative deflection $s(t)$ and the relative velocity $\dot{s}(t)$ is always implied. The behavior of the actuator in time domain depends mainly on the control signal $u(t)$. If the system is controlled, a classification in two control concepts is possible depending on the sensor location. If sensor location and manipulation location are the same, the controlled closed-loop operator system has a structure similar to the passive system of Eq. (3.11). If the sensor location and the actuator location differ additional control laws may be designed to feedback sensor signals from arbitrary locations. This feature is not available with passive elements.

3.3 Comparison of Passive and Active Elements

The relations of the force equations of passive and active elements are shown in the previous sections. A comparison of both systems underlines the differences.

Passive elements show the following properties:

- There is no external energy source. The energy can only be stored in springs or dissipated in dampers or friction elements, respectively.
- The simple design results in high reliability and robustness, less maintenance, low cost for manufacturing.
- The force generation depends on the relative movement of the mounting point of the elements on the related bodies.
- A soft connection of contiguous bodies necessary for comfort reasons in the case of secondary springs leads to large static displacements.

The last property follows from the rule of thumb for the eigenfrequency f_0 of a linear spring-mass-system based on the static deflection s_0,

$$f_0 \approx \frac{5}{\sqrt{s_0}} , \quad f_0 \text{ in Hz} , \quad s_0 \text{ in cm} . \tag{3.39}$$

A static deflection of the spring of $s_0 = 25\,\text{cm}$ is necessary for the typical eigenfrequency of the vehicle body of $f_0 = 1\,\text{Hz}$. Thus, additional static loads cause big and undesired relative displacements. Equation (3.39) follows for a spring-mass-system (mass m, spring constant k) with $f_0 = \omega_0/2\pi$, $\omega_0^2 = k/m$ from the static balance of gravitational force and preload force of the spring, $mg = ks_0$. This balance results in $k/m = g/s_0$ and $f_0 = \sqrt{g/(4\pi^2 s_0)}$.

Active elements have the following properties:

- The necessary energy source causes a continuous energy consumption.
- The complex design results in lower reliability, higher manufacturing and maintenance cost.
- All characteristics of passive systems can be achieved. The force generation is independent of local relative displacements, it is controllable by any relative or absolute measured signal even if the sensor location is off the actuator location. Depending on the measurements the actuator can response soft or stiff.

- A soft secondary suspension of the vehicle body, and, therefore, a low eigenfrequency is achievable. A continuously adjustment of the equilibrium condition is possible. The displacement due to additional static load can be adjusted.
- A short reaction time of the actuator allows a fast compensation of external disturbances.

To raise the reliability, sometimes active elements and passive elements are used in parallel.

3.4 Contact Forces between Wheel and Guideway

Wheels represent the most important passive primary suspensions. Wheels are robust and reliable. They have a simple design with low costs of production and maintenance. Wheels combine ideally the tasks of supporting, guiding and driving a vehicle. In comparison to alternative solutions with active components like magnetically levitated vehicles wheels are more economical.

Roughly classified, wheels can be distinguished as rigid and deformable wheels, the latter can be divided into linear elastic wheels (e. g. steel wheels) and wheels with a nonlinear elastic behavior (e. g. rubber wheels, tires). For rigid wheels the contact forces can be determined in a simple manner. Due to the complicated effects in the contact area this is not true for deformable wheels. Hence, many theories describing the rolling contact have been developed, e. g. Kalker (1979), Sperling (1977).

In the following section rolling of rigid and deformable wheels is described. Then, the contact forces for elastic wheels are evaluated (steel wheels), followed by the results for nonlinear elastic wheels (rubber tires).

3.4.1 Rolling of Rigid and Deformable Wheels

In Problems 2.6 and 2.7 of Sect. 2.2 idealized *rigid wheels* on rigid guideways are considered. The main characteristic of rigid contact is the point or line contact, respectively, between the contact partners. However, there are two states of motion which can appear alternatively:

a) pure kinematic rolling and
b) combined rolling and sliding.

These two states of motion are treated now more precisely for a plane motion of a driven homogeneous wheel (mass m, moment of inertia I_C, radius r, drive torque M_C), see Fig. 3.6. In case a) the material point P of the wheel, which corresponds to the contact point, is not moving instantaneously, it sticks to the guideway, $v_P = 0$. The contact point is the instantaneous center of rotation Q of the wheel. Therefore, the kinematic rolling condition

$$v_C = \omega r \tag{3.40}$$

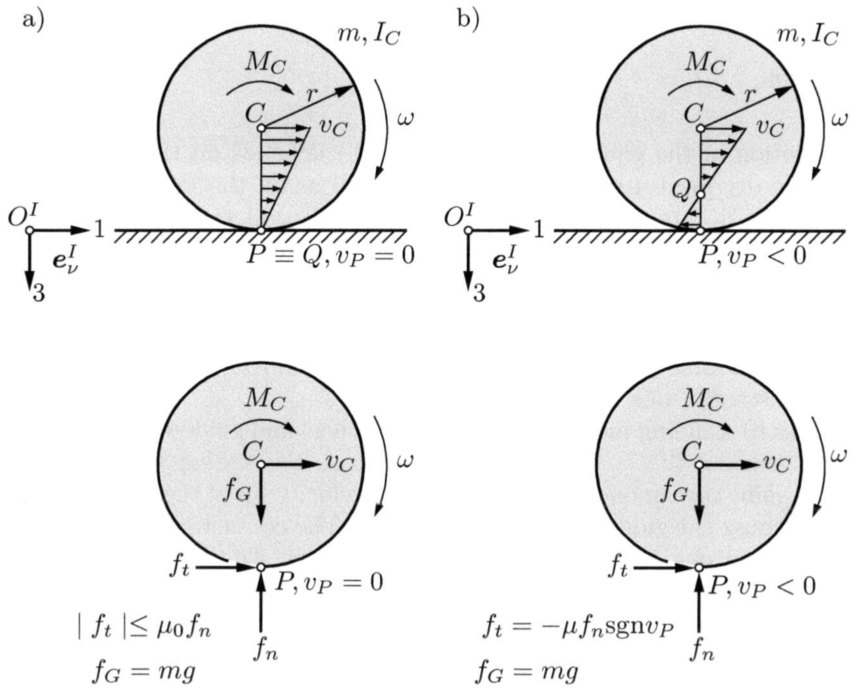

Fig. 3.6. States of motion of a free rigid wheel on a rigid guideway: a) kinematic rolling; b) combination of rolling and sliding

is valid. This condition represents a fixed relation between the velocity of the center of gravity v_C and the rotational velocity ω. The forces f_n, f_t at point P are reaction forces shown in Fig. 3.6 a) in the free body diagram. According to the Coulomb's friction law in the case of sticking it yields

$$|f_t| \leq \mu_0 f_n \ , \quad f_n \geq 0 \ , \tag{3.41}$$

with the coefficient of static friction μ_0. The special case $f_n = 0$ represents a vanishing contact what means a unilateral constraint. Newton's law of motion and Euler's law of motion yield

$$0 = mg - f_n \ , \tag{3.42}$$

$$m\dot{v}_C = f_t \ , \tag{3.43}$$

$$I_C\dot{\omega} = M_C - f_t r \ . \tag{3.44}$$

Together with Eq. (3.40) four equations for the four unknown variables v_C, ω, f_n, f_t are available. The evaluation results in

$$\dot{v}_C = \frac{M_C/r}{m + I_C/r^2} , \quad \omega = v_C/r , \tag{3.45}$$

$$f_n = mg , \quad f_t = \frac{mM_C/r}{m + I_C/r^2} . \tag{3.46}$$

The solution of the equation of motion (3.45) depends on the time behavior of the driving torque $M_C = M_C(t)$ as well as on the initial conditions. Pure rolling occurs as long as the condition of Eq. (3.41) is fulfilled. Thus, Eq. (3.46) yields a condition for the drive torque M_C,

$$|M_C| \leq \mu_0(m + I_C/r^2)gr . \tag{3.47}$$

The reaction force f_t does not perform any mechanical work, it acts in the wheel's instantaneous center of rotation.

In case b) a sliding motion exists between wheel and guideway. This occurs if the condition (3.47) is violated. In this case the sliding velocity in the contact point corresponds with the absolute velocity v_P of the point P of the wheel because the guideway is inertially fixed. The contact point is no longer the instantaneous center of rotation of the wheel. The following equation describes the velocity $\boldsymbol{v}_P$ in the inertial frame,

$$\boldsymbol{v}_P^I = \boldsymbol{v}_C^I + \tilde{\boldsymbol{\omega}}^I \boldsymbol{r}_{CP}^I = \begin{bmatrix} v_C \\ 0 \\ 0 \end{bmatrix} + \begin{bmatrix} 0 & 0 & -\omega \\ 0 & 0 & 0 \\ \omega & 0 & 0 \end{bmatrix} \begin{bmatrix} 0 \\ 0 \\ r \end{bmatrix} = \begin{bmatrix} v_C - r\omega \\ 0 \\ 0 \end{bmatrix} . \tag{3.48}$$

The ratio ν of the sliding velocity $v_P = v_C - r\omega$ to the rolling velocity v_C is denoted as rigid body slip

$$\nu = \frac{v_P}{v_C} = \frac{v_C - r\omega}{v_C} . \tag{3.49}$$

The force f_t in point P shown in the free body diagram, Fig. 3.6 b), is an applied force, which follows from the Coulomb's friction law in the case of sliding,

$$f_t = -\mu f_n \frac{v_P}{|v_P|} = -\mu f_n \, \mathrm{sgn} v_P , \quad f_n \geq 0 , \quad v_P = v_C - r\omega , \tag{3.50}$$

with the kinematic coefficient of friction ($\mu < \mu_0$). The applied force f_t depends on the reaction force f_n. Newton's law of motion and Euler's law of motion, Eqs. (3.42)-(3.44), hold in this case, too. In combination with Eq. (3.50) four equations for the four unknown variables v_C, ω, f_n, f_t are available again. Now the evaluation results in

$$\dot{v}_C = -\mu g \, \mathrm{sgn}(v_C - r\omega) , \quad \dot{\omega} = \frac{1}{I_C}(M_C - rm\dot{v}_C) , \tag{3.51}$$

$$f_n = mg , \quad f_t = -\mu mg \, \mathrm{sgn}(v_C - r\omega) . \tag{3.52}$$

The equations of motion (3.51) are coupled. To solve the problem a direction of $v_P = v_C - r\omega$ is assumed, e. g. $v_P < 0$, then $\dot{v}_P$ is determined and the differential equation for ω can be solved in dependency of $M_C = M_C(t)$ and the initial conditions. The validity of the assumption must be checked afterwards. If the assumption is wrong, the calculation must be repeated with the contrary assumption. The force f_t is dissipative. It performs the mechanical work

$$W = \int_0^t f_t v_P \mathrm{d}t = -\int_0^t \mu m g \left| v_P \right| \mathrm{d}t \,. \tag{3.53}$$

A comparison of both cases shows that in case a) the contact point is the instantaneous center of rotation of the wheel but not in case b). In case a) the force f_t is a reaction force and in case b) the force is an applied, dissipative force. The always existing friction reacts in different manner. In case a) the static friction force is used to accelerate the wheel center where a particular limit value must not be exceeded. In case b) the friction results in a sliding force. In both cases a sufficient number of equations are at hand to solve the dynamic problem completely.

Now *deformable wheels* on real non-rigid guideways are described. The main characteristic of a deformable rolling contact is a finite contact area between the contact partners.

In the following it is assumed that the eigenfrequencies of the contact partners and of the guideway are much higher than the frequencies of deformation in the contact patch. This is equivalent to the assumption of low frequency motions and perturbations. The assumption is true for vehicle performance problems like stability of motion, driving safety and driving comfort but not for structural dynamic problems like structure-borne noise radiation, noise development etc. For the vehicle dynamic problems handled here, it is assumed that the contact partners show elastic behavior only in the contact zone and the rest of the body is considered to be rigid. In this combination the models consist of rigid bodies with an elastic contact pair. In the following the focus is constricted to linear elastic contact pairs.

For static elastic contact pairs the contact area and the dedicated contact pressure in dependency of normal forces, principal curvatures and material parameters were calculated by Hertz (1895). His results are discussed later in more detail.

The effects of a rolling elastic contact pair with friction was treated by Carter (1926) and Fromm (1927), cp. also Krause and Poll (1980) using the Hertzian theory for discoidal wheels and plane motion (two-dimensional problem). Based on that, an illustrative physical explanation of the complicated processes in the contact area is given.

First, the rolling process is described using a brush model, see Fig. 3.7, for better understanding. If a circular brush is rotated and linear moved over a plane under normal load, it can be seen, that the brush hairs stick at the

Fig. 3.7. Deformations of the brush hairs

run-in of the contact area and show an increasing bending until the brush hairs slide back near the run-out edge. Such a brush model can be used as a discretized mechanical model for an elastic wheel. If elastic leaf springs are used instead of the brush hairs, qualitatively the forces shown in Fig. 3.8 a) are obtained where the contact area EA is divided into a sticking zone EG and a sliding zone GA. Due to the elastic deformation tangential static friction forces occur in the sticking zone while in the sliding zone kinematic friction forces emerge. The sum of all the different tangential forces result in the total tangential force. Therefore, the rolling elastic contact is much more complex as in thc casc of rigid contact due to the simultaneous existence of sticking and sliding in the contact area. This is also true for the motion because generally there is a small relative velocity v_C between the contact partners. This relative motion related to the rolling velocity is called slip. Two contributions to the slip can be distinguished:

a) deformation or micro slip caused by elastic deformation in the contact area,
b) rigid body slip ν as a result of the violation of the rolling condition which may be also found for a rigid contact pair, cp. Eq. (3.49).

The sum of both slip contributions is called total slip or true slip.

Using a two-dimensional continuous model shown in Fig. 3.8 b) in the case of $\mu = \mu_0$, the principle of rolling elastic contact is explained. The normal pressure distribution p is shown for the contact line EA corresponding to the Hertzian theory. According to the Coulomb's friction law, the shear stress distribution τ is proportional to the normal pressure distribution in the sliding

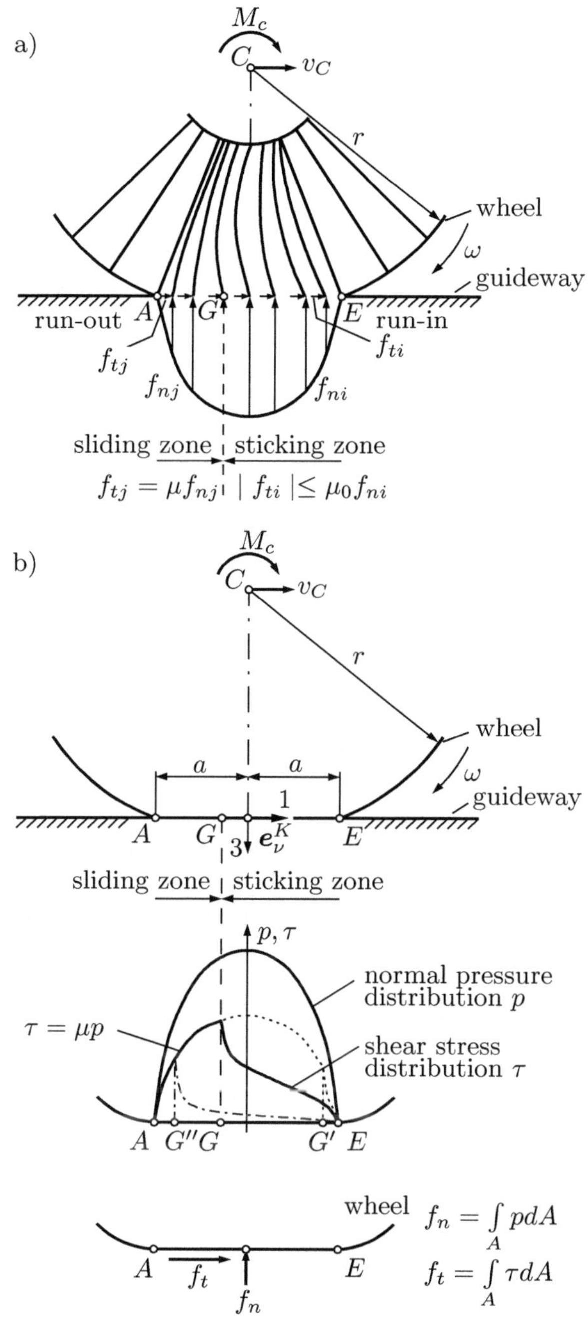

Fig. 3.8. Rolling elastic contact with friction: a) plane discrete mechanical model (brush model); b) plane continuous mechanical model

zone GA, $\tau = \mu p$. In the sticking zone GE the shear stresses are smaller. The integral over the different shear stresses τ results in the total tangential force $f_t = \int\limits_A \tau \mathrm{d}A$. The normal pressure distribution p provides the total normal force $f_n = \int\limits_A p\mathrm{d}A$.

The location of the boundary point G between the sticking and sliding zone is affected fundamentally by the rigid body slip ν. With large slip the marginal point G is closely situated to the run-in E (G' in Fig. 3.8 b); the shear stresses τ and the transmitted force f_t are high. With smaller slip ν the marginal point G moves close to the run-out A (G'' in Fig. 3.8 b). Here, the shear stresses τ and the force f_t are small. In the extreme case with a vanishing slip ($\nu = 0$) the sticking zone covers the full contact area ($G \equiv A$). Then the shear stresses τ disappear and it holds $f_t = 0$. The resulting force f_t increases with increasing slip due to a smaller sticking zone and a larger sliding zone. If the sliding zone is equal to the total contact zone ($G \equiv E$) a maximum tangential force occurs. This is the case for a certain value of rigid body slip ν. A larger slip results only in a higher sliding velocity. Similar to the rigid sliding contact the force f_t remains constant.

The main variable for the generation of forces in rolling contact is the slip ν. Therefore, the force f_t is described in dependency of the slip ν. The tangential force f_t is an applied force like the Coulomb sliding friction force and acts in the counter direction of the sliding velocity and the slip whereas the normal force f_n is a reaction force. The f_t-ν-characteristic is odd, $f_t(\nu) = -f_t(-\nu)$.

The ratio of the tangential and normal force $\bar{\varphi} = f_t/|f_n|$, the so-called rolling contact coefficient is generally plotted in dependency of the absolute value $|\nu|$ of the slip.

The contribution of Carter (1926) does not distinguish between the coefficient of sticking and sliding friction, $\mu = \mu_0$, the coefficient of friction $\mu = \text{const}$ is assumed which is independent of the sliding velocity. Then, the characteristic $\bar{\varphi} = \bar{\varphi}(\nu)$ shown in Fig. 3.9 a) is obtained. A description with a normalized rolling contact coefficient $\varphi = f_t/(\mu\,|f_n|) = \varphi(\nu)$ with $\varphi \le 1$ is also common. In practise the coefficient of kinematic or sliding friction, respectively, depends on the sliding velocity and, therefore, on the slip, $\mu = \mu(|\nu|)$, with a falling characteristic, cp. e. g. Kraft (1976). Then, a characteristic as shown in Fig. 3.9 b) holds. The occurring maximum $\bar{\varphi}_{max}$ is named maximum rolling contact coefficient.

The explanation given by Carter (1926), describing the two-dimensional rolling of a discoidal wheel with elastic contact and friction, is qualitatively also valid for an arbitrarily shaped wheel and a general motion (three-dimensional problem). The explanation is valid for pairing between steel wheel and rail as well as with some restrictions between tire and road. For the three-dimensional problem there is no preferred direction of the sliding velocity what results in different slip variables. It has to be distinguished between longitudinal, laterallateral and aligning slip. Correspondingly longitudinal and lateral force

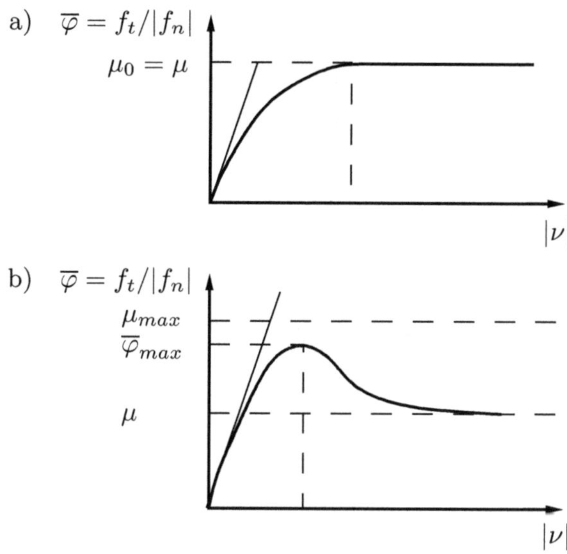

Fig. 3.9. The rolling contact coefficient $\bar{\varphi}$ in dependency of the slip ν: a) coefficients of friction $\mu_0 = \mu = \text{const}$; b) coefficient of kinematic friction $\mu = \mu(\nu)$, $\mu_0 > \mu$

tangential in the contact patch and an aligning torque perpendicular to the contact patch appears, too. These forces and the torque are applied ones but they depend on the normal force in the contact area which is a reaction force.

3.4.2 Definition of the Rigid Body Slip

The consideration of the planar motion of a rolling wheel so far shows the central role of the rigid body slip for the evaluation of the contact forces. Therefore, the slip between the contact partners must be defined in the case of a spatial motion, too. In Fig. 3.10, the contact partners wheel R and guideway S are separated from each other and a plane through the contact point P is displayed. The wheel plane is perpendicular to the wheel axis. Point P denotes the contact point, where the two contact partners R and S, which are assumed here to be rigid with regard of their contact, touch each other. The points P_R and P_S denote material points of R and S, respectively, which coincide instantaneously with the contact point P. For an elastic contact, the contact area is characterized by the plane K. In standard applications, the orientation of this plane corresponds to a tangential plane through P for locally rigid contact partners. For the kinematic description the following Cartesian coordinate frames are introduced:

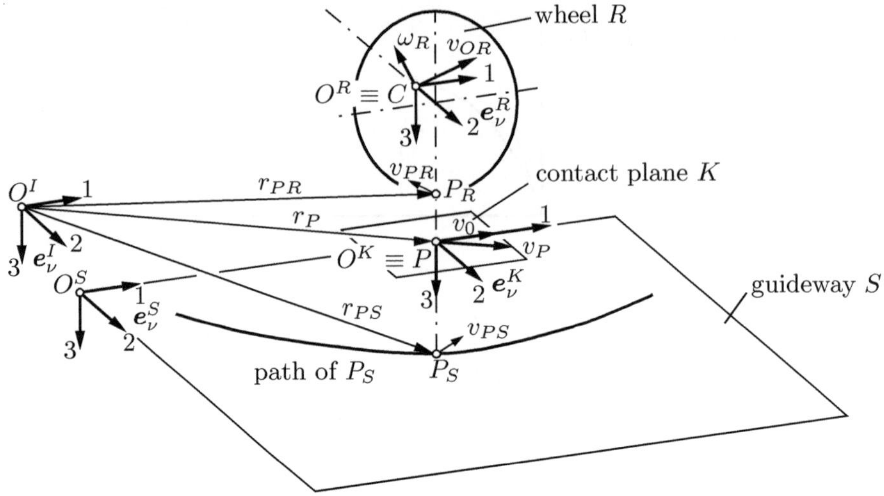

Fig. 3.10. Description of contact pairs for the calculation of rigid body slip

Inertial frame	I	$\{O^I, \mathbf{e}_\nu^I\}$,	
Wheel fixed frame	R	$\{O^R, \mathbf{e}_\nu^R\}$,	$O^R \equiv C$,
Guideway tangential frame	S	$\{O^S, \mathbf{e}_\nu^S\}$,	
Contact plane frame	K	$\{O^K, \mathbf{e}_\nu^K\}$,	$O^K \equiv P$.

For rotationally symmetric wheels an axle-fixed frame $\{O^Z, \mathbf{e}_\nu^Z\}$, which does not rotate around the symmetry axis, is often used instead of the body-fixed frame R. In Fig. 3.10, all $\mathbf{e}_3$-base vectors are assumed to be parallel, although this is not generally the case. The additionally introduced frame K with its origin P is fixed to the moving contact patch. The contact plane is defined by the base vectors $\mathbf{e}_1^K$ and $\mathbf{e}_2^K$ where $\mathbf{e}_1^K$ is perpendicular to the wheel axis and heads in driving direction, $\mathbf{e}_2^K$ points to the right with regard to the driving direction, while $\mathbf{e}_3^K$ is perpendicular to the contact plane and points downwards, cp. Fig. 3.10. In the following, all rigid body slip variables as well as the contact force variables are related to the contact plane frame K.

The motions of the contact partners relative to each other are restricted by constraints. The relative velocity does only occur in the contact plane, with the relative rotation perpendicular to it. The slip variables

$$\boldsymbol{\nu}^K = \begin{bmatrix} \nu_1 \\ \nu_2 \\ \nu_3 \end{bmatrix} = \frac{1}{v_0} \begin{bmatrix} \bar{v}_1^K \\ \bar{v}_2^K \\ \bar{\omega}_3^K \end{bmatrix} \tag{3.54}$$

can be derived from the coordinates of the sliding velocity $\bar{\boldsymbol{v}}_P^K$ at P and the sliding angular velocity $\bar{\boldsymbol{\omega}}^K$ described in frame K. In general, the slip variables are related to the wheel velocity v_0 in driving direction while in the case of a wheelset the mean value of the velocity at the left and right wheel is used. As it is shown in Fig. 3.10, this speed v_0 is equal to the absolute velocity of the contact point P in $\boldsymbol{e}_1^K$-direction.

The sliding velocity vectors $\{\bar{\boldsymbol{v}}_P, \bar{\boldsymbol{\omega}}\}$ result from the difference between the absolute velocities $\{\boldsymbol{v}_{PR}, \boldsymbol{\omega}_R\}$ and $\{\boldsymbol{v}_{PS}, \boldsymbol{\omega}_S\}$ of wheel R and guideway S, respectively, evaluated at P_R and P_S, which are coinciding with the contact point P. Alternatively, the sliding velocity can be determined from the difference between the corresponding relative velocities $\{\boldsymbol{v}_{PR,rel}, \boldsymbol{\omega}_{R,rel}\}$ and $\{\boldsymbol{v}_{PS,rel}, \boldsymbol{\omega}_{S,rel}\}$ with respect to the contact point P. This yields

$$\begin{aligned} \bar{\boldsymbol{v}}_P^K &= \boldsymbol{v}_{PR}^K - \boldsymbol{v}_{PS}^K = \boldsymbol{v}_{PR,rel}^K - \boldsymbol{v}_{PS,rel}^K , \\ \bar{\boldsymbol{\omega}}^K &= \boldsymbol{\omega}_R^K - \boldsymbol{\omega}_S^K \;\;= \boldsymbol{\omega}_{R,rel}^K - \boldsymbol{\omega}_{S,rel}^K \end{aligned} \tag{3.55}$$

in the contact plane frame K. Thus, motions of the guideway are also admitted which may be due to structural vibrations. Considering Eq. (3.55), the slip can be defined as

$$\boldsymbol{\nu}^K = \begin{bmatrix} \nu_1 \\ \nu_2 \\ \nu_3 \end{bmatrix} = \frac{1}{v_0} \begin{bmatrix} v_{R1}^K - v_{S1}^K \\ v_{R2}^K - v_{S2}^K \\ \omega_{R3}^K - \omega_{S3}^K \end{bmatrix} = \frac{1}{v_0} \begin{bmatrix} v_{R1,rel}^K - v_{S1,rel}^K \\ v_{R2,rel}^K - v_{S2,rel}^K \\ \omega_{R3,rel}^K - \omega_{S3,rel}^K \end{bmatrix} . \tag{3.56}$$

The slip process represents a plane motion, which is described by two translational and one rotational kinematic quantities. The longitudinal slip ν_1 as well as the lateral slip ν_2 are derived from the difference between the absolute or relative velocities at the points P_R and P_S with regard to the orientation of $\boldsymbol{e}_1^K$ and $\boldsymbol{e}_2^K$, respectively. Whereas the aligning slip ν_3 results from the difference between the angular velocities in $\boldsymbol{e}_3^K$-direction perpendicular to the contact plane. The aligning slip is also called spin creepage in railway engineering. The longitudinal slip ν_1 and lateral slip ν_2 are non-dimensional variables, they are commonly represented in percent or one tenth of a percent ($\nu_i \,\widehat{=}\, \nu_i \cdot 100\% \,\widehat{=}\, \nu_i \cdot 1000‰$, $i = 1, 2$); the aligning slip has the unit 1/length.

In general, the calculation of slip is difficult. It depends on the current contact geometry and motion which differ from application to application. Hence, only some general hints can be given. It is appropriate to evaluate the absolute velocity $\boldsymbol{v}_{Pi}$ at the material points P_i, $i = R, S$, by means of the kinematic relations of rigid bodies described in the inertial frame I, cp. Eq. (2.21),

$$\boldsymbol{v}_{Pi}^I = \boldsymbol{v}_{Oi}^I + \tilde{\boldsymbol{\omega}}_i^I \boldsymbol{\rho}_{Oi,Pi}^I , \tag{3.57}$$

$$i = R, S,$$

$$\boldsymbol{v}_{Pi}^K = \boldsymbol{S}^{KI} \boldsymbol{v}_{Pi}^I , \tag{3.58}$$

or in the moving body-fixed reference frame $i = R$ or $i = S$, respectively, cp. Eq. (2.26),

$$\boldsymbol{v}_{Pi}^{i} = \boldsymbol{v}_{Oi}^{i} + \tilde{\boldsymbol{\omega}}_{i}^{i}\boldsymbol{\rho}_{Oi,Pi}^{i} + \dot{\boldsymbol{\rho}}_{Oi,Pi}^{i} \,, \qquad i = R, S, \tag{3.59}$$

$$\boldsymbol{v}_{Pi}^{K} = \boldsymbol{S}^{Ki}\boldsymbol{v}_{Pi}^{i} \,, \tag{3.60}$$

followed by a transformation into the contact plane frame K. Because P_i is a material point of the rigid body, the relative velocity vanishes, $\dot{\boldsymbol{\rho}}_{Oi,Pi} \equiv \mathbf{0}$. Therefore, Eq. (3.59) is simplified to

$$\boldsymbol{v}_{Pi}^{i} = \boldsymbol{v}_{Oi}^{i} + \tilde{\boldsymbol{\omega}}_{i}^{i}\boldsymbol{\rho}_{Oi,Pi}^{i} \,, \quad i = R, S \,. \tag{3.61}$$

If an axle-fixed frame $\{O^Z \equiv C, \boldsymbol{e}_\nu^Z\}$ is used instead of R, Eq. (3.59) is still valid for $i = Z$ and $\dot{\boldsymbol{\rho}}_{C,PR}^{Z} = \tilde{\boldsymbol{\omega}}_{ZR}^{Z}\boldsymbol{\rho}_{C,PR}^{Z}$, which leads formally to Eq. (3.61), regarding $\boldsymbol{\omega}_Z \equiv \boldsymbol{\omega}_R = \boldsymbol{\omega}_{IZ} + \boldsymbol{\omega}_{ZR}$. If the contact points P_i change relatively with regard to the bodies i, $i = R, S$, the current position vectors $\boldsymbol{\rho}_{Oi,Pi}$ just have to be applied to Eq. (3.57) and Eq. (3.59). In many applications the guideway is fixed, which implies $\boldsymbol{v}_{PS}^{K} \equiv \mathbf{0}$ and $\boldsymbol{\omega}_S^K \equiv \mathbf{0}$.

The evaluation of the relative velocities $\boldsymbol{v}_{Pi,rel}$ for $i = R, S$ is appropriately done in the contact plane frame K. Therefore, it is useful to consider the contact plane K as motionless while the bodies R and S move above or below K, respectively. This yields to negative values $v_{i1,rel}^{K}$, $i = R, S$, of the longitudinal relative velocities.

3.4.3 Contact Forces for Elastic Wheels on Elastic Rails

This section describes some results of Kalker's theory, cp. Kalker (1979), for the three-dimensional problem of an rolling elastic contact with friction. In the case of ideal test conditions, Kalker's theory has been confirmed by experiments. Therefore, it is accepted worldwide in railway engineering. First, a single wheel is considered. Then, the conclusions are transferred to a wheelset which is an important component of railway vehicles.

Initially, the slip of a rigid body, introduced in Sect. 3.4.2, is considered. When calculating the slip of real wheel/rail pairs, a difficulty arise from the evaluation of the contact point position, depending on gauge and profile geometry as well as the location of the wheelset related to the track. In this context, it is advisable to use special algorithms for the contact geometry or to refer to measured data, cp. e. g. Garg and Dukkipati (1984). Often the evaluation is simplified by approximations with regard to contact geometry, e. g. conical wheels and a rail profile based on a combination of circular arcs of different radii. If the yaw angle of the wheel axis is small, the lateral displacement of the rail's and wheel's contact point are of the same size, therefore, no lateral slip occurs, cp. Sect. 3.4.2. Hence, the lateral slip can be evaluated by the assumption of the contact point lying in the middle of the rail.

■ Problem 3.2 Slip for a conical wheel

Evaluate the slip variables considering a conical wheel (rolling radius r_0, apex angle 2δ, $\delta \ll 1$), which is running straight ahead on a fixed rail with center velocity v_C and angular velocity ω, cp. Fig. 3.11, without any lateral motion.

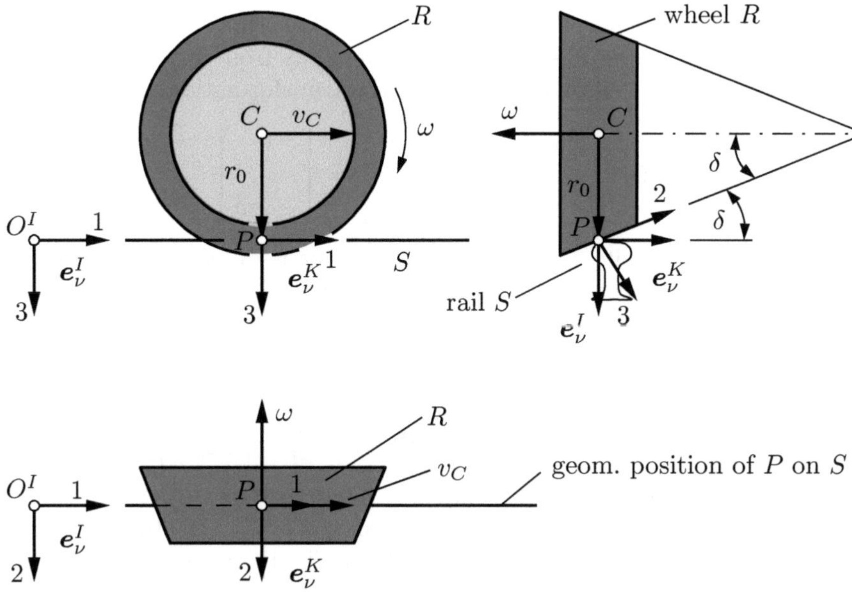

Fig. 3.11. Conical wheel with slip on a rail

Solution

The evaluation of the sliding velocity $\bar{\boldsymbol{v}}_P^K$ between wheel and rail can be carried out in two manners. In the contact plane frame K, which is moving with velocity v_C along the rail, the sliding velocity $\bar{\boldsymbol{v}}_P^K$ directly results from the difference of the relative velocities at the material points of wheel and rail coinciding with the contact point P,

$$\bar{\boldsymbol{v}}_P^K = \boldsymbol{v}_{PR,rel}^K - \boldsymbol{v}_{PS,rel}^K = \begin{bmatrix} -r_0\omega \\ 0 \\ 0 \end{bmatrix} - \begin{bmatrix} -v_C \\ 0 \\ 0 \end{bmatrix} = \begin{bmatrix} v_C - r_0\omega \\ 0 \\ 0 \end{bmatrix} . \tag{1}$$

When using the inertial frame I, the sliding velocity $\boldsymbol{v}_{PR}^I$ of the wheel's point currently corresponding with the contact point P arise from the rigid body relation given in Eq. (3.57). The absolute velocity of rail point $\boldsymbol{v}_{PS}^I$ coinciding with P is vanishing. The transformation into the contact plane frame K by

$\boldsymbol{S}^{KI} = \boldsymbol{\delta}_1$ does not change the coordinates in longitudinal direction, therefore, the result

$$\bar{\boldsymbol{v}}_P^K = \boldsymbol{S}^{KI}\left(\boldsymbol{v}_{PR}^I - \boldsymbol{0}\right) = \begin{bmatrix} 1 & 0 & 0 \\ 0 & \mathrm{c}\delta & -\mathrm{s}\delta \\ 0 & \mathrm{s}\delta & \mathrm{c}\delta \end{bmatrix} \begin{bmatrix} v_C - r_0\omega \\ 0 \\ 0 \end{bmatrix} = \begin{bmatrix} v_C - r_0\omega \\ 0 \\ 0 \end{bmatrix} \tag{2}$$

is identical to Eq. (1). To evaluate the sliding angular velocity $\bar{\boldsymbol{\omega}}^K$ its description in the inertial frame I is appropriate. One gets from $\boldsymbol{\omega}_R^I = [0 \; -\omega \; 0]^{\mathrm{T}}$, cp. Fig. 3.11, and $\boldsymbol{\omega}_S^I \equiv \boldsymbol{0}$ by analogy with the transformation Eq. (2)

$$\bar{\boldsymbol{\omega}}^K = \boldsymbol{S}^{KI}\left(\boldsymbol{\omega}_R^I - \boldsymbol{0}\right) = \begin{bmatrix} 1 & 0 & 0 \\ 0 & \mathrm{c}\delta & -\mathrm{s}\delta \\ 0 & \mathrm{s}\delta & \mathrm{c}\delta \end{bmatrix} \begin{bmatrix} 0 \\ -\omega \\ 0 \end{bmatrix} = \begin{bmatrix} 0 \\ -\omega\mathrm{c}\delta \\ -\omega\mathrm{s}\delta \end{bmatrix} . \tag{3}$$

Using the related velocity $v_0 = v_C$, the slip vector is obtained according to Eq. (3.56). It can be linearized for small angles $\delta \ll 1$:

$$\boldsymbol{\nu}^K \equiv \begin{bmatrix} \nu_1 \\ \nu_2 \\ \nu_3 \end{bmatrix} = \frac{1}{v_0} \begin{bmatrix} \bar{v}_1^K \\ \bar{v}_2^K \\ \bar{\omega}_3^K \end{bmatrix} = \frac{1}{v_C} \begin{bmatrix} v_C - r_0\omega \\ 0 \\ -\omega\mathrm{s}\delta \end{bmatrix} \approx \begin{bmatrix} 1 - r_0\dfrac{\omega}{v_C} \\ 0 \\ -\dfrac{\omega}{v_C}\delta \end{bmatrix} . \tag{4}$$

Assuming for instance $r_0\omega \approx 1.001 v_C$, a slip vector of $\boldsymbol{\nu}^K \approx \left[-0.001 \; 0 \; -\delta/r_0\right]^{\mathrm{T}}$ is obtained. In this example, the longitudinal slip is given by the related difference between the wheel's centroid and circumferential velocity. Whereas the aligning slip ν_3 results from the projection of the wheel's angular velocity on the $\boldsymbol{e}_3^K$ base vector of the contact plane. ■

Using the slip variables, the rigid body sliding velocity $\boldsymbol{v}_Q^K$ can be specified in each point Q of the contact plane K. Sliding means a plane motion, hence, under consideration of the position vector $\boldsymbol{r}_{PQ}^K = [x \; y \; 0]^{\mathrm{T}}$, cp. Fig. 3.12, it yields

$$\begin{aligned} \boldsymbol{v}_Q^K &= \boldsymbol{v}_P^K + \tilde{\bar{\boldsymbol{\omega}}}^K \boldsymbol{r}_{PQ}^K \\ &= \begin{bmatrix} \bar{v}_1^K \\ \bar{v}_2^K \\ 0 \end{bmatrix} + \begin{bmatrix} 0 & -\bar{\omega}_3^K & 0 \\ \bar{\omega}_3^K & 0 & 0 \\ 0 & 0 & 0 \end{bmatrix} \begin{bmatrix} x \\ y \\ 0 \end{bmatrix} = \begin{bmatrix} \bar{v}_1^K - \bar{\omega}_3^K y \\ \bar{v}_2^K + \bar{\omega}_3^K x \\ 0 \end{bmatrix} . \end{aligned} \tag{3.62}$$

Together with Eq. (3.54) one gets

$$\boldsymbol{v}_Q^K \equiv \boldsymbol{v}(x, y) = v_0 \begin{bmatrix} \nu_1 - \nu_3 y \\ \nu_2 + \nu_3 x \\ 0 \end{bmatrix} . \tag{3.63}$$

Due to the elastic deformation of wheel and rail in the contact plane as well as the material point's motion relative to the coordinate frame K approximated

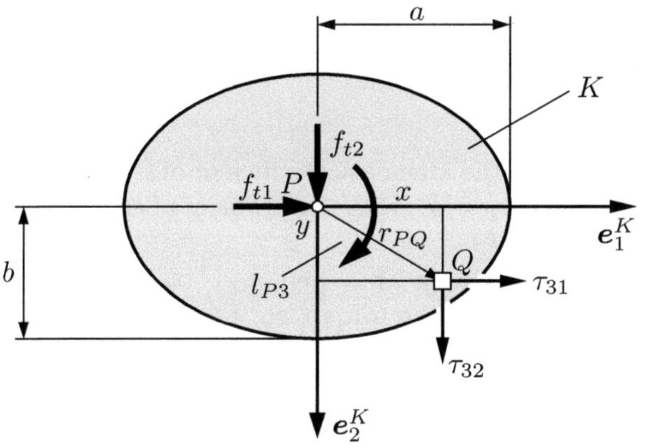

Fig. 3.12. Contact plane K with forces acting on the wheel

by the velocity $-v_0 \boldsymbol{e}_1^K$, the true sliding velocity $\boldsymbol{w}(x,y)$ at point Q depends on the rigid body's sliding velocity $\boldsymbol{v}(x,y)$ and the material derivative $\dot{\boldsymbol{u}}(x,y,t)$ of the elastic surface displacements $\boldsymbol{u}^K = \boldsymbol{u}_R^K - \boldsymbol{u}_S^K = [u_1\ u_2\ 0]^{\mathrm{T}}$. That is, cp. Kalker (1979),

$$\boldsymbol{w}_Q^K \equiv \boldsymbol{w}(x,y) = \boldsymbol{v}(x,y) + \dot{\boldsymbol{u}}(x,y,t)$$
$$= v_0 \begin{bmatrix} \nu_1 - \nu_3 y \\ \nu_2 + \nu_3 x \\ 0 \end{bmatrix} + \begin{bmatrix} \partial u_1/\partial t - v_0 \partial u_1/\partial x \\ \partial u_2/\partial t - v_0 \partial u_2/\partial x \\ 0 \end{bmatrix} . \tag{3.64}$$

The shape of the contact area K and surface pressure $p(x,y) = |\sigma_{33}(x,y)|$ are assumed according to the static elastic contact described by Hertz (1895). Hence, the contact area is an ellipse with the semi-axes a and b. The surface pressure at point Q with its coordinates x, y reads as

$$p(x,y) = \frac{3|f_n|}{2\pi ab}\sqrt{1 - \left(\frac{x}{a}\right)^2 - \left(\frac{y}{b}\right)^2} , \tag{3.65}$$

with $|f_n|$ representing the absolute value of the normal force in $\boldsymbol{e}_3^K$-direction. Sections which are parallel to either of the semi-axis, i.e. $x = \text{const}$ or $y = \text{const}$ exhibit elliptic pressure distribution. The friction characteristic at the contact patch is described by Coulomb's laws of friction. If the tangential stresses τ_{31} and τ_{32} acting on the wheel are combined to the vector $\boldsymbol{\tau} = \boldsymbol{\tau}(x,y) = [\tau_{31}\ \tau_{32}]^{\mathrm{T}}$, the sticking or sliding zone, respectively, are characterized by

$$|\boldsymbol{\tau}| \leq \mu p \quad \text{for} \quad \boldsymbol{w} = \boldsymbol{0} , \tag{3.66}$$

or

$$\left.\begin{aligned} \boldsymbol{\tau} &= -\mu p \frac{\boldsymbol{w}}{|\boldsymbol{w}|} \\ |\boldsymbol{\tau}| &= \mu p \end{aligned}\right\} \quad \text{for} \quad \boldsymbol{w} \neq \mathbf{0} \ , \tag{3.67}$$

where $\mu = \text{const}$ denotes the kinematic coefficient of friction; the static coefficient of friction is set to $\mu_0 = \mu$. Now, the problem of contact force evaluation can be expressed as follows.

Firstly, the tangential stresses $\boldsymbol{\tau} = \left[\tau_{31}\ \tau_{32}\right]^{\mathrm{T}}$ are determined in order to derive the contact forces and the torque, cp. Fig. 3.12,

$$\boldsymbol{f}^K = \left[f_{t1}\ f_{t2}\ l_{P3}\right]^{\mathrm{T}} \ , \tag{3.68}$$

$$f_{t1} = \int_K \tau_{31} \mathrm{d}x\mathrm{d}y \ , \tag{3.69}$$

$$f_{t2} = \int_K \tau_{32} \mathrm{d}x\mathrm{d}y \ , \tag{3.70}$$

$$l_{P3} = \int_K (\tau_{32} x - \tau_{31} y) \mathrm{d}x\mathrm{d}y \ , \tag{3.71}$$

under consideration of Eqs. (3.66) and (3.67) with the real sliding velocity $\boldsymbol{w}$ given by Eq. (3.64). The vectors of surface displacement $\boldsymbol{u}$ and tangential stress $\boldsymbol{\tau}$ are linked together by the constitutive relations of an elastic half-space. Now, the border between sticking and sliding zone is evaluated for given values ν_1, ν_2, ν_3, v_0, $|f_n|$, a, b. This reveals a boundary value problem with a free boundary, which in general has to be solved numerically. The relation

$$\boldsymbol{f}^K = \boldsymbol{f}^K(\boldsymbol{\nu}^K) \tag{3.72}$$

is referred to as contact force to slip relation or contact force law, respectively. In the case of a rolling elastic contact between wheel and rail the underlying theories can be distinguished according to Kalker (1979) by their complexity.

1) Simplified theory with the material law specified by

$$\boldsymbol{u} = \begin{bmatrix} u_1 \\ u_2 \end{bmatrix} = \begin{bmatrix} \gamma_1 \tau_{31} \\ \gamma_2 \tau_{32} \end{bmatrix} \ . \tag{3.73}$$

2) Exact theory with the material law derived from the basic equations of elasticity theory where the contact partners are assumed as isotropic elastic half-spaces.

3) Linear theory for small sliding velocities. According to Eq. (3.64) it is assumed that $\boldsymbol{w}(x,y) \to \mathbf{0}$. This yields to a linear contact force law

$$\boldsymbol{f}^K = -\boldsymbol{F}\boldsymbol{\nu}^K \,, \quad \boldsymbol{F} = \mathbf{const}\,. \tag{3.74}$$

where the 3×3 -contact matrix $\boldsymbol{F}$ appears. The linear theory represents an approximation, because the friction law is violated at the run-out. It's value consists in an approximation of the exact theory for $\boldsymbol{\nu} \to \mathbf{0}$. This is depicted by Fig. 3.9. The tangents at the origin can be correctly determined by means of the linear theory.

A theory is denoted dynamic or quasistatic, depending on wether the inertia effects are considered or not. Additionally, the rolling contact is called steady or unsteady, if the terms $\partial u_\nu/\partial t$, $\nu = 1,2$, in Eq. (3.64) are vanishing or not, respectively. For most applications it is sufficient to use quasistatic and steady theories.

In terms of tangential forces the following symmetry conditions are generally valid:

$$f_{t1}(\nu_1,\nu_2,\nu_3) = -f_{t1}(-\nu_1,\nu_2,\nu_3) = f_{t1}(\nu_1,-\nu_2,-\nu_3)\,, \tag{3.75}$$

$$f_{t2}(\nu_1,\nu_2,\nu_3) = f_{t2}(-\nu_1,\nu_2,\nu_3) \quad = -f_{t2}(\nu_1,-\nu_2,-\nu_3)\,. \tag{3.76}$$

3.4.3.1 Linear Law of Contact Forces

In the following the half-axes a, b of the contact ellipse are determined by Hertz's theory and the results of the linear theory of Kalker are presented. It is assumed that the $\boldsymbol{e}_1^K, \boldsymbol{e}_3^K$- and the $\boldsymbol{e}_2^K, \boldsymbol{e}_3^K$- plane are the principal curvature planes of wheel and rail track, respectively, in the contact point P, what is in real applications not exactly the case. The deviations can be neglected if there are only small angles between the principal curvature planes and the coordinate plane. The starting point for the calculation of the half-axes of the contact ellipse are the principal curvature radii $R_{R1}, R_{R2}, R_{S1}, R_{S2}$, with reference to the contact point P for wheel and rail in the undeformed state, see Fig. 3.13. The principal curvature radii R_{R1}, R_{S1} can be determined in the $\boldsymbol{e}_1^K, \boldsymbol{e}_3^K$- plane. For a straight rail track $R_{S1} \to \infty$ holds while R_{R1} is given by the rolling radius r_0, $R_{R1} = r_0$. The principal curvature radii R_{R2}, R_{S2} are visible in the $\boldsymbol{e}_2^K, \boldsymbol{e}_3^K$- plane. A principal curvature radius is positive if the body considered is convex in P, and the radius is negative if the body is concave. In Fig. 3.13 all principal curvature radii are positive. The half-axes a and b of the ellipse are for contacting bodies of the same material

$$\boldsymbol{e}_1^K : a = m\sqrt[3]{\frac{3(1-\bar{\nu}^2)\,|f_n|}{E(A+B)}}\,, \tag{3.77}$$

$$\boldsymbol{e}_2^K : b = n\sqrt[3]{\frac{3(1-\bar{\nu}^2)\,|f_n|}{E(A+B)}} \tag{3.78}$$

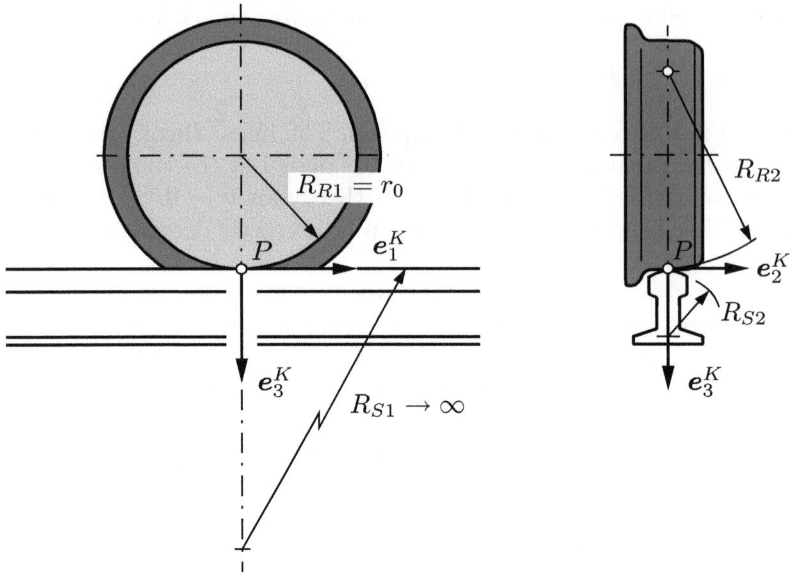

Fig. 3.13. Principal curvature radii of wheel and rail

where:

m, n	parameters,
$\lvert f_n \rvert$	absolute value of the normal force,
$E, \bar{\nu}$	Young's modulus and Poisson's ratio of the material of the contact pair (e. g. steel: $E \approx 210\,\mathrm{kN/mm^2}$, $\bar{\nu} \approx 0.3$)
A, B	curvatures of the contact partners in both principal curvature planes,

$$A = \frac{1}{R_{R2}} + \frac{1}{R_{S2}}\,, \quad B = \frac{1}{R_{R1}} + \frac{1}{R_{S1}} = \frac{1}{r_0}\,. \tag{3.79}$$

The parameters m, n result from Table 3.4 by means of the angle ϑ ($0° \le \vartheta \le 180°$) where

$$\vartheta = \arccos \frac{A - B}{A + B}\,, \tag{3.80}$$

$$A - B = \frac{1}{R_{R2}} + \frac{1}{R_{S2}} - \frac{1}{r_0}\,, \tag{3.81}$$

$$A + B = \frac{1}{R_{R2}} + \frac{1}{R_{S2}} + \frac{1}{r_0}\,. \tag{3.82}$$

For the relation of the half-axes a, b of the contact ellipse holds in dependency of the angle ϑ:

Table 3.4. Coefficients m, n for evaluation of the semi-axes of contact ellipse a, b and axis ratio $g = \min(a/b,\ b/a)$, from Kalker (1967b)

ϑ [°]	m	n	$g = \frac{b}{a} = \frac{n}{m}$	ϑ [°]	m	n	$g = \frac{a}{b} = \frac{m}{n}$
0	∞	0	0	90	1	1	1
0.5	61.40	0.1018	0.00166	95	0.944	1.061	0.890
1	36.89	0.1314	0.00356	100	0.893	1.128	0.792
1.5	27.48	0.1522	0.00554	105	0.846	1.202	0.704
2	22.26	0.1691	0.00760	110	0.802	1.284	0.625
3	16.50	0.1964	0.0119	115	0.759	1.378	0.551
4	13.31	0.2188	0.0164	120	0.717	1.486	0.483
6	9.79	0.2552	0.0261	125	0.678	1.611	0.421
8	7.86	0.2850	0.0363	130	0.641	1.754	0.365
10	6.604	0.3112	0.0471	135	0.604	1.926	0.314
20	3.813	0.4123	0.108	140	0.567	2.136	0.265
30	2.731	0.493	0.181	145	0.530	2.397	0.221
35	2.397	0.530	0.221	150	0.493	2.731	0.181
40	2.136	0.567	0.265	160	0.4123	3.813	0.108
45	1.926	0.604	0.314	170	0.3112	6.604	0.0471
50	1.754	0.641	0.365	172	0.2850	7.86	0.0363
55	1.611	0.678	0.421	174	0.2552	9.79	0.0261
60	1.486	0.717	0.483	176	0.2188	13.31	0.0164
65	1.378	0.759	0.551	177	0.1964	16.50	0.0119
70	1.284	0.802	0.625	178	0.1691	22.26	0.00760
75	1.202	0.846	0.704	178.5	0.1522	27.48	0.00554
80	1.128	0.893	0.792	179.0	0.1314	36.89	0.00365
85	1.061	0.944	0.890	179.5	0.1018	61.40	0.00166
90	1.00	1.00	1	180	0	∞	0

$$
\begin{aligned}
0° \le \vartheta < 90° &\Leftrightarrow a > b\,,\\
\vartheta = 90° &\Leftrightarrow a = b\,,\\
90° < \vartheta \le 180° &\Leftrightarrow a < b\,.
\end{aligned}
\tag{3.83}
$$

With Eqs. (3.80) and (3.79) the relation between a, b in dependency of A and B follows

$$
A \lesseqgtr B \Leftrightarrow a \lesseqgtr b\,. \tag{3.84}
$$

Using the linear theory of Kalker the contact forces can be calculated by means of the half-axes a, b of the contact ellipse and the ratio g

$$
g = \min(a/b,\ b/a)\,. \tag{3.85}
$$

For the contact forces acting on the wheel the following relation is valid, cp. Eq. (3.74),

$$\underbrace{\begin{bmatrix} f_{t1} \\ f_{t2} \\ l_{P3} \end{bmatrix}}_{\boldsymbol{f}^K} \underset{=-}{=-} \underbrace{\begin{bmatrix} f_{11} & 0 & 0 \\ 0 & f_{22} & f_{23} \\ 0 & -f_{23} & f_{33} \end{bmatrix}}_{\boldsymbol{F}} \underbrace{\begin{bmatrix} \nu_1 \\ \nu_2 \\ \nu_3 \end{bmatrix}}_{\boldsymbol{\nu}^K} , \tag{3.86}$$

$$f_{11} = abGC_{11} , \quad f_{22} = abGC_{22} , \quad f_{33} = (ab)^2 GC_{33} , \quad f_{23} = (ab)^{\frac{3}{2}} GC_{23} . \tag{3.87}$$

The equations contain the shear modulus G (steel: $G \approx 80\,\mathrm{kN/mm^2}$) and the four Kalker coefficients C_{ij}. These coefficients depend on the ratio g and the Poisson's ratio $\bar{\nu}$. They are listed in Table 3.5, cp. Kalker (1967b). Equations (3.86) and (3.87) result in

$$f_{t1} = -c^2 GC_{11}\nu_1 , \tag{3.88}$$

$$f_{t2} = -c^2 G(C_{22}\nu_2 + cC_{23}\nu_3) , \tag{3.89}$$

$$l_{P3} = -c^3 G(-C_{23}\nu_2 + cC_{33}\nu_3) , \tag{3.90}$$

where $c = \sqrt{ab}$ is the geometric mean radius of the contact ellipse. The tangential force f_{t1} acting in longitudinal direction of the rail track only depends on the longitudinal slip ν_1 while the tangential force f_{t2} acting in lateral direction depends on the lateral slip ν_2 and on the aligning slip ν_3. This is consistent to the general laws of symmetry of Eqs. (3.75) and (3.76). The contact forces acting on the rail track are in opposite to the forces acting on the wheel.

The following relationship between the material constants $E, \bar{\nu}$ and G is valid for linear elastic materials:

$$G = \frac{E}{2(1+\bar{\nu})} . \tag{3.91}$$

In the case of different materials for wheel $\{G_R, \bar{\nu}_R\}$ and rail track $\{G_S, \bar{\nu}_S\}$ Kalker (1967b) uses substituted parameters to evaluate Eqs. (3.87)-(3.90) and also Eqs. (3.77) and (3.78),

$$\frac{1}{G} = \frac{1}{2}\left(\frac{1}{G_R} + \frac{1}{G_S}\right) , \quad G = \frac{2G_R G_S}{G_R + G_S} ,$$

$$\frac{\bar{\nu}}{G} = \frac{1}{2}\left(\frac{\bar{\nu}_R}{G_R} + \frac{\bar{\nu}_S}{G_S}\right) , \quad \frac{1-\bar{\nu}^2}{E} = \frac{1}{4}\left(\frac{1-\bar{\nu}_R}{G_R} + \frac{1-\bar{\nu}_S}{G_S}\right) . \tag{3.92}$$

Table 3.5. Kalker's coefficients depending on Poisson's ratio $\bar{\nu}$ and the axis-ratio g of the contact ellipse, from Kalker (1967b) ($\Lambda = \ln(16/g^2)$, $\ln 4 = 1.386$)

		C_{11}			C_{22}			C_{23}			C_{33}		
	g	$\bar{\nu}=0$	0.25	0.5	0	0.25	0.5	0	0.25	0.5	0	0.25	0.5
$\frac{a}{b}$	0.0	$\pi^2/(4(1-\bar{\nu}))$			$\pi^2/4$			$\pi\sqrt{g}/(3(1-\bar{\nu}))\cdot[1+\bar{\nu}(\Lambda/2+\ln 4-5)]$			$\pi^2/[16(1-\bar{\nu})g]$		
	0.1	2.51	3.31	4.85	2.51	2.52	2.53	0.33	0.473	0.73	6.42	8.28	11.7
	0.2	2.59	3.37	4.81	2.59	2.63	2.66	0.48	0.603	0.81	3.46	4.27	5.66
	0.3	2.68	3.44	4.80	2.68	2.75	2.81	0.61	0.715	0.89	2.49	2.96	3.72
	0.4	2.78	3.53	4.82	2.78	2.88	2.98	0.72	0.823	0.98	2.02	2.32	2.77
	0.5	2.88	3.62	4.83	2.88	3.01	3.14	0.83	0.929	1.07	1.74	1.93	2.22
	0.6	2.98	3.72	4.91	2.98	3.14	3.31	0.93	1.03	1.18	1.56	1.68	1.86
	0.7	3.09	3.81	4.97	3.09	3.28	3.48	1.03	1.14	1.29	1.43	1.50	1.60
	0.8	3.19	3.91	5.05	3.19	3.41	3.65	1.13	1.25	1.40	1.34	1.37	1.42
	0.9	3.29	4.01	5.12	3.29	3.54	3.82	1.23	1.36	1.51	1.27	1.27	1.27
$\frac{b}{a}$	1.0	3.40	4.12	5.20	3.40	3.67	3.98	1.33	1.47	1.63	1.21	1.19	1.16
	0.9	3.51	4.22	5.30	3.51	3.81	4.16	1.44	1.59	1.77	1.16	1.11	1.06
	0.8	3.65	4.36	5.42	3.65	3.99	4.39	1.58	1.75	1.94	1.10	1.04	0.95
	0.7	3.82	4.54	5.58	3.82	4.21	4.67	1.76	1.95	2.18	1.05	0.97	0.85
	0.6	4.06	4.78	5.80	4.06	4.50	5.04	2.01	2.23	2.50	1.01	0.90	0.75
	0.5	4.37	5.10	6.11	4.37	4.90	5.56	2.35	2.62	2.96	0.96	0.82	0.65
	0.4	4.84	5.57	6.57	4.84	5.48	6.31	2.88	3.24	3.70	0.91	0.75	0.55
	0.3	5.57	6.34	7.34	5.57	6.40	7.51	3.79	4.32	5.01	0.87	0.67	0.45
	0.2	6.96	7.78	8.82	6.96	8.14	9.79	5.72	6.63	7.89	0.83	0.60	0.34
	0.1	10.7	11.7	12.9	10.7	12.8	16.0	12.2	14.6	18.0	0.80	0.53	0.23

■ **Problem 3.3 Contact area and contact forces for wheel-rail contact**

In the following, the wheel-rail contact of Problem 3.2 for a given cone angle $\delta = 2.9^\circ$ ($\tan\delta \approx \delta = 1/20 = 0.05$) and a rolling radius $r_0 = 500$ mm is considered. The track is characterized by a straight line and a transversal main curvature radius $R_{S2} = 300$ mm. Wheel and rail consist of steel ($E \approx 210$ kN/mm^2, $G \approx 80$ kN/mm^2, $\bar{\nu} \approx 0.3$). Evaluate for given wheel loads of $f_R = 100$ kN and $f_R = 50$ kN, respectively,
a) the size and location of the contact ellipse,
b) the size and orientation of the tangential forces and the aligning torque due to the slip variable presented in Problem 3.2.

Solution

a) Considering $\delta \ll 1$, the wheel's main curvature radius simplifies to $R_{R1} = r_0$ as well as the normal force is almost equal to the wheel load, i.e. $f_n = f_R$. For the main curvature radii

$$R_{R1} = r_0 = 500 \text{ mm} , \quad R_{R2} \to \infty , \quad R_{S1} \to \infty , \quad R_{S2} = 300 \text{ mm} , \tag{1}$$

the values of A, B and ϑ result from Eqs. (3.79) - (3.82),

$$A = \frac{1}{R_{R2}} + \frac{1}{R_{S2}} = \frac{1}{300} \text{ mm}^{-1} , \quad B = \frac{1}{R_{R1}} + \frac{1}{R_{S1}} = \frac{1}{500} \text{ mm}^{-1} , \tag{2}$$

$$A - B = \frac{2}{1500} \text{ mm}^{-1} , \quad A + B = \frac{8}{1500} \text{ mm}^{-1} , \tag{3}$$

$$\vartheta = \arccos \frac{A-B}{A+B} = \arccos 0.25 \approx 75.5^{\circ} . \tag{4}$$

A linear interpolation, using the auxiliary variable ϑ, yields the parameters

$$m = 1.19 , \quad n = 0.85 . \tag{5}$$

For the given numerical numbers, the semi-axes of the ellipse are obtained by Eq. (3.77), Eq. (3.78) for $f_n = 100$ kN as

$$a = 7.43 \text{ mm} , \quad b = 5.31 \text{ mm} . \tag{6}$$

In this case, the semi-axis a in $\mathbf{e}_1^K$-direction along the rail, is greater than the semi-axis b orientated across the rail. The relation $a > b$ immediately results from $A > B$, cp. Eq. (2). The area $\bar{A}$ of the ellipse and the axis ratio $g = \min(a/b, b/a)$ are

$$\bar{A} = \pi ab = 124 \text{ mm}^2 = 1.24 \text{ cm}^2 , \quad g = \frac{b}{a} = \frac{n}{m} = 0.714 . \tag{7}$$

The normal force $f_n = 50$ kN leads to different absolute values of $a = 5.90$ mm, $b = 4.21$ mm, $\bar{A} = 78$ mm^2, whereas the axis ratio $g = b/a = 0.175$ remains unchanged.
b) In the case of a constant axis ratio $g =$ const, Kalker's coefficients C_{ij} for $\bar{\nu} = 0.3$ are determined by quadratic interpolation of the values $\bar{\nu} = 0$, $\bar{\nu} = 0.25$, $\bar{\nu} = 0.5$,

$$\begin{aligned} C_{ij}(\bar{\nu} = 0.3) = C_{ij}(\bar{\nu} = 0.25) &+ 0.08\,[C_{ij}(\bar{\nu} = 0.25) - C_{ij}(\bar{\nu} = 0)] \\ &+ 0.12\,[C_{ij}(\bar{\nu} = 0.5) - C_{ij}(\bar{\nu} = 0.25)] . \end{aligned} \tag{8}$$

Table 3.6. Values of Eq. (8)

$g = \frac{b}{a}$	$C_{11}(\bar{\nu} = 0.3)$	$C_{22}(\bar{\nu} = 0.3)$	$C_{23}(\bar{\nu} = 0.3)$	$C_{33}(\bar{\nu} = 0.3)$
0.8	4.544	4.065	1.786	1.025
0.7	4.722	4.296	1.987	0.9446
0.714	4.697	4.264	1.959	0.9559

The results are shown in Table 3.6 for $g = 0.8$ and $g = 0.7$. Then, Kalker's coefficients for $g = 0.714$ result from a linear interpolation of the values for $g = 0.7$ and $g = 0.8$. Finally, the contact forces are calculated by means of Eqs. (3.88)-(3.90), with $c = \sqrt{ab} = 6.28$ mm according to Eq. (6) as well as the slip values $\boldsymbol{\nu}^K \approx [-0.001 \; 0 \; -\delta/r_0]^{\mathrm{T}}$ taken from Problem 3.2 with $\delta/r_0 = 10^{-4}$ mm^{-1},

$$f_{t1} = -c^2 G \, C_{11} \, \nu_1 = 14.82 \text{ kN} \,, \tag{9}$$

$$f_{t2} = -c^2 G \, (C_{22} \, \nu_2 + c \, C_{23} \, \nu_3) = 3.88 \text{ kN} \,, \tag{10}$$

$$l_{P3} = -c^3 G \, (-C_{23} \, \nu_2 + c \, C_{33} \, \nu_3) = 11.9 \text{ kNmm} = 0.019 \text{ kNm} \,. \tag{11}$$

Although no transversal slip occurs in this example, i.e. $\nu_2 = 0$, the aligning slip ν_3 causes a tangential force f_{t2} transversal to the rail. However, this transversal component is smaller than the longitudinal component f_{t1}. Both tangential forces and the aligning torque act on the wheel in positive direction.

For the normal force $f_n = 50$ kN, Kalker's coefficients are equal to those of $f_n = 100$ kN because of the same ratio $g = b/a$. But the changed value of $c = \sqrt{ab} = 4.98$ mm results in force values $f_{t1} = 9.34$ kN, $f_{t2} = 1.94$ kN, $l_{P3} = 4.72$ kNmm. ∎

3.4.3.2 Contact Forces Considering Saturation

The advantage of Kalker's linear theory is the straightforward evaluation of the contact forces $\boldsymbol{f}^K$ as mentioned in Sect. 3.4.3.1. In contrary, the simplified as well as the exact theory can only be applied numerically. The corresponding computer programs Rollen, Usetab and Contact based on the theory of Kalker (1979) and Kalker (1967a) are now commercially available, see VORtech Computing (2009).

The limitation of the linear theory is that only initial gradients of the contact forces $f^K(\boldsymbol{\nu}^K \to \mathbf{0})$ can be determined, cp. Fig. 3.9. To determine the contact forces and to verify the theoretical approaches, a lot of experimental tests have been made. It turned out that the initial gradients of contact forces vary between 50 % - 100 % of the values according to Kalker's theory. This discrepancy is caused by a thin film of contamination in the contact patch. If the thin film of contamination is removed from the wheel and the rail, the experimental results show a good correlation with Kalker's exact theory and the initial gradients with the linear theory as well. Therefore, the design of railway vehicles has to be insensitive to changes of the contact forces. Moreover, for the determination of contact forces the accuracy demanded may be reduced. The last statement leads to the many heuristic approaches such as the nonlinear saturation characteristics of the rolling contact coefficient, Fig. 3.9, starting from the initial gradients which are simple to calculate.

All nonlinear approaches have in common that the resulting tangential force f_t^* and the corresponding normalized rolling contact coefficient φ^* are limited as a result of Coulomb's law,

$$f_t^* = \sqrt{f_{t1}^{*2} + f_{t2}^{*2}} \le \mu\,|f_n| \;, \quad \varphi^* = \frac{f_t^*}{\mu\,|f_n|} \le 1 \,. \tag{3.93}$$

The corresponding values for $f_t \ll \mu\,|f_n|$, respectively $\varphi \ll 1$,

$$f_t = \sqrt{f_{t1}^2 + f_{t2}^2} \;, \quad \varphi = \frac{f_t}{\mu\,|f_n|} \tag{3.94}$$

are known from Kalker's linear theory. But they can not be used for increasing slip. At first, approximations for the nonlinear saturation behavior are searched in the form $f_t^* = f_t^*(f_t)$, respectively $\varphi^* = \varphi^*(\varphi)$, with the following properties

$$f_t^* = \begin{cases} f_t & f_t \ll \mu\,|f_n| \\ & \text{for} \\ \mu\,|f_n| & f_t \ge \mu\,|f_n| \end{cases} \quad \text{or} \quad \varphi* = \begin{cases} \varphi & \varphi \ll 1 \,, \\ & \text{for} \\ 1 & \varphi \ge 1 \,. \end{cases} \tag{3.95}$$

The dimensionless representation of $\varphi^* = \varphi^*(\varphi)$ offers advantages for calculations. The following equations show three simple approximations $\varphi_i^*(\varphi)$, $i = 1, 2, 3$, which satisfy the required properties. The approximations get better with increasing index i, cp. Fig. 3.14,

$$i = 1 : \varphi_1^* = \begin{cases} \varphi & 0 \le \varphi \le \varphi_{G1} \\ & \text{for} \\ 1 & \varphi \ge \varphi_{G1} \end{cases} \;, \quad \varphi_{G1} = 1 \,; \tag{3.96}$$

$$i = 2 : \varphi_2^* = 1 - \mathrm{e}^{-\varphi} \;, \quad 0 \le \varphi < \infty \,; \tag{3.97}$$

$$i = 3 : \varphi_3^* = \begin{cases} \varphi - \frac{1}{3}\varphi^2 + \frac{1}{27}\varphi^3 & 0 \le \varphi \le \varphi_{G3} \\ & \text{for} \\ 1 & \varphi \ge \varphi_{G3} \end{cases} \;, \quad \varphi_{G3} = 3 \,. \tag{3.98}$$

The linear/constant characteristic of Eq. (3.96) is a rough approximation while the cubic/constant characteristic of Eq. (3.98), which is based on investigations of Johnson and Vermeulen (1964), approximates the exact characteristic of saturation well, cp. Garg and Dukkipati (1984). The exponential characteristic of Eq. (3.97) is also a rough approximation again, but it does not require partitioning of the slip domain.

The aim is to get the coordinates f_{t1}, f_{t2} of the tangential force $\boldsymbol{f}_t$ using the developed approximation. Therefore, an information about the directions of the forces is necessary. Useful relations are known for two special cases, cp. Garg and Dukkipati (1984).

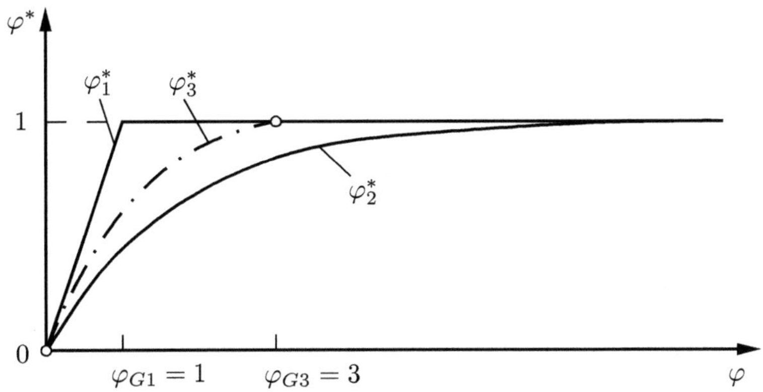

Fig. 3.14. Saturation characteristics of the normalized rolling contact coefficient: a) φ_1^* linear/constant; b) φ_2^* exponential; c) φ_1^* cubic/constant

Limit Case I. If the slip is small ($\varphi \ll 1$), the coordinates f_{t1}, f_{t2} of the tangential force follow from Kalker's linear theory. The direction of the tangential force $f_t = \sqrt{f_{t1}^2 + f_{t2}^2}$ is described by the angle $\alpha = \alpha_{\mathrm{I}}$ measured from longitudinal direction,

$$\begin{aligned} \sin\alpha_{\mathrm{I}} &= f_{t2}/f_t \\ \cos\alpha_{\mathrm{I}} &= f_{t1}/f_t \end{aligned} \quad , \qquad \alpha_{\mathrm{I}} = \arctan(f_{t2}/f_{t1}) \; . \tag{3.99}$$

The saturation behavior of the tangential force f_t^* is transferred to the coordinates f_{t1}^* and f_{t2}^* proportionally,

$$\begin{aligned} f_{t1}^* &= f_t^* f_{t1}/f_t = f_t^* \cos\alpha_{\mathrm{I}} \; , \\ f_{t2}^* &= f_t^* f_{t2}/f_t = f_t^* \sin\alpha_{\mathrm{I}} \; , \\ f_t^* &= \varphi^*(\varphi)\mu\,|f_n| \; , \quad \varphi \ll 1 \; . \end{aligned} \tag{3.100}$$

Limit Case II. If the slip is big enough only kinematic friction occurs ($\varphi \gtrsim 3$), and the resulting tangential force $f_t^* = \sqrt{f_{t1}^{*2} + f_{t2}^{*2}} = \mu\,|f_n|$ points into the direction contrary to the resulting slip $\nu = \sqrt{\nu_1^2 + \nu_2^2}$. This direction is described with the angle $\alpha = \alpha_{\mathrm{II}}$ measured from longitudinal direction,

$$\begin{aligned} \sin\alpha_{\mathrm{II}} &= -\nu_2/\nu \\ \cos\alpha_{\mathrm{II}} &= -\nu_1/\nu \end{aligned} \quad , \qquad \alpha_{\mathrm{II}} = \arctan(\nu_2/\nu_1) \; . \tag{3.101}$$

This yields the tangential forces f_{t1}^*, f_{t2}^*

$$
\begin{aligned}
f_{t1}^* &= -f_t^* \nu_1/\nu = f_t^* \cos\alpha_{II} \,, \\
f_{t2}^* &= -f_t^* \nu_2/\nu = f_t^* \sin\alpha_{II} \,, \\
f_t^* &= \mu\,|f_n| = \varphi^*(\varphi)\mu\,|f_n| \,, \quad \varphi^* > 3 \,.
\end{aligned}
\tag{3.102}
$$

Equations (3.100) and (3.102) are also the basis for a more general way to calculate the components f_{t1}^*, f_{t2}^* of the tangential forces between the limit cases I ($\varphi \ll 1$, $\alpha = \alpha_I$) and II ($\varphi \gtrsim 3$, $\alpha = \alpha_{II}$). The angle $\alpha = \alpha(\varphi)$ is described as a function of φ similar to the normalized rolling contact coefficient $\varphi^*(\varphi)$. According to Garg and Dukkipati (1984), a linear characteristic $\alpha = \alpha(\varphi)$ for $\alpha_I \leq \alpha \leq \alpha_{II}$ may be chosen. This yields in conjunction with the approximation $i = 3$

$$
i = 3 : \alpha_3(\varphi) = \begin{cases} \alpha_I + (\alpha_{II} - \alpha_I)\varphi/\varphi_{G3} & \text{for } 0 \leq \varphi \leq \varphi_{G3} \\ \alpha_{II} & \phantom{\text{for }} \varphi \geq \varphi_{G3} \end{cases} , \quad \varphi_{G3} = 3 \,.
\tag{3.103}
$$

Thus, for the calculation of approximated tangential forces f_{t1}^* and f_{t2}^* it holds considering the saturation

$$
\begin{aligned}
&f_{t1}^* = f_t^* \cos\alpha(\varphi) \,, \quad f_{t2}^* = f_t^* \sin\alpha(\varphi) \,, \quad f_t^* = \varphi^*(\varphi)\mu\,|f_n| \\
&\alpha_I \leq \alpha(\varphi) \leq \alpha_{II} \,, \quad 0 \leq \varphi < \infty \,, \quad 0 \leq \varphi^*(\varphi) \leq 1 \,.
\end{aligned}
\tag{3.104}
$$

With the combination of different approximations for $\varphi^*(\varphi)$ and $\alpha(\varphi)$ many approximations for the tangential forces are possible. For coarse calculations the approximations $\varphi^* = \varphi_i^*$, $i = 1, 2$ or 3, with $\alpha = \alpha_I$ for $\varphi < 1$ respectively $\alpha = \alpha_{II}$ for $\varphi > 1$ may be used. Better approximations are achieved for $\varphi^* = \varphi_3^*$, $\alpha = \alpha_3$.

■ **Problem 3.4 Contact forces considering approximated saturation**
Calculate starting from Problem 3.3 with normal force $f_n = 100\,\text{kN}$ and coefficient of kinematic friction $\mu = 0.4$:

a) The approximated value for the tangential forces f_{t1}^* and f_{t2}^* using different saturation approaches φ_i^*, see Eqs. (3.96)-(3.98), and direction angles α_i, see Eqs. (3.99), (3.101) and (3.103).
b) The achievable maximum acceleration a_{1max} of the railway vehicle.
c) The longitudinal slip ν_1 during constant acceleration $a_1 = \text{const}$ of the train within $0 \leq a_1 \leq a_{1max}$.

Solution
a) Starting point are the results of Problem 3.3 for $f_n = 100\,\text{kN}$:

$$f_{t1} = 14.82\,\mathrm{kN}\,, \quad f_{t2} = 3.88\,\mathrm{kN}\,, \quad \nu_1 = -0.001\,, \quad \nu_2 = 0\,. \tag{1}$$

This yields

$$f_t = \sqrt{f_{t1}^2 + f_{t2}^2} = 15.32\,\mathrm{kN}\,, \quad \nu = \sqrt{\nu_1^2 + \nu_2^2} = 0.001\,. \tag{2}$$

The normalized rolling contact coefficient φ^* is

$$\varphi^* = \frac{f_t}{\mu\,|f_n|} = 0.383\,. \tag{3}$$

Taking into account that the lateral slip ν_2 is zero due to driving straightforward even if the longitudinal slip ν_1 is very high, the boundary angles α_{I} and α_{II} read

$$\alpha_{\mathrm{I}} = \arctan(f_{t2}/f_{t1}) = 14.67^\circ\,, \quad \alpha_{\mathrm{II}} = \arctan(\nu_2/\nu_1) = 0\,. \tag{4}$$

The angle α_3 is calculated with Eq. (3.103)

$$\alpha_3(\varphi) = \alpha_{\mathrm{I}} + (\alpha_{\mathrm{II}} - \alpha_{\mathrm{I}})\varphi/3 = \alpha_{\mathrm{I}}\left(1 - \frac{\varphi}{3}\right) = 12.8^\circ\,. \tag{5}$$

With these results the tangential forces f_{t1}^*, f_{t2}^* have the values given in Table 3.7 for different combinations $\{\varphi_i^*, \alpha_j\}$ of the approximations. The differences of the results are only small. The result in the last column using the combination $\{\varphi_3^*, \alpha_3\}$ is considered to be the best approximation.

b) Simplified assumptions are applied for the railway vehicle:

- The motion of the vehicle is straight on along a horizontal rail track.
- The normal forces and the tangential forces in the rail contact are equal for each wheel $f_{t1,i}^* = f_{t1}^*$, $|f_n|_i = m_i g$, $m_i = m$.
- The cone angle is small, $\delta < 1$, so the wheel loads f_R are approximated by the normal forces, $f_R = mg \approx |f_N|$.
- Forces from air resistance are neglected.

Table 3.7. Tangential forces f_{t1}^* and f_{t2}^* for different combinations of $\{\varphi_i^*, \alpha_j\}$

	Combinations of the approximation methods			
	φ_1^*, α_{I}	φ_2^*, α_{I}	φ_3^*, α_{I}	φ_3^*, α_3
$\varphi_i^*(\varphi)$, $\varphi = 0.383$	0.383	0.318	0.336	0.336
$f_t^* = \varphi_i^*(\varphi)\mu\|f_n\|$	15.32 kN	12.72 kN	13.44 kN	13.44 kN
α	14.67°	14.67°	14.67°	12.80°
$f_{t1}^* = f_t^* \cos\alpha$	14.82 kN	12.31 kN	13.00 kN	13.11 kN
$f_{t2}^* = f_t^* \sin\alpha$	3.88 kN	3.22 kN	3.40 kN	2.98 kN

The achievable maximum acceleration while driving straight on depends on the maximum transmissible tangential force $f^*_{t1,max} = \mu\,|f_n|$ per wheel. From Newton's law of motion in $\boldsymbol{e}^I_1$-direction for the whole vehicle with p rail contact points follows

$$a_{1max}\sum_{i=1}^{p} m_i = \sum_{i=1}^{p}(f^*_{t1,max})_i = \sum_{i=1}^{p}\mu\,|f_n|_i = \mu g\sum_{i=1}^{p} m_i\,. \tag{6}$$

This yields with $\mu = 0.4$

$$a_{1max} = \mu g = 3.92\,\mathrm{m/s^2}\,. \tag{7}$$

c) If the acceleration $a_1 = \text{const}$ $(0 \le a_1 \le a_{1max})$ of the railway vehicle is known, the necessary tangential force f^*_{t1} per wheel follows from Eq. (6),

$$f^*_{t1} = m a_1 = \frac{|f_n|}{g} a_1\,. \tag{8}$$

Using of the saturation characteristic, in a first step the tangential force f_{t1} and in a second step using Kalker's linear theory the corresponding longitudinal slip ν_1 is determined. To simplify matters here the approximations $\{\varphi^*_2, \alpha_{\mathrm{II}}\}$ are used. So a simple but qualitative correct solution for ν_1 is found. Initially from Eqs. (3.104) and (3.97) with $\alpha_{\mathrm{II}} = 0$ follows

$$f^*_{t1} = \varphi^*_2(\varphi)\mu\,|f_n|\cos\alpha_{\mathrm{II}} = (1-\mathrm{e}^{-\varphi})\mu\,|f_n|\,, \tag{9}$$

$$f^*_{t2} = \varphi^*_2(\varphi)\mu\,|f_n|\sin\alpha_{\mathrm{II}} = 0\,. \tag{10}$$

Comparing of Eq. (9) and Eq. (8) and solving for φ, cp. Eq. (3.94), yields

$$\varphi \equiv \frac{f_t}{\mu\,|f_n|} = -\ln\left(1-\frac{a_1}{\mu g}\right)\,, \quad f_t = \sqrt{f^2_{t1}+f^2_{t2}}\,. \tag{11}$$

If f^2_{t2} is neglected compared to f^2_{t1} what is consistent with the approximation $\alpha = \alpha_{\mathrm{II}} = 0$, cp. Eq. (10), it follows

$$f_{t1} \approx f_t = -\mu\,|f_n|\ln\left(1-\frac{a_1}{\mu g}\right)\,. \tag{12}$$

On the other hand from Kalker's linear theory, see Eq. (3.88), with the numerical values from Problem 3.3 it follows

$$f_{t1} = -c^2 G C_{11}\nu_1 = -14.82\cdot 10^3 \nu_1\,\mathrm{kN}\,. \tag{13}$$

A comparison of Eqs. (12) and (13), and solving for the longitudinal slip ν_1, yields

$$\nu_1 = \frac{\mu\,|f_n|}{c^2 G C_{11}}\ln\left(1-\frac{a_1}{\mu g}\right)\,, \tag{14}$$

$$= 2.70\ln\left(1-\frac{a_1}{\mu g}\right)\cdot 10^{-3}\,. \tag{15}$$

Table 3.8. Results for the longitudinal slip using Eq. (15)

$\frac{a_1}{\mu g} = \frac{a_1}{a_{1max}}$	0	0.1	0.2	0.3	0.4	0.5
$-\nu_1 [‰]$	0	0.28	0.60	0.96	1.38	1.87
$\frac{a_1}{\mu g} = \frac{a_1}{a_{1max}}$	0.6	0.7	0.8	0.9	0.95	1.0
$-\nu_1 [‰]$	2.47	3.25	4.34	6.21	8.09	∞

The evaluation of Eq. (15) results in the longitudinal slip ν_1 in dependence of $a_1/\mu g = a_1/a_{1max}$, see Table 3.8. Due to the acceleration of the railway vehicle only negative values of longitudinal slip occur. They are as usual denoted in one-tenth of a percent, $\nu_1\,[‰] \mathrel{\widehat{=}} 10^3\nu_1$. On the non-accelerated wheel no longitudinal slip occurs. With increasing numbers of constant acceleration the longitudinal slip increases. For maximum acceleration the longitudinal slip converges to infinity what means wheel spin. ■

■ Problem 3.5 Contact forces and linear equations of motion for a railway wheelset

For the free wheelset from Problem 2.9 using the contact patch systems $\{P_i, \boldsymbol{e}_\nu^i\}$, $i = l, r$ with $l \mathrel{\widehat{=}}$ left, $r \mathrel{\widehat{=}}$ right determine:

a) the slip

$$\boldsymbol{\nu}_i^i = \begin{bmatrix} \nu_{i1} & \nu_{i2} & \nu_{i3} \end{bmatrix}^{\mathrm{T}} , \tag{1}$$

b) the contact forces

$$\vec{\boldsymbol{f}}_i^i = -\begin{bmatrix} f_{ti,1} & f_{ti,2} & l_i \end{bmatrix}^{\mathrm{T}} , \tag{2}$$

according to Kalker's linear theory,

c) the linear equations of motion for small lateral and yaw motions using the results from Problem 2.9.

Solution

The notations from Problem 2.9 are used where the contact forces $f_{ti,j}, l_i$, $j = 1, 2$, on both wheels are defined pointing in opposite direction, see Fig. 2.23. This is considered with the negative sign in Eq. (2).

a) The calculation of the slip is carried out in three steps:

1) Calculation of the absolute velocities of the wheels' and the rails' material points (wheel R and rail S) coinciding with the contact points P_i and calculation of the absolute angular velocities in the intermediate system Z.

2) Transformation of the relative velocities and the relative angular velocities into the contact patch systems $\boldsymbol{e}_\nu^i$, $i = l, r$.
3) Linearization of the slip for small deviations from the nominal motion with respect to translation and rotation.

Step 1: The absolute velocity of the point coinciding with P_i on the wheel results from the rigid body relation,

$$\boldsymbol{v}_{Pi}^Z = \boldsymbol{v}_C^Z + \tilde{\boldsymbol{\omega}}_{IK}^Z \boldsymbol{\rho}_{Ci}^Z , \quad \boldsymbol{v}_C^Z = \boldsymbol{S}^{ZI} \boldsymbol{v}_C^I , \quad i = l, r . \tag{3}$$

With the values defined in Problem 2.9,

$$\boldsymbol{S}^{ZI} = \begin{bmatrix} \mathrm{c}\gamma & \mathrm{s}\gamma & 0 \\ -\mathrm{c}\alpha\mathrm{s}\gamma & \mathrm{c}\alpha\mathrm{c}\gamma & \mathrm{s}\alpha \\ \mathrm{s}\alpha\mathrm{s}\gamma & -\mathrm{s}\alpha\mathrm{c}\gamma & \mathrm{c}\alpha \end{bmatrix} , \quad \boldsymbol{v}_C^I = \begin{bmatrix} \dot{x} \\ \dot{y} \\ \dot{z} \end{bmatrix} ,$$

$$\boldsymbol{\omega}_{IK}^Z = \begin{bmatrix} \dot{\alpha} \\ \dot{\gamma}\mathrm{s}\alpha - \dot{\beta} \\ \dot{\gamma}\mathrm{c}\alpha \end{bmatrix} , \qquad \boldsymbol{\rho}_{Ci}^Z = \begin{bmatrix} 0 \\ \mp a + \Delta a_i \\ r_i \end{bmatrix} , \tag{4}$$

it follows

$$\boldsymbol{v}_{Pi}^Z = \begin{bmatrix} \dot{x}\mathrm{c}\gamma + \dot{y}\mathrm{s}\gamma + (\dot{\gamma}\mathrm{s}\alpha - \dot{\beta})r_i - \dot{\gamma}(\pm a + \Delta a_i)\mathrm{c}\alpha \\ -\dot{x}\mathrm{c}\alpha\mathrm{s}\gamma + \dot{y}\mathrm{c}\alpha\mathrm{c}\gamma + \dot{z}\mathrm{s}\alpha - \dot{\alpha}r_i \\ \dot{x}\mathrm{s}\alpha\mathrm{s}\gamma - \dot{y}\mathrm{s}\alpha\mathrm{c}\gamma + \dot{z}\mathrm{c}\alpha + \dot{\alpha}(\pm a + \Delta a_i) \end{bmatrix} , \quad i = l, r . \tag{5}$$

Here and in the following the upper algebraic sign holds for $i = l$ and the lower one for $i = r$. The absolute velocity of the point coinciding with P_i on the rail is equal to zero since the rail is fixed. Therefore, Eq. (5) already denotes the relative velocity $\bar{\boldsymbol{v}}_{Pi}^Z = \boldsymbol{v}_{Pi}^Z$. For the same reason the angular velocity $\boldsymbol{\omega}_{IK}^Z$ of the wheelset, cp. Eq. (4), already denotes the relative angular velocity, $\bar{\boldsymbol{\omega}}^Z = \boldsymbol{\omega}_{IK}^Z$.

Step 2: The transformation of the relative velocity into the contact area systems $\boldsymbol{e}_\nu^i$ is carried out with the transformation matrices $\boldsymbol{S}^{iZ}$, cp. Problem 2.9,

$$\bar{\boldsymbol{v}}_{Pi}^i = \boldsymbol{S}^{iZ} \bar{\boldsymbol{v}}_{Pi}^Z , \quad \bar{\boldsymbol{\omega}}^i = \boldsymbol{S}^{iZ} \bar{\boldsymbol{\omega}}^Z , \quad \boldsymbol{S}^{iZ} = \begin{bmatrix} 1 & 0 & 0 \\ 0 & \mathrm{c}\delta_i & \pm\mathrm{s}\delta_i \\ 0 & \mp\mathrm{s}\delta_i & \mathrm{c}\delta_i \end{bmatrix} , \quad i = l, r . \tag{6}$$

Then, it follows

$$\begin{aligned} \bar{v}_{Pi,1}^i &= \dot{x}\mathrm{c}\gamma + \dot{y}\mathrm{s}\gamma + (\dot{\gamma}\mathrm{s}\alpha - \dot{\beta})r_i + \dot{\gamma}(\pm a - \Delta a_i)\mathrm{c}\alpha , \\ \bar{v}_{Pi,2}^i &= -\dot{x}\mathrm{s}\gamma\mathrm{c}(\alpha \pm \delta_i) + \dot{y}\mathrm{c}\gamma\mathrm{c}(\alpha \pm \delta_i) + \dot{z}\mathrm{s}(\alpha \pm \delta_i) \\ &\quad - \dot{\alpha}r_i\mathrm{c}\delta_i + \dot{\alpha}(-a \pm \Delta a_i)\mathrm{s}\delta_i , \\ \bar{\omega}_3^i &= \dot{\gamma}\mathrm{c}(\alpha \pm \delta_i) \pm \dot{\beta}\mathrm{s}\delta_i , \quad i = l, r . \end{aligned} \tag{7}$$

Step 3: The slip according to Eq. (3.54) follows from Eq. (7) by relation to the medium speed v_0. Assuming that the wheelset in its nominal state moves along the rail track with $\dot{x} = \dot{x}_0 = v_0$, for pure rolling the angular velocity $\beta = \beta_0 = \Omega = v_0/r_0$ would arise. With respect to the nominal state, considering the constraint equations

$$z \approx 0 \,, \quad \alpha = \alpha(y) \,, \quad \dot{\alpha} = \alpha' \dot{y} \,, \quad (\cdot)' \equiv \mathrm{d}(\cdot)/\mathrm{d}y \,, \tag{8}$$

for small deviations of position and angle, the velocities Eq. (7) can be linearized, cp. Eqs. (24) and (25) of Problem 2.9. With

$$\begin{aligned} &\dot{x} = v_0 + \dot{\tilde{x}} \,, && \dot{\tilde{x}} \ll v_0 \,, \ \dot{y} \ll v_0 \,, && \dot{z} = z'\dot{y} \,, \ z' \ll 1 \,, \\ &\dot{\beta} = \Omega + \dot{\tilde{\beta}} \,, && \dot{\tilde{\beta}} \ll \Omega \,, \ \Omega = v_0/r_0 \,, && \\ &\alpha \ll 1 \,, && \dot{\alpha} \ll \Omega \,, \ \Delta a_i \ll a \,, && \\ &\gamma \ll 1 \,, && \dot{\gamma} \ll \Omega \,, && \end{aligned} \tag{9}$$

one gets from (7) for the slip

$$\boldsymbol{\nu}_i^i = \frac{1}{v_0} \begin{bmatrix} \bar{v}_{Pi,1}^i \\ \bar{v}_{Pi,2}^i \\ \bar{\omega}_3^i \end{bmatrix} = \frac{1}{v_0} \begin{bmatrix} v_0(1 - r_i/r_0) \pm a\dot{\gamma} \\ (\dot{y} - v_0\gamma)\mathrm{c}(\alpha \pm \delta_i) - \dot{\alpha}(r_i \mathrm{c}\delta_i + a\mathrm{s}\delta_i) \\ \dot{\gamma}\mathrm{c}(\alpha \pm \delta_i) \pm \Omega \mathrm{s}\delta_i \end{bmatrix} \,, \ i = l, r \,. \tag{10}$$

Considering small wheel conicities, $\delta_i \ll 1$, $i = l, r$, from Eq. (10) it remains

$$\boldsymbol{\nu}_i^i = \frac{1}{v_0} \begin{bmatrix} v_0(1 - r_i/r_0) \pm a\dot{\gamma} \\ \dot{y} - v_0\gamma - \dot{\alpha} r_i \\ \dot{\gamma} \pm \Omega\delta_i \end{bmatrix} \,, \quad i = l, r \,. \tag{11}$$

b) The contact forces

$$\bar{\boldsymbol{f}}_i^i = -\boldsymbol{F}_i \boldsymbol{\nu}_i^i \tag{12}$$

follow from Kalker's linear theory with the slip vector $\boldsymbol{\nu}_i^i$, cp. Eqs. (3.86) and (3.87). The calculation of Kalker coefficients required for the matrices $\boldsymbol{F}_i$, is difficult for real wheel/rail contacts. The reason is that the contact points P_i are shifted into lateral direction on the profiles of wheel and rail, and the main radii of curvature depend highly on the wheelset position. For small deviations from the nominal position the Kalker coefficients are calculated with the average main radii of curvature, then the deviations due to the different main curvature planes of wheel and rail are considered to be negligible. As normal forces the static values f_{n0} are used, cp. Eq. (28) in Problem 2.9. Assuming constant and similar Kalker coefficients for both contact points P_i, $i = l, r$, it follows from Eq. (12) with Eq. (2) using the slip of Eq. (11)

$$\bar{\boldsymbol{f}}^i_i = -\begin{bmatrix} f_{ti,1} \\ f_{ti,2} \\ l_i \end{bmatrix} = -\begin{bmatrix} f_{11} & 0 & 0 \\ 0 & f_{22} & +f_{23} \\ 0 & -f_{23} & f_{33} \end{bmatrix} \begin{bmatrix} \nu_{i1} \\ \nu_{i2} \\ \nu_{i3} \end{bmatrix} ,$$

$$\begin{aligned} f_{ti,1} &= \frac{f_{11}}{v_0}\left[v_0(1 - r_i/r_0) \pm a\dot{\gamma}\right] , \\ f_{ti,2} &= \frac{f_{22}}{v_0}\left[\dot{y} - v_0\gamma - \dot{\alpha}r_i\right] + \frac{f_{23}}{v_0}\left[\dot{\gamma} \pm \Omega\delta_i\right] , \\ l_i &= -\frac{f_{23}}{v_0}\left[\dot{y} - v_0\gamma - \dot{\alpha}r_i\right] + \frac{f_{33}}{v_0}\left[\dot{\gamma} \pm \Omega\delta_i\right] , \quad i = l, r . \end{aligned} \tag{13}$$

Due to different slip the contact forces on the left and right wheel are different.

c) With Eqs. (9) and (13), the equations of motion found in Problem 2.9 can be linearized and described explicitly by generalized coordinates. Usually the inertial system is used. In Problem 2.9 the translatory equations of motion (10)-(12) are given in the inertial frame I,

$$\begin{aligned} m\ddot{x} &= f_{l1} + f_{r1} , \\ m\ddot{y} &= f_{l2} + f_{r2} , \\ m\ddot{z} &= f_{l3} + f_{r3} + mg . \end{aligned} \tag{14}$$

Using the rotational matrix $\boldsymbol{S}^{IZ}$ and the linearized values, a transformation of the rotational equations of motion (21)-(23) into the inertial frame I leads to

$$\begin{aligned} I_1\ddot{\alpha} + I_2\dot{\gamma}\Omega &= (l_{l1} + l_{r1}) - \gamma(l_{l2} + l_{r2}) , \\ -I_2\ddot{\beta} &= \gamma(l_{l1} + l_{r1}) + (l_{l2} + l_{r2}) - \alpha(l_{l3} + l_{r3}) , \\ I_1\ddot{\gamma} - I_2\dot{\alpha}\Omega &= \alpha(l_{l2} + l_{r2}) + (l_{l3} + l_{r3}) . \end{aligned} \tag{15}$$

The coordinates $f_{i\nu}$, $l_{i\nu}$, $i = l, r$, of the forces and torques are given in Problem 2.9 by Eqs. (8), (9), (16) and (17) depending on the contact forces. Substitution, linearization and neglect of small values lead to the equations relevant to the lateral and yaw motion

$$m\ddot{y} = -\gamma(f_{tl,1} + f_{tr,1}) - (f_{tl,2} + f_{tr,2}) + (\alpha + \delta_l)f_{nl} + (\alpha - \delta_r)f_{nr} , \tag{16}$$

$$m\ddot{z} = -(f_{nl} + f_{nr}) + mg , \tag{17}$$

$$I_1\ddot{\alpha} + I_2\dot{\gamma}\Omega = a(f_{nl} - f_{nr}) + r_l f_{tl,2} + r_r f_{tr,2} , \tag{18}$$

$$I_1\ddot{\gamma} - I_2\dot{\alpha}\Omega = -a(f_{tl,1} - f_{tr,1}) - (l_l + l_r) . \tag{19}$$

Due to the constraint equations (8) it holds $z \approx 0$ and $\alpha = \alpha(y)$. Therewith, the unknown reaction forces f_{nl} and f_{nr} can be calculated from Eqs. (17) and (18), where the inertia terms due to $m\ddot{z} \ll mg$, $I_1\ddot{\alpha} \ll mg\alpha$ are neglected,

$$f_{nl} = \frac{1}{2a}I_2\dot{\gamma}\Omega - \frac{1}{2a}(r_l f_{tl,2} + r_r f_{tr,2}) + \frac{1}{2}mg , \tag{20}$$

$$f_{nr} = -\frac{1}{2a}I_2\dot{\gamma}\Omega + \frac{1}{2a}(r_l f_{tl,2} + r_r f_{tr,2}) + \frac{1}{2}mg . \tag{21}$$

For the lateral and the yaw motion follows from Eqs. (16) and (19) with the contact forces of Eq. (13) and the normal forces given in Eqs. (20) and (21)

$$\begin{aligned}
m\ddot{y} &+ 2\frac{f_{22}}{v_0}\left(\dot{y} - v_0\gamma - \dot{\alpha}\frac{r_l + r_r}{2}\right) + 2\frac{f_{23}}{v_0}\left(\dot{\gamma} + \Omega\frac{\delta_l - \delta_r}{2}\right) \\
&+ 2\gamma f_{11}\left(1 - \frac{r_l + r_r}{2r_0}\right) - mg\left(\alpha + \frac{\delta_l - \delta_r}{2}\right) - \frac{\delta_l + \delta_r}{2a} I_2\dot{\gamma}\Omega \\
&+ \frac{\delta_l + \delta_r}{2a}\left[\frac{f_{22}}{v_0}(r_l + r_r)(\dot{y} - v_0\gamma) - \frac{f_{22}}{v_0}(r_l^2 + r_r^2)\dot{\alpha}\right. \\
&\left. + \frac{f_{23}}{v_0}(r_l + r_r)\dot{\gamma} + \frac{f_{23}}{v_0}(r_l\delta_l - r_r\delta_r)\Omega\right] = 0\,, \qquad (22)
\end{aligned}$$

$$\begin{aligned}
I_1\ddot{\gamma} - I_2\dot{\alpha}\Omega &- 2\frac{f_{23}}{v_0}\left(\dot{y} - v_0\gamma - \dot{\alpha}\frac{r_l + r_r}{2}\right) + 2\frac{f_{33}}{v_0}\left(\dot{\gamma} + \Omega\frac{\delta_l - \delta_r}{2}\right) \\
&+ 2a\frac{f_{11}}{v_0}\left(-\frac{v_0}{r_0}\frac{r_l - r_r}{2} + a\dot{\gamma}\right) = 0\,. \qquad (23)
\end{aligned}$$

The term $mg\,(\alpha + (\delta_l - \delta_r)/2)$ in Eq. (22) is called lateral gravitational stiffness due to the fact that it represents a restoring force in lateral direction. In literature sometimes a small gravitational term can also be found in yaw equation (23), which is due to a different orientation of the contact area frames. Considering the constraint equations (8), $\alpha = \alpha(y), \dot{\alpha} = \alpha'\dot{y}$, in Eqs. (22) and (23) it turns out that the lateral and yaw motion are decoupled from the other motions. The resulting differential equations are nonlinear, although the coordinates y, γ and their derivations occur only linear. The reason is that the running radii r_i and the angles of inclination δ_i as well as $\alpha = \alpha(y)$ are nonlinear functions of the lateral displacement y. They depend on the profile shapes of the contact partners and on the locations of the contact points, cp. Garg and Dukkipati (1984). Linear equations of motion arise from conic wheels ($\delta_l = \delta_r = \delta_0$) and rectangular rails, cp. Problem 2.7. With Eqs. (17)-(19) and Eq. (23) from Problem 2.7 one gets

$$\begin{aligned}
\delta_l = \delta_r &= \delta_0 && \Rightarrow \frac{1}{2}(\delta_l + \delta_r) = \delta_0\,, \quad \frac{1}{2}(\delta_l - \delta_r) = 0\,, \\
z = r_0 &- \frac{1}{2}(r_l + r_r) \approx 0 && \Rightarrow \frac{1}{2}(r_l + r_r) \approx r_0\,, \\
-y \approx &\left[q - \frac{1}{2}(r_l + r_r)\right]\alpha && = \left[\frac{a}{\tan\delta_0} - r_0\right]\alpha \approx \frac{a}{\delta_0}\alpha \Rightarrow -\dot{\alpha} \approx \frac{\delta_0}{a}\dot{y}\,, \\
\alpha \approx &\frac{r_l - r_r}{2a} \approx -\frac{\delta_0}{a}y && \Rightarrow \frac{1}{2}(r_l - r_r) \approx -\delta_0 y\,. \qquad (24)
\end{aligned}$$

With Eq. (24) it follows from Eqs. (22) and (23)

$$m\ddot{y} + 2\frac{f_{22}}{v_0}\left(\dot{y} + \frac{\delta_0 r_0}{a}\dot{y} - v_0\gamma\right) + 2\frac{f_{23}}{v_0}\dot{\gamma} + mg\frac{\delta_0}{a}y - I_2\frac{\delta_0 v_0}{a r_0}\dot{\gamma}$$
$$+ \frac{\delta_0}{a}\left[\frac{f_{22}}{v_0}2r_0(\dot{y} - v_0\gamma) + \frac{f_{23}}{v_0}2r_0\dot{\gamma}\right] = 0\,, \tag{25}$$

$$I_1\ddot{\gamma} + I_2\frac{\delta_0 v_0}{a r_0}\dot{y} - 2\frac{f_{23}}{v_0}\left(\dot{y} + \frac{\delta_0 r_0}{a}\dot{y} - v_0\gamma\right) + 2\frac{f_{33}}{v_0}\dot{\gamma}$$
$$+ 2a\frac{f_{11}}{v_0}\left(\frac{\delta_0 v_0}{r_0}y + a\dot{\gamma}\right) = 0\,. \tag{26}$$

For small values ($\delta_0 r_0/a \ll 1$) the desired equations of motion read in matrix notation

$$\begin{bmatrix} m & 0 \\ 0 & I_1 \end{bmatrix}\begin{bmatrix} \ddot{y} \\ \ddot{\gamma} \end{bmatrix} + \left\{\frac{2}{v_0}\begin{bmatrix} f_{22} & f_{23} \\ -f_{23} & a^2 f_{11} + f_{33} \end{bmatrix} + \begin{bmatrix} 0 & -I_2\dfrac{\delta_0 v_0}{a r_0} \\ I_2\dfrac{\delta_0 v_0}{a r_0} & 0 \end{bmatrix}\right\}\begin{bmatrix} \dot{y} \\ \dot{\gamma} \end{bmatrix}$$
$$+ \begin{bmatrix} mg\dfrac{\delta_0}{a} & -2f_{22} \\ 2f_{11}\dfrac{a\delta_0}{r_0} & 2f_{23} \end{bmatrix}\begin{bmatrix} y \\ \gamma \end{bmatrix} = \begin{bmatrix} 0 \\ 0 \end{bmatrix}. \tag{27}$$

These are the differential equations of an ordinary mechanical system with all kinds of forces, cp. Eq. (2.99),

$$\boldsymbol{M}\ddot{\boldsymbol{y}} + (\boldsymbol{D} + \boldsymbol{G})\dot{\boldsymbol{y}} + (\boldsymbol{K} + \boldsymbol{N})\boldsymbol{y} = \boldsymbol{0}\,,$$
$$\boldsymbol{M} = \boldsymbol{M}^{\mathrm{T}} > 0\,,\quad \boldsymbol{D} = \boldsymbol{D}^{\mathrm{T}}\,,\quad \boldsymbol{G} = -\boldsymbol{G}^{\mathrm{T}}\,,\quad \boldsymbol{K} = \boldsymbol{K}^{\mathrm{T}}\,,\quad \boldsymbol{N} = -\boldsymbol{N}^{\mathrm{T}}\,, \tag{28}$$

with

$$\boldsymbol{y} = \begin{bmatrix} y \\ \gamma \end{bmatrix},\quad \boldsymbol{M} = \begin{bmatrix} m & 0 \\ 0 & I_1 \end{bmatrix},\quad \boldsymbol{D} = \frac{2}{v_0}\begin{bmatrix} f_{22} & 0 \\ 0 & a^2 f_{11} + f_{33} \end{bmatrix},$$

$$\boldsymbol{G} = \begin{bmatrix} 0 & \dfrac{2f_{23}}{v_0} - I_2\dfrac{\delta_0 v_0}{a r_0} \\ -\dfrac{2f_{23}}{v_0} + I_2\dfrac{\delta_0 v_0}{a r_0} & 0 \end{bmatrix},$$

$$\boldsymbol{K} = \begin{bmatrix} mg\dfrac{\delta_0}{a} & f_{11}\dfrac{a\delta_0}{r_0} - f_{22} \\ f_{11}\dfrac{a\delta_0}{r_0} - f_{22} & 2f_{23} \end{bmatrix},$$

$$\boldsymbol{N} = \begin{bmatrix} 0 & -f_{11}\dfrac{a\delta_0}{r_0} - f_{22} \\ f_{11}\dfrac{a\delta_0}{r_0} + f_{22} & 0 \end{bmatrix}. \tag{29}$$

Particular properties of the equations of motion of the wheelset are

a) the occurrence of nonconservative position forces,
b) decreasing damping forces with increasing speed.

Both properties result from the contact forces and have an essential impact on the stability of the wheelset. On pure kinematic rolling, cp. Problem 2.7, a deviation from the nominal position leads to a periodic lateral motion of the wheelset called hunting motion. Unlike to this, in the general case considered here initial perturbations may decrease below a critical speed and increase above it. From the equations of motion (27)-(29) the corresponding critical speed can be evaluated as shown in Sect. 9.2.2. ■

3.4.4 Contact Forces of Elastic Tires on a Rigid Road

The general considerations on the planar rolling of a wheel including frictional rolling contact, see Sect. 3.4.1, are also valid for the tire/road contact. Nevertheless, the results for the expansion to the spatial problem of the wheel/rail contact, see Sect. 3.4.3, can not directly be transferred to the tire/road contact. The reason is the fiber-reinforced tire structure which does not behave like an isotropic elastic half space around the contact area. Thus, the linear relations between stresses and strains are not valid anymore, which are the basis of Kalker's theory. Further differences to the wheel/rail contact are the significant larger elastic deformations within the tread and the side wall or carcass, respectively, of the tire, as well as a strongly slip dependent coefficient of friction of the tire/road contact. The length $2a$ of the contact area or contact patch, respectively, of a tire amounts to around 150 mm. It is 10 times larger than a typical length of a wheel/rail contact.

Although the results of Sect. 3.4.3 can not be transferred completely, the proceeding and the structure of the equations will be retained in the following. The goal remains to give a law for the contact forces (3.72) in dependance of the slip. Concerning the tire/road contact problem, almost all developed theories are based on material laws like Eq. (3.73), and usually discoidal tires are assumed. Especially the cornering behavior of the tire is considered which plays a major role for holding the driving direction of road vehicles. Regarding the consideration of the elastic tire deformation within the contact patch, three different modeling approaches are distinguished:

- brush or spring model,
- model of a pre-stressed rope on an elastic foundation,
- model of a pre-stressed beam on an elastic foundation.

A comparison of the steady-state rolling elastic contact theories based on these models is given in Sperling (1977). Other contributions deal with the development of unsteady rolling theories, see Weber (1981), Boehm (1985), Pacejka (2002).

3.4.4.1 The Brush Model

In the following, a simplified steady state theory from Pacejka (2002) is presented, which is based on the brush model by Fromm (1927). The advantage of this theory is the explicit description of the contact forces, and, furthermore, it allows an easy enhancement and adjustment to measured data. The tread of the tire is assumed to be elastic, the carcass and the wheel rim to be rigid.

First, the model input quantities are defined. A rigid discoidal wheel rolling straight ahead with an elastic rolling contact on a planar horizontal road is considered. In Fig. 3.15 a rolling wheel without slip (velocity v_C, angular velocity $\omega = \overset{*}{\omega}$) is shown. Due to the elastic contact, it is subject to a static deflection,

$$\rho = r_0 - r \,, \tag{3.105}$$

where r_0 and r denote the wheel radii without and with load, respectively. The point P_R in Fig. 3.15 coincides with the instantaneous center of rotation of the rolling wheel without slip. The center is located slightly below the road surface due to the elastic deflection. The rolling condition results in

$$v_C = \overset{*}{\omega} r_e \,, \quad r_e \approx r_0 - \rho/3 \,, \tag{3.106}$$

where r_e denotes the effective or dynamical rolling radius. For the driven or the braked wheel with $\omega \neq \overset{*}{\omega}$ the point P_R, see Fig. 3.15, is no longer identical

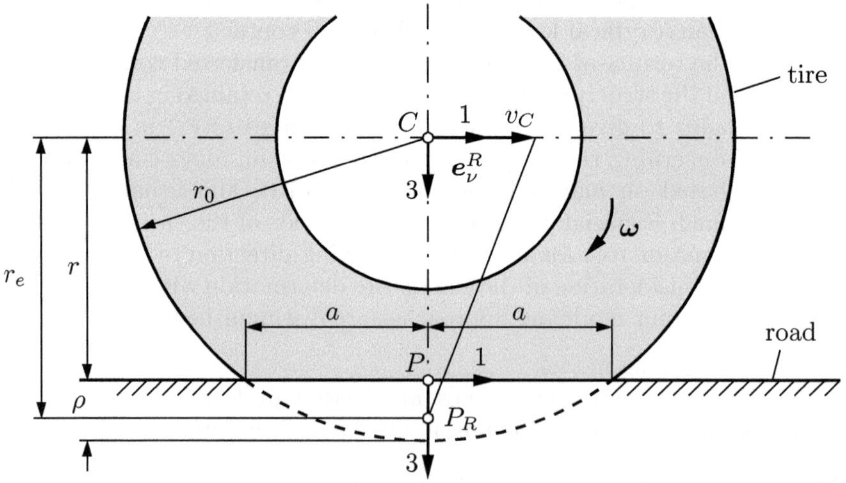

Fig. 3.15. Tire conditions for the elastic rolling contact with static deflection ρ, and effective rolling radius r_e

to the instantaneous center of rotation of the wheel. Its velocity written in the contact patch frame $K\{P, \mathbf{e}_\nu^K\}$ amounts to

$$v_{P1}^K = v_C - \omega r_e \,. \tag{3.107}$$

Since the road is rigid, this velocity is equal to the relative velocity $\bar{v}_{P1}^K = v_{P1}^K$. According to the slip definition (3.54) the longitudinal slip follows as

$$\nu_1^K = 1 - \frac{\omega r_e}{v_C} \,. \tag{3.108}$$

In vehicle engineering other slip definitions, presented in Table 3.9, are used, too. In the following, the slip definition (3.49) will be used, where in specific applications it might be beneficial to distinguish between driving and braking slip, according to Mitschke and Wallentowitz (2004). The point P_R in Fig. 3.15 is considered to be fixed to the wheel, with the distance $\overline{CP_R} = r_e$ within the wheel plane, see also ??. The velocity of the wheel within the wheel plane serves as the reference velocity v_0. It is equal to the absolute velocity of the origin P of the contact plane frame K along the $\mathbf{e}_1^K$-direction, cp. Sect. 3.4.2.

After these preparations, a rigid, cornering, discoidal wheel with elastic rolling contact is presented, cp. Pacejka (2002). Let the wheel plane be vertical. Then, the moving axle-fixed reference frame $R\{C, \mathbf{e}_\nu^R\}$ and the contact area frame $K\{P, \mathbf{e}_\nu^K\}$ have the same orientation, see Fig. 3.16 a). The absolute velocities of the wheel (index R) $\{\boldsymbol{v}_{CR}, \boldsymbol{\omega}_R\}$ amount to

Table 3.9. Comparison of different slip definitions, considering longitudinal slip as an example, for a wheel rolling within a vertical plane

Slip for driving / braking	$\frac{v_C - \omega r_e}{v_C}$	$\frac{\omega r_e - v_C}{v_C}$	$\frac{\omega r_e - v_C}{\omega r_e}$	driving: $\frac{\omega r_e - v_C}{\omega r_e}$; braking: $\frac{v_C - \omega r_e}{v_C}$
Source	This book	Pacejka (1986)	Pacejka (1975)	Mitschke and Wallentowitz (2004)
Slip value for wheelspin, $\omega \to \infty$	$-\infty$	∞	1	1
Slip value for blocking wheel, $\omega = 0$	1	-1	$-\infty$	1
Slip value for free rolling wheel, $v_C = \omega r_e$	0	0	0	0

v_C translative velocity of the wheel center along the wheel plane
ω angular velocity of the wheel around the wheel axis
r_e effective rolling radius

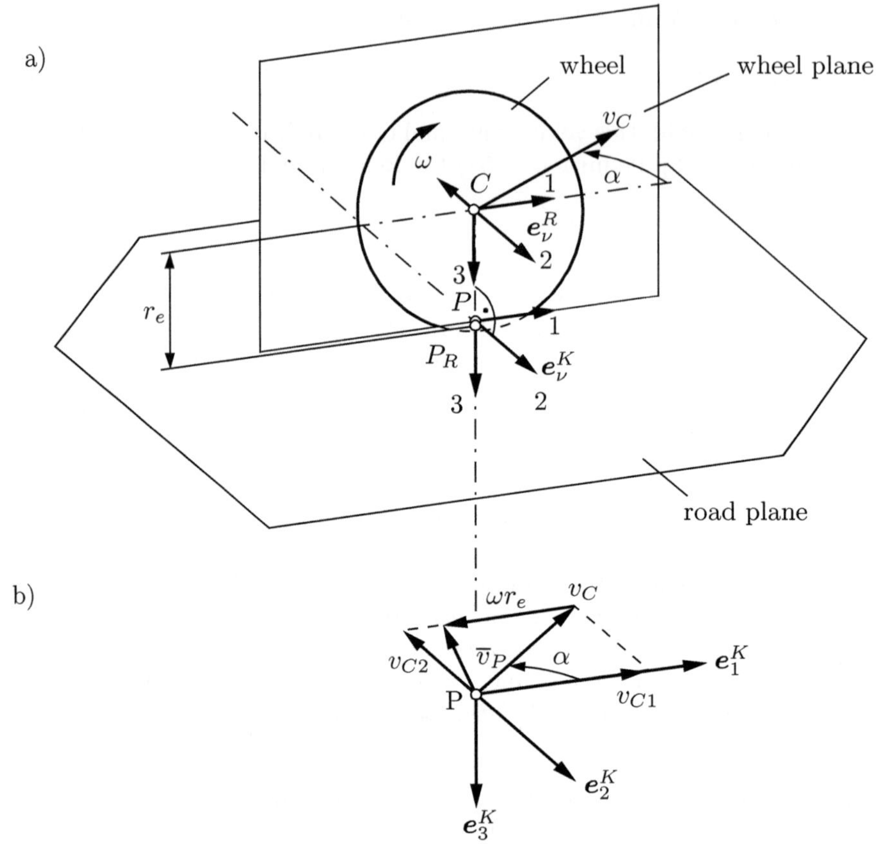

Fig. 3.16. Illustration of the conditions of a skewing or cornering, respectively, wheel: a) wheel and road plane; b) projection on the road plane

$$\boldsymbol{v}_{CR}^K = \boldsymbol{v}_{CR}^R = \begin{bmatrix} v_C \cos\alpha \\ -v_C \sin\alpha \\ 0 \end{bmatrix}, \quad \boldsymbol{\omega}_R^K = \boldsymbol{\omega}_R^R = \begin{bmatrix} 0 \\ -\omega \\ 0 \end{bmatrix}, \tag{3.109}$$

where v_C, ω denote the velocity and the angular velocity of the wheel, respectively. The slip angle, in accordance with vehicle engineering terminologies, is referred to as α. The absolute velocity $\boldsymbol{v}_{PR}$ of the wheel point P results from the rigid body relation

$$\boldsymbol{v}_{PR}^K = \boldsymbol{v}_{PR}^R = \boldsymbol{v}_{CR}^R + \tilde{\boldsymbol{\omega}}_R^R \boldsymbol{r}_{CP}^R = \begin{bmatrix} v_C \cos\alpha - \omega r_e \\ -v_C \sin\alpha \\ 0 \end{bmatrix}. \tag{3.110}$$

Therein $\boldsymbol{r}^R_{CP} = [0\ 0\ r_e]^{\mathrm{T}}$ with the effective rolling radius r_e is used instead of r, cp. Fig. 3.16. The absolute velocities $\{\boldsymbol{v}_{PS}, \boldsymbol{\omega}_S\}$ of the road (index S) are vanishing.

Consequently, the sliding velocities $\{\bar{\boldsymbol{v}}_P, \bar{\boldsymbol{\omega}}\}$ agree with $\{\boldsymbol{v}_{PR}, \boldsymbol{\omega}_R\}$ from Eqs. (3.110), (3.109), cp. Fig. 3.16 b). The slip according to definition (3.54) reads as

$$\boldsymbol{\nu}^K \equiv \begin{bmatrix} \nu_1 \\ \nu_2 \\ \nu_3 \end{bmatrix} = \frac{1}{v_0} \begin{bmatrix} \bar{v}_1^K \\ \bar{v}_2^K \\ \bar{\omega}_3^K \end{bmatrix} = \frac{1}{v_0} \begin{bmatrix} v_C \cos\alpha - \omega r_e \\ -v_C \sin\alpha \\ 0 \end{bmatrix} . \tag{3.111}$$

The reference velocity v_0 agrees with $v^R_{CR,1} = v_C \cos\alpha$, cp. Eq. (3.109). Thus, for the slip it remains

$$\begin{aligned} \nu_1 &= \frac{v_C \cos\alpha - \omega r_e}{v_C \cos\alpha} = 1 - \frac{\omega r_e}{v_C \cos\alpha} , \\ \nu_2 &= -\frac{v_C \sin\alpha}{v_C \cos\alpha} = -\tan\alpha , \\ \nu_3 &= 0 . \end{aligned} \tag{3.112}$$

Next, the contact forces are calculated separately for lateral slip and then for the longitudinal slip based on the brush model.

3.4.4.2 Contact Forces for Pure Lateral Slip

For pure lateral slip ($\nu_1 = 0$ or $r_e\omega = v_c \cos\alpha$) Fig. 3.17 shows the conditions within the contact patch during the steady rolling process. The continuous load in normal direction is simplified by a parabolic function instead of the elliptical shaped characteristic following from the Hertzian theory, cp. Fig. 3.17 a),

$$q_n(x) = \frac{3\,|f_n|}{4a} \left[1 - \left(\frac{x}{a}\right)^2\right] \geq 0 , \quad -a \leq x \leq a , \tag{3.113}$$

where $|f_n|$ corresponds to the wheel load and a denotes half the length of the patch. The deformation of the wheel within the contact area is characterized by the lateral displacements u_2 of the brush hairs, which are considered to be leaf springs, see Fig. 3.17 b). Within the sticking zone the line of contact runs straight and parallel to the velocity v_C starting from the run-in boundary E to the limit point G,

$$u_2(x) = (a - x) \tan\alpha , \quad x_G \leq x \leq a , \tag{3.114}$$

while within the sliding zone it is decreasing according to the sliding friction level, and it joins the wheel plane again at the run-out A. For the special case of small slip angles $\alpha \to 0$ and thus for small lateral slip $\nu_2 \to 0$ the

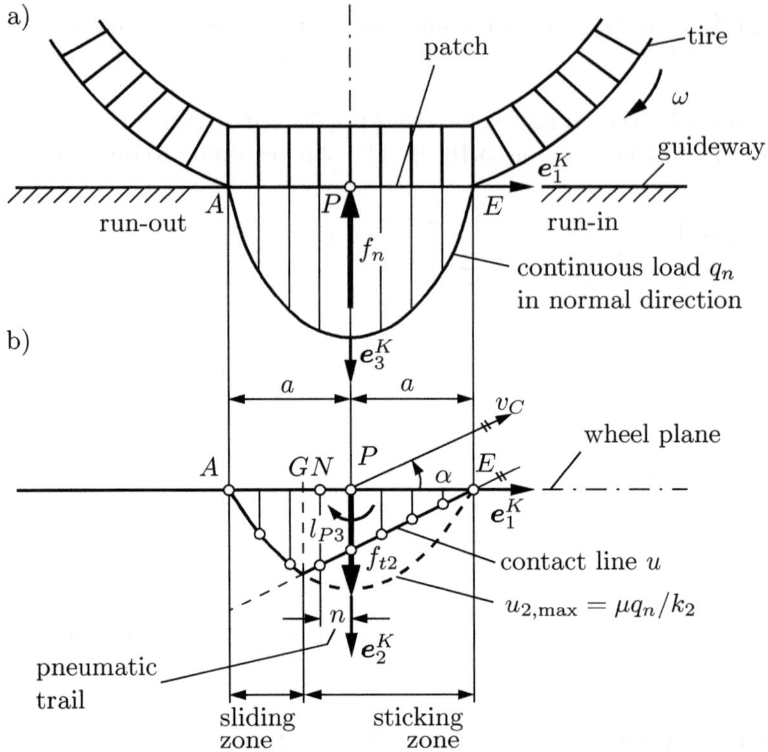

Fig. 3.17. Conditions within the contact area according to the brush model: a) side view; b) top view

limit point G moves to the run-out A. Then, Eq. (3.114) with $\tan\alpha \approx \alpha$ is valid within the entire contact patch $-a \le x \le a$. The lateral tread stiffness $k_2 = \text{const}$, which is related to the patch length, can be extracted from static measurements. It corresponds to the lateral stiffness of the leaf springs for constant patch width. From the displacements $u_2(x)$ follows the lateral force f_{t2} as well as the aligning torque l_{P3},

$$f_{t2} = k_2 \int_{-a}^{a} u_2(x) \mathrm{d}x = 2k_2 a^2 \alpha = -2k_2 a^2 \nu_2 \,, \tag{3.115}$$

$$l_{P3} = k_2 \int_{-a}^{a} u_2(x) x \mathrm{d}x = -\frac{2}{3} k_2 a^3 \alpha = \frac{2}{3} k_2 a^3 \nu_2 \,. \tag{3.116}$$

Therefore, the initial gradient of the contact force to slip relation for pure lateral slip complies with the linear theory of Kalker for the wheel/rail contact, cp. Eq. (3.86),

$$\boldsymbol{f}^K = -\boldsymbol{F}\boldsymbol{\nu}^K ,$$

$$\begin{bmatrix} f_{t2} \\ l_{P3} \end{bmatrix} = - \begin{bmatrix} f_{22} & * \\ -f_{23} & * \end{bmatrix} \begin{bmatrix} \nu_2 \\ 0 \end{bmatrix} , \tag{3.117}$$

$$f_{22} = - \left. \frac{\partial f_{t2}}{\partial \nu_2} \right|_{\nu_2 \to 0} = 2k_2 a^2 , \quad f_{23} = \left. \frac{\partial l_{P3}}{\partial \nu_2} \right|_{\nu_2 \to 0} = \frac{2}{3} k_2 a^3 . \tag{3.118}$$

The coefficients f_{22}, f_{23} are referred to as cornering stiffness and aligning torque stiffness, the lateral force f_{t2} as cornering force and the moment l_{P3} as alignment torque, too.

At next, the special case of large slip angles $\alpha \to \pi/2$ is considered. This results in a very large lateral slip $\nu_2 = -\tan\alpha \to -\infty$ and the limit point G moves to the run-in E. In this case the entire contact area corresponds to the sliding zone, the continuous load q_{t2} in the lateral direction results from Eq. (3.113) and Coulomb's sliding friction law as

$$|q_{t2}(x)| = \mu q_n(x) = \frac{3}{4} \frac{\mu}{a} |f_n| \left[1 - \left(\frac{x}{a} \right)^2 \right] , \quad -a \leq x \leq a . \tag{3.119}$$

From this follows the maximum possible lateral displacement $u_{2,max}(x)$, see Fig. 3.17 b),

$$|u_{2,max}(x)| = |q_{t2}(x)| / k_2 = \mu q_n(x) / k_2 . \tag{3.120}$$

The coordinate x_G of the limit point G can be derived by equating the corresponding relations given by Eqs. (3.120) and (3.114),

$$|u_{2,max}(x_G)| = |u_2(x_G)| \Rightarrow \mu q_n(x_G)/k_2 = (a - x_G) |\tan\alpha| . \tag{3.121}$$

If the coordinate x is described as

$$x = a - 2a\lambda , \quad 0 \leq \lambda \leq 1 , \tag{3.122}$$

then, from Eqs. (3.121) and (3.113) it follows with $x_G = a - 2a\lambda_G$

$$\lambda_G = 1 - \theta_2 |\tan\alpha| , \quad \theta_2 = \frac{2}{3} \frac{k_2 a^2}{\mu |f_n|} . \tag{3.123}$$

For $\lambda_G = 0$ one gets $x_G = a$, i. e. the limit point coincides with the run-in E. The corresponding slip angle $\alpha = \alpha_G$, resulting in a complete sliding within the entire contact zone, follows directly from Eq. (3.123) as

$$|\tan\alpha_G| = \frac{1}{\theta_2} . \tag{3.124}$$

Then, the contact forces f_{t2} and l_{P3} can be calculated explicitly.

For $|\alpha| \geq \alpha_G$ (complete sliding) the continuous load $q_{t2}(x)$ in the lateral direction, cp. Eq. (3.119), is integrated along the patch length, resulting in

$$f_{t2} = \int_{-a}^{a} q_{t2}(x)\mathrm{d}x \ , \quad l_{P3} = \int_{-a}^{a} q_{t2}(x)x\mathrm{d}x = 0 \ , \tag{3.125}$$

where the torque l_{P3} disappears due to the even function $q_{t2}(x) = q_{t2}(-x)$.

For $|\alpha| \leq \alpha_G$ the contact forces are derived from the sum of two integrals along the sliding area ($-a \leq x \leq x_G$ or $1 \geq \lambda \geq \lambda_G$) using Eq. (3.125) and along the sticking area ($x_G \leq x \leq a$ or $\lambda_G \geq \lambda \geq 0$),using Eqs. (3.115), (3.116), respectively. After introducing the normalized rolling contact coefficients φ_2, φ_3 and the normalized lateral slip ν_2^*, the results read as

$$\begin{aligned} &\varphi_2 = \frac{f_{t2}}{\mu\,|f_n|} \ , \quad \varphi_3 = \frac{l_{P3}}{\mu a\,|f_n|} \ , \quad \nu_2^* = \theta_2\,|\nu_2| = \theta_2\,|\tan\alpha| \ , \\ &\lambda_G = 1 - \nu_2^* \ , \quad |\tan\alpha_G| = \frac{1}{\theta_2} \ , \quad \nu_{2G}^* = 1 \ , \quad \theta_2 = \frac{2}{3}\frac{k_2 a^2}{\mu\,|f_n|} \ , \\ &-\frac{\pi}{2} \leq \alpha \leq \frac{\pi}{2} \ , \end{aligned} \tag{3.126}$$

$$\varphi_2 = \begin{cases} (1-\lambda_G^3)\mathrm{sgn}\alpha = (3\nu_2^* - 3\nu_2^{*2} + \nu_2^{*3})\mathrm{sgn}\alpha & |\alpha| \leq \alpha_G \ , \ \nu_2^* \leq 1 \ , \\ & \text{for} \\ \mathrm{sgn}\alpha & |\alpha| \geq \alpha_G \ , \ \nu_2^* \geq 1 \ , \end{cases} \tag{3.127}$$

$$-\varphi_3 = \begin{cases} \lambda_G^3(1-\lambda_G)\mathrm{sgn}\alpha & \\ = (\nu_2^* - 3\nu_2^{*2} + 3\nu_2^{*3} - \nu_2^{*4})\mathrm{sgn}\alpha & |\alpha| \leq \alpha_G \ , \ \nu_2^* \leq 1 \ , \\ & \text{for} \\ 0 & |\alpha| \geq \alpha_G \ , \ \nu_2^* \geq 1 \ . \end{cases} \tag{3.128}$$

The rolling contact coefficients are uneven functions of the slip angle and consequently of the lateral slip, $\varphi_2(\alpha) = -\varphi_2(-\alpha)$, $\varphi_3(\alpha) = -\varphi_3(-\alpha)$. The corresponding functions are displayed graphically in Fig. 3.18. By moving the reference point from P to the reference point N the contact quantities $\{f_{t2}, l_{P3}\}$ are reduced to one resulting force f_{t2} only, cp. Fig. 3.17. From the equilibrium of torques, $l_{N3} = l_{P3} - x_N f_{t2} = 0$, arises the coordinate x_N of the point N,

$$x_N = \frac{l_{P3}}{f_{t2}} = a\frac{\varphi_3}{\varphi_2} = \begin{cases} -\dfrac{a}{3}\dfrac{1 - 3\nu_2^* + 3\nu_2^{*2} - \nu_2^{*3}}{1 - \nu_2^* + \nu_2^{*2}/3} \leq 0 & |\alpha| \leq \alpha_G \ , \ \nu_2^* \leq 1 \ , \\ & \text{for} \\ 0 & |\alpha| \geq \alpha_G \ , \ \nu_2^* \geq 1 \ . \end{cases} \tag{3.129}$$

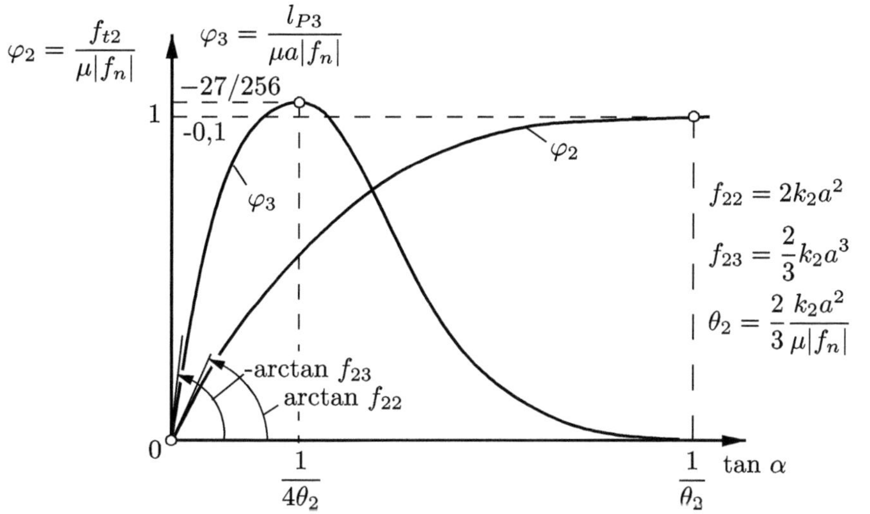

Fig. 3.18. Normalized rolling contact coefficient φ_2, φ_3 for pure lateral slip (slip angle α)

For complete sliding ($|\alpha| \geq \alpha_G$) one gets $x_N = 0$, $N \equiv P$, while for lateral slip ($|\alpha| \leq \alpha_G$) it follows $x_N \leq 0$. Then, the point N is displaced backwards and is located between the point P and the run-out A, see Fig. 3.17. Therefore, the length of the line $\overline{PN} = |x_N|$ is called pneumatic trail n. For disappearing slip, $\nu_2^* \to 0$, Eq. (3.129) results in the limit n_0 of the pneumatic trail,

$$n_0 = |x_n(\nu_2^* \to 0)| = -a\frac{l_{P3}}{f_{t2}}\bigg|_{\nu_2^* \to 0} = \frac{a}{3}\,. \tag{3.130}$$

The value $n_0 = a/3$ is smaller than the real trail. Also, the dependency of the contact forces f_{t2} and l_{P3} on the normal force $|f_n|$ according to the brush model does not completely agree with the experimental results. An improvement can be achieved by taking the elasticity of the carcass into account by means of a rope model, cp. Pacejka (2002). Nevertheless, the given theoretical results fit qualitatively well to the experimental data. This also holds for the contact force due to longitudinal slip, which is described using the brush model in the follow section.

3.4.4.3 Contact Force for Pure Longitudinal Slip

Again the conditions within the contact patch are considered, see Fig. 3.19. Here, the contact patch frame K is separated from the road to get a better overview. Figure 3.19 a) shows the conditions within the road-fixed inertial frame I. The sliding velocity $\bar{v}_1^K$ results from the difference of the absolute velocities of the wheel and the road, cp. Eq. (3.110),

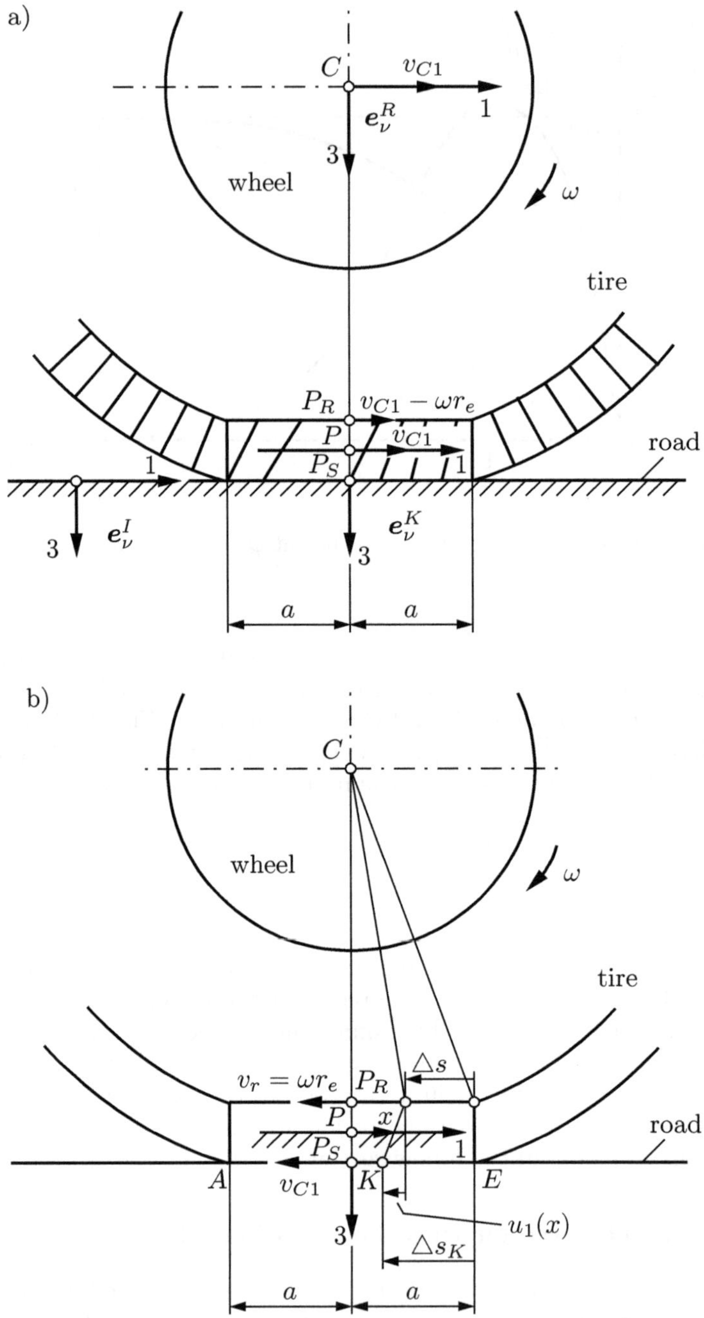

Fig. 3.19. Conditions within the contact zone for pure longitudinal slip according to the brush model: a) absolute velocities; b) relative velocities

$$\bar{v}_1^K = v_{C1} - \omega r_e \ , \quad v_{C1} = v_C \cos\alpha \ , \tag{3.131}$$

where the effective rolling radius r_e is used again. If the contact area frame K is considered as not moving, see Fig. 3.19 b), the result is the same. Here $\bar{v}_1^K$ results from the difference of the corresponding relative velocities where v_{C1} denotes the relative velocity of the road and $v_r = \omega r_e$ the relative velocity of the wheel.

The deformations of the brush hairs, which are considered to be leaf springs, are evaluated conveniently within the frame K depending on the longitudinal coordinate x ($\boldsymbol{e}_1^K$-direction). Observing one leaf spring on its way through the contact area, starting at the run-in $E(x = a)$ to an arbitrary position indicated by x, see Fig. 3.19b, then the distance $\Delta s = a - x$ is passed in a time $\Delta t = \Delta s / v_r$, whereas $v_r = \omega r_e$ denotes the relative velocity of the wheel within the contact zone with respect to frame K. The distance Δs_K passed by the contact point K of the leaf spring in the time Δt within the sticking area amounts to

$$\Delta s_K = v_{C1} \Delta t \ , \quad \Delta t = \frac{\Delta s}{v_r} \ , \quad \Delta s = a - x \ , \quad v_r = \omega r_e \ , \tag{3.132}$$

where v_{C1} denotes the relative velocity of the road in respect to the frame K. The longitudinal displacement $u_1(x)$ of the leaf spring itself follows from the distance difference. In the contact area frame K one gets

$$-u_1(x) = \Delta s_K - \Delta s = \left(\frac{v_{C1}}{v_r} - 1 \right) \Delta s = \frac{v_{C1} - \omega r_e}{\omega r_e} (a - x) \ . \tag{3.133}$$

This relation applies to the sticking zone $x_G \leq x \leq a$. The same results are derived if the process is observed within the inertial frame, see Fig. 3.19 a). The displacement u_1 follows from the sliding velocity $v_{C1} - \omega r_e$ multiplied with the time $\Delta t = \Delta s / v_r$ according to Eq. (3.132). Using the longitudinal slip ν_1, Eq. (3.112) then it follows from Eq. (3.133)

$$u_1(x) = -(a - x) \frac{\nu_1}{1 - \nu_1} \ , \quad \nu_1 = \frac{v_{C1} - \omega r_e}{v_{C1}} \ , \quad x_G \leq x \leq a \ . \tag{3.134}$$

If for the slip definition the velocity $v_r = \omega r_e$ of the wheel is selected as the reference velocity, a differently defined longitudinal slip σ_1 is obtained. With this slip σ_1 it follows from Eq. (3.133) directly

$$u_1(x) = -(a - x)\sigma_1 \ , \quad \sigma_1 = \frac{v_{C1} - \omega r_e}{\omega r_e} \ , \quad x_G \leq x \leq a \ . \tag{3.135}$$

The relation between the different definitions of slip is simply

$$\sigma_1 = \frac{\nu_1}{1 - \nu_1} \ , \quad \nu_1 = \frac{\sigma_1}{1 + \sigma_1} \ . \tag{3.136}$$

If Eq. (3.135) is compared to the corresponding relation (3.114) for the lateral displacements $u_2(x)$ for pure lateral slip a formal analogy, $u_1(x) = u_2(x)$, is found for

$$\sigma_1 = -\tan\alpha = \nu_2 \, . \tag{3.137}$$

Thus, the results valid for pure lateral slip can be transferred directly to the case of longitudinal slip. Identical coefficients of friction $\mu_1 = \mu_2 = \mu$, $\mu = \text{const}$, and identical stiffnesses $k_1 = k_2$ in longitudinal and lateral direction are assumed. Then, for the contact force it follows from Eqs. (3.137) and (3.136) the identity

$$f_{t1}\left(\sigma_1 = \frac{\nu_1}{1-\nu_1}\right) \equiv f_{t2}(\nu_2) \, , \tag{3.138}$$

cp. Fig. 3.20. By adapting the abscissa scale, the normalized lateral rolling contact coefficient $\varphi_2 = \varphi_2(\alpha)$ is identical to the normalized longitudinal rolling contact coefficient $\varphi_1(\sigma_1)$. With Eq. (3.137) the identity reads as

$$\varphi_1(-\arctan\sigma_1) \equiv \varphi_2(\alpha) \, . \tag{3.139}$$

As a result of the analogy of longitudinal and lateral displacements within the contact patch, further results can be transferred also for $k_1 \neq k_2$. Full longitudinal sliding begins analogous to Eq. (3.124) for $\tan\alpha_G = \pm 1/\theta_1$. From Eqs. (3.137) and (3.136) follow the corresponding limit slips σ_{1G} and ν_{1G}, respectively,

$$\sigma_{1G} = \pm\frac{1}{\theta_1} \, , \qquad \nu_{1G} = \frac{1}{1 \pm \theta_1} \, , \qquad \theta_1 = \frac{2}{3}\frac{k_1 a^2}{\mu\,|f_n|} \, . \tag{3.140}$$

For small longitudinal slips $|\nu_1| \ll 1$ the linearized characteristic of the longitudinal displacements, $u_1(x) = -(a-x)\,\nu_1$, results from Eq. (3.134). Analogously to Eq. (3.115) the corresponding contact force f_{t1} is obtained,

$$f_{t1} = -2k_1 a^2 \nu_1 \, , \qquad |\nu_1| \ll 1 \, . \tag{3.141}$$

Hereby the coefficient f_{11} of the contact force slip relation for longitudinal slip reads as, cp. Eq. (3.118)

$$f_{11} = -\left.\frac{\partial f_{t1}}{\partial \nu_1}\right|_{\nu_1 \to 0} = 2k_1 a^2 \, . \tag{3.142}$$

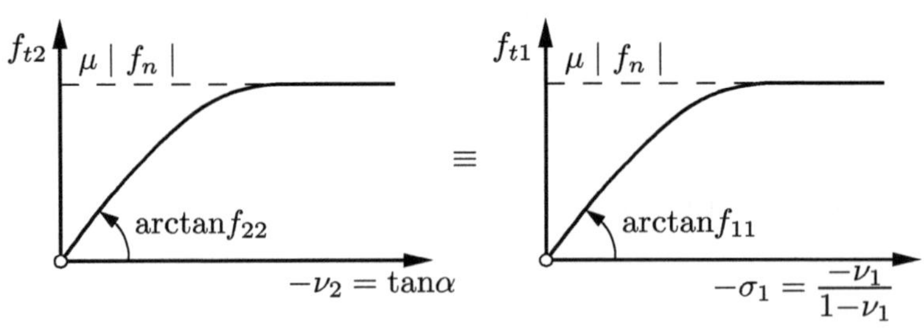

Fig. 3.20. Analogy of contact force characteristics for lateral and longitudinal slip, $\mu_1 = \mu_2 = \mu$, $k_1 = k_2$

3.4.4.4 Linear Contact Force Law

From Eq. (3.141) together with Eq. (3.117), Eq. (3.118) follows the linear dependency between contact forces and slip for longitudinal and lateral slip in analogy to the linear theory of Kalker for the wheel/rail contact, cp. Eq. (3.86),

$$\boldsymbol{f}^K = -\boldsymbol{F}\boldsymbol{\nu}^K \,, \tag{3.143}$$

$$\begin{bmatrix} f_{t1} \\ f_{t2} \\ l_{P3} \end{bmatrix} = -\begin{bmatrix} f_{11} & 0 & * \\ 0 & f_{22} & * \\ 0 & -f_{23} & * \end{bmatrix} \begin{bmatrix} \nu_1 \\ \nu_2 \\ 0 \end{bmatrix} ,$$

$$f_{11} = 2k_1 a^2 \,, \quad f_{22} = 2k_2 a^2 \,, \quad f_{23} = \frac{2}{3} k_2 a^3 \,, \quad (* \text{ arbitrary}) \,. \tag{3.144}$$

Thus, a description is found for the linear relation of the contact force and the slip in the elastic rolling contact between tire and road as well as between wheel and rail, see Eq. (3.86). However, the influence of the aligning slip is neglected in Eq. (3.144). As already mentioned, only the starting slope of the contact forces for $\boldsymbol{\nu}^K \to \boldsymbol{0}$ can be obtained from the linear theory. An advantage of such a standardized description is the possibility to add heuristical approaches of the saturation. The saturation behaviour of the wheel/rail contact forces at simultaneous longitudinal and lateral slip can be adopted directly to the tire/road contact, cp. Eqs. (3.93)-(3.104). Because of the analogy of the local deformation caused by pure longitudinal or pure lateral slip, respectively, in the tire/road contact even if both types of slip occur simultaneously, further relations can be found based on the brush model. This will be shown in the following section. Again, equal coefficients of friction $\mu_1 = \mu_2 = \mu$, $\mu =$ const, and equal tread stiffness values $k_1 = k_2 = k$ in longitudinal and lateral direction are assumed.

3.4.4.5 Contact Forces for Simultaneous Longitudinal and Lateral Slip

The hairs of the brush model are regarded as leaf springs. Due to the isotropic stiffness and friction their displacement is directly opposite to the sliding direction $\bar{\boldsymbol{v}}_P$. This leads in the sticking zone as well as in the sliding zone to a corresponding longitudinal and lateral slip. Hence, the directions of the forces and displacements are defined just like for the special cases considered so far in contradiction to the wheel/rail contact. The displacement vector $\boldsymbol{u}$ of the leaf springs in the sticking zone reads in the contact area frame K

$$\boldsymbol{u}^K = -\Delta t \, \bar{\boldsymbol{v}}_P^K \,, \quad \Delta t = \frac{\Delta s}{v_r} = \frac{a-x}{\omega r_e} \,,$$

$$\begin{bmatrix} u_1 \\ u_2 \end{bmatrix} = -\frac{a-x}{\omega r_e} \begin{bmatrix} v_C \cos\alpha - \omega r_e \\ -v_C \sin\alpha \end{bmatrix} , \tag{3.145}$$

with Δt corresponding to Eq. (3.132) and $\bar{\boldsymbol{v}}_P^K$ from Eq. (3.110) in two-dimensional vector notation. If the slip definition $\boldsymbol{\sigma}^K$ with the circumferential velocity $v_r = \omega r_e$ as reference velocity is used instead of $\boldsymbol{\nu}^K$, the following equations are obtained, cp. Eqs. (3.111), (3.112),

$$\boldsymbol{\sigma}^K = \frac{1}{v_r}\bar{\boldsymbol{v}}_P^K = \frac{v_0}{v_r}\frac{1}{v_0}\bar{\boldsymbol{v}}_P^K = \frac{v_0}{v_r}\boldsymbol{\nu}^K \,, \quad \frac{v_0}{v_r} = \frac{v_C\cos\alpha}{\omega r_e} = \frac{1}{1-\nu_1} \,,$$

$$\begin{bmatrix}\sigma_1\\ \sigma_2\end{bmatrix} = \frac{1}{1-\nu_1}\begin{bmatrix}\nu_1\\ \nu_2\end{bmatrix} \,, \quad \nu_2 = -\tan\alpha \,. \tag{3.146}$$

With Eq. (3.146) it follows from Eq. (3.145)

$$\boldsymbol{u}^K(x) = -(a-x)\boldsymbol{\sigma}^K \,, \quad x_G \le x \le a \,. \tag{3.147}$$

The advantage of this notation is that the displacements u_i depend exclusively on the slip $\sigma_i, i = 1, 2$, independent from each other. Therefore, the contact forces at simultaneous longitudinal and lateral slip are calculated accordingly to the case of pure lateral slip. The results of Eqs. (3.113)-(3.129) can be transferred directly. The tangential force $\boldsymbol{f}_t^*$ reads in the frame K, cp. also Eq. (3.102),

$$\boldsymbol{f}_t^{*K} = -\frac{f_t^*}{\sigma}\boldsymbol{\sigma}^K \,. \tag{3.148}$$

In analogy to Eqs. (3.126) and Eq. (3.127) one gets

$$f_t^* = \sqrt{f_{t1}^{*2} + f_{t2}^{*2}} \,, \quad \varphi^* = \frac{f_t^*}{\mu\,|f_n|} \le 1 \,, \tag{3.149}$$

$$\sigma = \sqrt{\sigma_1^2 + \sigma_2^2} \,, \quad \sigma^* = \theta\sigma \,, \tag{3.150}$$

$$\lambda_G^* = 1 - \sigma^* \,, \quad \sigma_G = \frac{1}{\theta} \,, \quad \sigma_G^* = 1 \,, \quad \theta = \frac{2}{3}\frac{ka^2}{\mu\,|f_n|} \,, \tag{3.151}$$

$$\varphi^* = \begin{cases} 1 - \lambda_G^{*3} = 3\sigma^* - 3\sigma^{*2} + \sigma^{*3} & \text{for } \sigma^* \le 1 \\ 1 & \text{for } \sigma^* \ge 1 \,. \end{cases} \tag{3.152}$$

The components f_{t1}^*, f_{t2}^* of the tangential force result from

$$\begin{aligned} f_{t1}^* &= -\varphi^*\mu\,|f_n|\,\sigma_1/\sigma \,, \\ f_{t2}^* &= -\varphi^*\mu\,|f_n|\,\sigma_2/\sigma \end{aligned} \tag{3.153}$$

by using the normalized rolling contact coefficient φ^* from Eq. (3.152). The aligning torque l_{P3}^* results in analogy to Eq. (3.129) as

$$l_{P3}^* = x_N f_{t2}^* \,, \tag{3.154}$$

$$x_N = \frac{l_{P3}^*}{f_{t2}^*} = \begin{cases} -\dfrac{a}{3}\dfrac{1-3\sigma^*+3\sigma^{*2}-\sigma^{*3}}{1-\sigma^*+\sigma^{*2}/3} < 0 & \text{for } \sigma^* \le 1 \\ 0 & \text{for } \sigma^* \ge 1. \end{cases} \tag{3.155}$$

Pure sliding occurs for $\sigma^* \geq 1$ or $\sigma \geq 1/\theta$. Figure 3.21 shows the typical f_{t1}^*, f_{t2}^*-diagram representing the longitudinal and lateral forces f_{t1}^* and f_{t2}^* in dependence on the longitudinal slip ν_1 and the slip angle α. The diagram is limited by the friction circle with the radius $\mu\,|f_n|$. As an example a driven cornering tire is shown in the diagram's origin. It has to be noticed that the contact area frame K is rotated by 180° for reasons of convenience. In this example the longitudinal slip is $\nu_1 = -0.05$ and the slip angle is $\alpha = 0.1\,\text{rad} \,\widehat{=}\, 5.73°$. The corresponding point E is located within the friction circle. The resulting tangential force vector $\boldsymbol{f}_t^*$ has the coordinates f_{t1}^*, f_{t2}^* and points opposite to the sliding velocity vector $\bar{\boldsymbol{v}}_P$. The left side of the diagram characterizes the driving case while the right side covers the braking case. If a tire is blocked, the circumferential velocity $v_r = \omega r_e$ disappears and the sliding velocity $\bar{\boldsymbol{v}}_P$ coincides with the velocity of the center of mass $\boldsymbol{v}_C$, $\bar{\boldsymbol{v}}_P = \boldsymbol{v}_C$. The resulting force $\boldsymbol{f}_{tB}^*$ points opposite to $\bar{\boldsymbol{v}}_P$, its absolute value is given by the sliding friction force, $|\boldsymbol{f}_{tB}^*| = \mu\,|f_n|$. The corresponding point B in the characteristic diagram is located on the friction circle, whereas the slip angle α arises between the axis of abscissa and the line $\overline{\Gamma B}$. The case of pure sliding is marked in Fig. 3.21, too. The corresponding points are located on the friction circle, namely for braking between B and G' and for driving between A and G. The points G and G' have unequal values of their ordinate which explains the asymmetry of the curves $\alpha = \text{const}$ in the characteristic diagram. The construction of the characteristic diagram by the tangential force-slip-characteristics $f_{t1}^* = f_{t1}^*(\nu_1)$ and $f_{t2}^* = f_{t2}^*(\alpha)$ is indicated in Fig. 3.21 by the corresponding plots. In general, the characteristic diagram of the longitudinal and lateral forces found in this way shows good agreement with experimental results. However, the aligning torque characteristic (3.154), (3.155) shows great discrepancies. Pacejka (1986) has proposed a correction term for the aligning torque with regard to the tangential and lateral stiffnesses c_1 and c_2 of the carcass, respectively. It is assumed that the point of application Q of the tangential forces is displaced in the contact area relative to P by $\boldsymbol{r}_{PQ}^K = \begin{bmatrix} w_1 & w_2 \end{bmatrix}^{\mathrm{T}}$ as a result of the elasticity of the carcass. Then, it follows for the aligning torque

$$
\begin{aligned}
l_{P3,korr}^* &= l_{P3}^* - w_2 f_{t1}^* + w_1 f_{t2}^* \,, \quad w_i = \frac{f_{ti}^*}{c_i} \,, \quad i = 1, 2 \,, \\
l_{P3,korr}^* &= x_N(\sigma^*) f_{t2}^* + f_{t1}^* f_{t2}^* \left(\frac{1}{c_1} - \frac{1}{c_2} \right) .
\end{aligned}
\tag{3.156}
$$

All these results obtained for the brush model allow several extensions, cp. Pacejka (1975), Pacejka (1986):

- Consideration of anisotropic stiffness and friction behavior, $k_1 \neq k_2$, $\mu_1 \neq \mu_2$.
- Implementation of slip-dependent and normal load-dependent coefficients of friction $\mu = \mu(f_n, \nu)$, occurring especially on wet roads, cp. Gaebel et al. (2008).

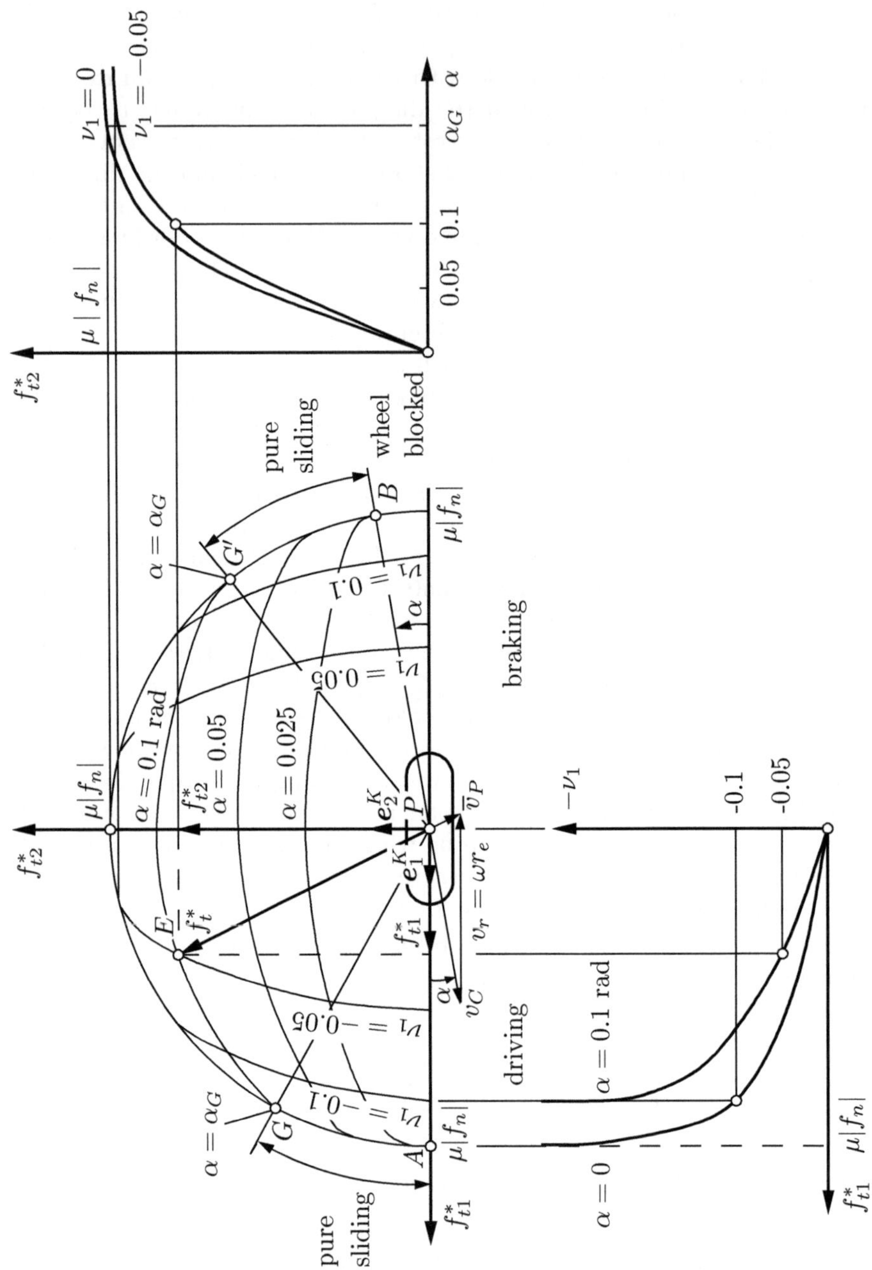

Fig. 3.21. Characteristic diagram of the longitudinal and lateral forces f_{t1}^* and f_{t2}^*, respectively, in dependence on the longitudinal slip ν_1 and the slip angle α for $\theta = 5$, from Pacejka (1986)

- Embedding of experimentally identified contact force slip relations in semi-empirical models. In this case characteristics are used, which are acquired at certain nominal conditions and transferred to other conditions by means of similarity relations, cp. Gaebel et al. (2008).
- Description of unsteady conditions, e. g. dynamical tire load or lateral force variations, cp. Weber (1981), Gutzeit et al. (2006).

■ Problem 3.6 Contact forces for a road vehicle

Figure 3.22 shows a simplified model of a road vehicle, as given by Riekert and Schunck (1940), also known as bicycle model. This model permits a rigorous analysis of the motion in the horizontal plane, if the forces and torques affecting the vehicle are known. The twist $\{\boldsymbol{v}_C, \dot{\psi} = \Omega\}$, the lengths $l_v = l_h = l$, the steering angle δ, the weight G of the vehicle, and the driving force $f_{t1,h}$ at the rear axle are given. The indices v and h refer to the virtual front and rear tire, respectively, placed in the middle of the vehicle. The velocity $\boldsymbol{v}_C$ of the center of mass diverges with the body slip angle $\beta \ll 1$ from the longitudinal axis of the vehicle. The angular velocity Ω shall be small, $\Omega l \ll v_C$. Both tires have the same characteristics. The length of the contact patch is $2a$, the coefficient of sliding friction is $\mu =$ const, the specific stiffnesses $k_1 = k_2 = k$ of the tread and the stiffnesses c_1, c_2 of the carcass are known. The following values are given: $a = 0.1\,\mathrm{m}$, $k = 900\,\mathrm{kN/m^2}$, $c_1 \to \infty$, $c_2 = 60\,\mathrm{kN/m}$, $\mu\,|f_n| = \mu G/2 = 2000\,\mathrm{N}$.

Evaluate the contact forces in dependence on the slip angle on a) the front tire, b) the rear tire and c) the resulting wrench $\{\boldsymbol{f}, \boldsymbol{l}_C\}$ referred to the center of mass.

Solution

The slip angle α_v, α_h can be taken from Fig. 3.22,

$$\alpha_v = \delta + \beta - \frac{\Omega l_v}{v_C} \, , \qquad \alpha_h = \beta + \frac{\Omega l_h}{v_C} \, . \tag{1}$$

Assuming small angles for the velocities v_v, v_h of the tire centers it yields

$$v_v \approx v_h \approx v_C \, . \tag{2}$$

Now, the slip is calculated according to definition (3.112).

a) The front wheel is not driven, thus

$$\nu_{1v} = 0 \, , \qquad \nu_{2v} = -\tan\alpha_v \tag{3}$$

holds. The contact forces according to Eq. (3.115), Eq. (3.116) or Eq. (3.118) are obtained for small slip angles as

$$\begin{aligned} f_{t1,v} &= 0 \, , \\ f_{t2,v} &= -f_{22}\nu_{2v} = 2ka^2 \tan\alpha_v = 18 \ \tan\alpha_v \,\mathrm{kN} \, , \\ l_{P3\,,v} &= f_{23}\nu_{2v} = -\frac{2}{3}ka^3 \tan\alpha_v = -0.6 \ \tan\alpha_v \,\mathrm{kNm} \, . \end{aligned} \tag{4}$$

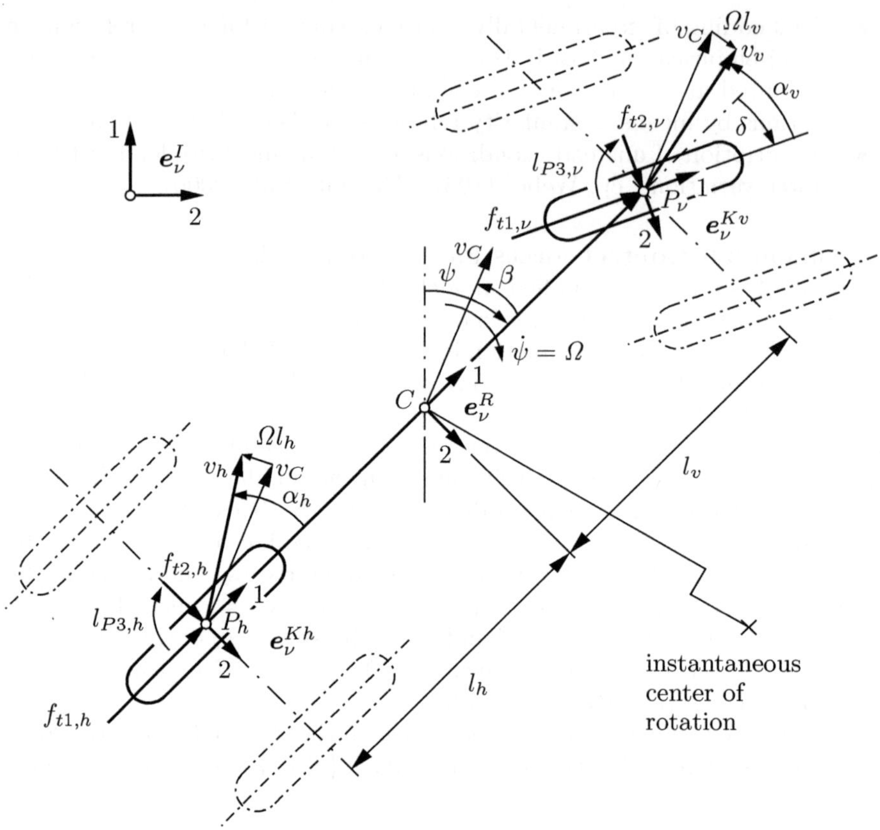

Fig. 3.22. Simplified vehicle model by Riekert and Schunck (1940) with contact forces

The slip angle α_v may take also larger values depending on the steering angle δ. The limit angle α_{vG}, above which pure sliding occurs, results from Eq. (3.124)

$$|\tan\alpha_{vG}| = \frac{1}{\theta} = \frac{3\mu\,|f_n|}{2ka^2} = \frac{1}{3}\,, \quad \alpha_{vG} = 18.43^\circ\,. \tag{5}$$

According to Eqs. (3.126)-(3.128) the corresponding contact forces yield for $\alpha_v > 0$ with $\nu_2^* = \theta\tan\alpha_v = 3\tan\alpha_v$

$$f_{t1,v} = 0 \,, \tag{6}$$

$$f_{t2\,,v} = \mu\,|f_n|\,\varphi_2 = \mu\,|f_n| \begin{cases} 3\nu_2^* - 3\nu_2^{*2} + \nu_2^{*3} & \nu_2^* \le 1 \\ & \text{for} \\ 1 & \nu_2^* \ge 1 \end{cases} , \tag{7}$$

$$l_{P3,v} = a\mu\,|f_n|\,\varphi_3 = -a\mu\,|f_n| \begin{cases} \nu_2^* - 3\nu_2^{*2} + 3\nu_2^{*3} - \nu_2^{*4} & \nu_2^* \le 1 \\ & \text{for} \\ 0 & \nu_2^* \ge 1 \end{cases} . \tag{8}$$

The evaluation of Eqs. (7) and (8) results in values, which are stated in Table 3.10. Figure 3.23 illustrates the corresponding curves.

b) The rear wheel is driven. From the given tangential force $f_{t1,h}$ results a longitudinal slip ν_{1h}, which is initially not known and assumed in an interval $-0.35 \le \nu_{1h} \le 0.35$ ($\nu_{1h} < 0$ represents driving, $\nu_{1h} > 0$ braking). The lateral slip yields

$$\nu_{2h} = -\tan\alpha_h \,. \tag{9}$$

The contact forces result for simultaneous longitudinal and lateral slip from Eqs. (3.149)-(3.156) with consideration of $\sigma_{1h} = \nu_{1h}/(1-\nu_{1h})$, $\sigma_{2h} = -\tan\alpha_h\,/(1-\nu_{1h})$, $\sigma = \sqrt{\sigma_{1h}^2 + \sigma_{2h}^2}$, $\sigma^* = \theta\sigma = 3\sigma$,

$$\begin{aligned} f_{t1,h}^* &= -\varphi^*\mu\,|f_n|\,\sigma_1/\sigma \,, \\ f_{t2\,,h}^* &= -\varphi^*\mu\,|f_n|\,\sigma_2/\sigma \,, \\ l_{P3,h}^* &= x_N f_{t2,h}^* \,, \quad l_{P3,korr}^* = l_{P3,h}^* - f_{t1,h}^* f_{t2,h}^* \frac{1}{c_2} \,, \end{aligned} \tag{10}$$

Table 3.10. Contact force and torque at front wheel

α_v [°]	$\tan\alpha_v$ [-]	$f_{t2,v}$ [N]	$l_{P3,v}$ [Nm]	$x_{Nv} = l_{P3,v}/f_{t2,v}$ [mm]
0	0	0	0	$-33.3\bar{3}$
2.86	0.05	772	-18.53	-24.0
5.71	0.10	1314	-20.58	-15.7
8.53	0.15	1667	-14.96	-9.0
11.31	0.20	1872	- 7.68	-4.10
14.04	0.25	1969	- 2.36	-1.20
16.70	0.30	1998	- 0.18	-0.09
18.43	$0.3\bar{3}$	2000	0	0

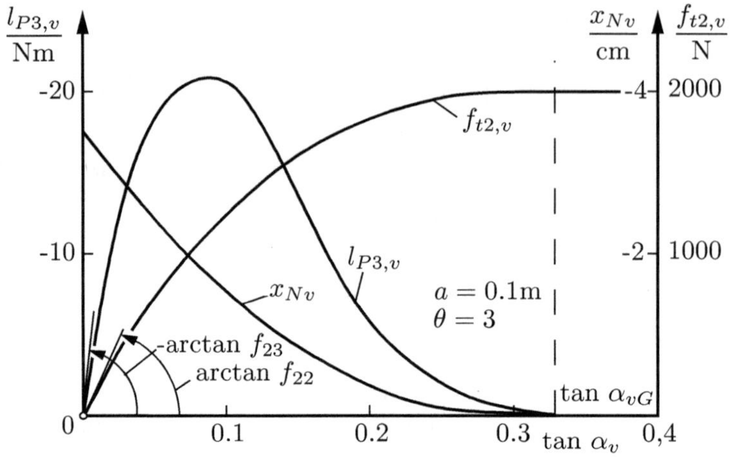

Fig. 3.23. Characteristics of the contact forces $f_{t2,v}$; $l_{P3,v}$ and of the coordinate x_{Nv} of the point of force incidence at the front wheel in dependence of the slip angle α_v

$$\varphi^* = \begin{cases} 3\sigma^* - 3\sigma^{*2} + \sigma^{*3} & \sigma^* \le 1 \\ & \text{for} \\ 1 & \sigma^* \ge 1 \end{cases}, \tag{11}$$

$$x_N = \begin{cases} -\frac{a}{3}\frac{1 - 3\sigma^* + 3\sigma^{*2} - \sigma^{*3}}{1 - \sigma^* + \sigma^{*2}/3} & \sigma^* \le 1 \\ & \text{for} \\ 0 & \sigma^* \ge 1 \end{cases}. \tag{12}$$

The results of the evaluation of Eqs. (10)-(12) for $\tan\alpha_h = 0.15$ are stated in Table 3.11. It turns out that the corrected aligning torques $l^*_{P3,korr}$ according

Table 3.11. Contact forces and torques at rear wheel for $\tan\alpha_h = 0.15$

ν_{1h}	σ_{1h}	σ_{2h}	σ	$f^*_{t1,h}$ [N]	$f^*_{t2,h}$ [N]	$l^*_{P3,h}$ [Nm]	$l^*_{P3,korr}$ [Nm]
0.35	0.5385	-0.2308	0.5859	-1838	788	0	24.139
0.30	0.4286	-0.2143	0.4792	-1728	1008	0	29.030
0.20	0.25	-0.1875	0.3125	-1599	1199	0	31.953
0.10	0.1111	-0.1667	0.2003	-1038	1559	-6.392	20.579
0	0	-0.1500	0.1500	0	1667	-14.96	-14.960
-0.10	-0.0909	-0.1364	0.1639	965	1448	-10.86	-34.149
-0.20	-0.1667	-0.125	0.2084	1504	1128	-4.06	-32.335
-0.30	-0.2308	-0.1154	0.2561	1775	888	-0.852	-27.122
-0.35	-0.2593	-0.1111	0.2821	1832	785	-0.243	-24.351

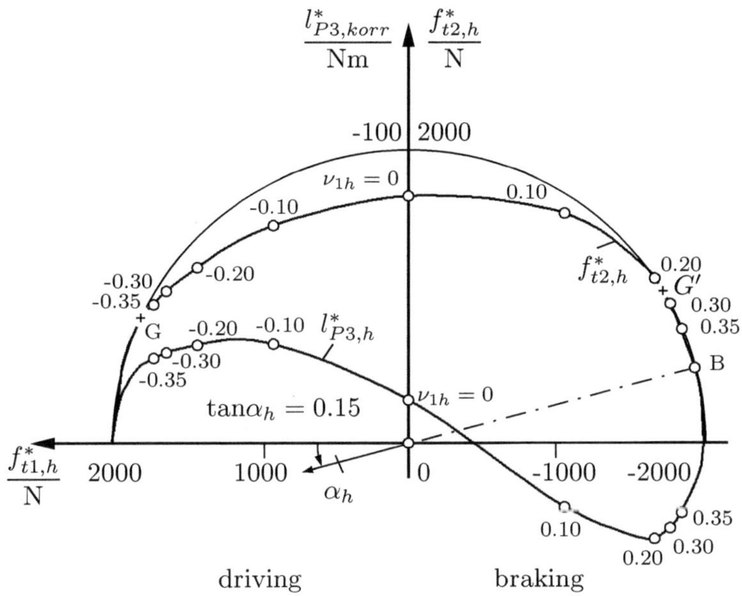

Fig. 3.24. Characteristics of the contact force $f^*_{t2,h}$ and torque $l^*_{P3,korr}$ at the rear tire in dependence on the longitudinal force $f^*_{t1,h}$ for $\tan\alpha_h = 0.15$, $\theta = 3$, $a = 0.1$ m

to Eq. (3.156) deviate significantly from the uncorrected values $l^*_{P3,h}$. The results are shown in Fig. 3.24 together with the friction circle. This $f^*_{t2,h}$, $f^*_{t1,h}$-diagram is designed corresponding to Fig. 3.21, where the boundary points with the friction circle are G and G'. The blocked wheel is characterized by the point B. From Fig. 3.24 it can be seen, that the lateral force $f^*_{t2,h}$ decreases for an increasing absolute value of the longitudinal force, $\left|f^*_{t1,h}\right| > 0$. Furthermore, the corresponding longitudinal slip ν_{1h} can be estimated for given longitudinal forces $f^*_{t1,h}$.

c) The resulting wrench $\{\boldsymbol{f}, l_C\}$ related to the center of mass yield in a vehicle-fixed reference frame $R\{C,\ \boldsymbol{e}^R_\nu\}$, cp. Fig. 3.22,

$$
\begin{aligned}
f^R_1 &= f^*_{t1,h} - f_{t2,v}\sin\delta\ , \\
f^R_2 &= f^*_{t2,h} + f_{t2,v}\cos\delta\ , \\
l^R_{C3} &= l^*_{P3,korr} + l_{P3,v} + f_{t2,v} l_v \cos\delta - f^*_{t2,h} l_h\ .
\end{aligned}
\tag{13}
$$

These contact forces are required for the equations of motion, from which the vehicle motion follows. The lateral motion of an automobile is discussed in more detail in Chap. 9. ■

■ Problem 3.7 Contact forces and linear equations of motion for a drawbar trailer

For the model of a trailer treated in Problem 2.13, the contact forces have to be determined and added to the equations of motion. Small motions relative to the reference position $y_R = 0$, $\gamma_R = 0$ can be assumed. For the case of a laterally fixed trailer joint, $y = y_R = 0$, the path of motions has to be discussed.

Solution

The linearized equations of motion (41) from Problem 2.13 read as

$$\begin{bmatrix} m & 0 & -\bar{c}m \\ 0 & I_{2,2} & 0 \\ -\bar{c}m & 0 & I_G \end{bmatrix} \begin{bmatrix} \ddot{\tilde{y}} \\ \ddot{\tilde{\beta}} \\ \ddot{\tilde{\gamma}} \end{bmatrix} + \begin{bmatrix} k_y & 0 & 0 \\ 0 & 0 & 0 \\ 0 & 0 & k_\gamma \end{bmatrix} \begin{bmatrix} \tilde{y} \\ \tilde{\beta} \\ \tilde{\gamma} \end{bmatrix} = \begin{bmatrix} -f_{t1}\gamma - f_{t2} \\ r f_{t1} \\ a f_{t2} - l_3 \end{bmatrix} . \tag{1}$$

The contact forces f_{t1}, f_{t2}, $l_{P3} \equiv l_3$ correspond with the free body diagram in Fig. 2.30 b), where their directions are introduced opposite to the positive coordinate directions. Under assumption of the brush model, cp. Eq. (3.144), the contact force law yields

$$\begin{bmatrix} f_{t1} \\ f_{t2} \\ l_3 \end{bmatrix} = \begin{bmatrix} f_{11} & 0 & * \\ 0 & f_{22} & * \\ 0 & -f_{23} & * \end{bmatrix} \begin{bmatrix} \nu_1 \\ \nu_2 \\ 0 \end{bmatrix} . \tag{2}$$

The wheel of the trailer is rolling freely, it is neither driven nor braked. Therefore, the longitudinal slip and the longitudinal force disappear, $\nu_1 = 0$, $f_{t1} = 0$. Then, the equation of the wheel rotation can be separated. From Eqs. (1) and (2) as well as (3.144) it follows

$$\begin{bmatrix} m & -\bar{c}m \\ -\bar{c}m & I_G \end{bmatrix} \begin{bmatrix} \ddot{\tilde{y}} \\ \ddot{\tilde{\gamma}} \end{bmatrix} + \begin{bmatrix} k_y & 0 \\ 0 & k_\gamma \end{bmatrix} \begin{bmatrix} \tilde{y} \\ \tilde{\gamma} \end{bmatrix} = \begin{bmatrix} -f_{22}\nu_2 \\ (af_{22} + f_{23})\nu_2 \end{bmatrix} , \tag{3}$$

$$f_{22} = 2k_2 a^{*2} , \quad f_{23} = \frac{2}{3} k_2 a^{*3} , \quad af_{22} + f_{23} = (a + a^*/3) f_{22} \approx af_{22} . \tag{4}$$

Here, k_2 denotes the lateral stiffness of the tire tread and a^* the length of the contact patch, where $a^* \ll a$ is assumed. The lateral slip ν_2 results according to definition (3.54) to

$$\nu_2 = \bar{v}_2^K / v_0 = v_{C2}^K / v_0 . \tag{5}$$

The velocity $\boldsymbol{v}_{C2}^K$ of the center of mass of the wheel in the contact area frame K results from the transformation of the velocities $\boldsymbol{v}_{C2}^I$ given in the inertial frame I, cp. Eq. (4) from Problem 2.13,

$$\boldsymbol{v}_{C2}^{K} = \boldsymbol{S}^{KI}\boldsymbol{v}_{C2}^{I} = (-\gamma)_3\ \boldsymbol{v}_{C2}^{I}$$

$$= \begin{bmatrix} c\gamma & s\gamma & 0 \\ -s\gamma & c\gamma & 0 \\ 0 & 0 & 1 \end{bmatrix} \begin{bmatrix} v_0 + a\dot{\gamma}s\gamma \\ \dot{y} - a\dot{\gamma}c\gamma \\ 0 \end{bmatrix} = \begin{bmatrix} v_0 c\gamma + \dot{y}s\gamma \\ -v_0 s\gamma + \dot{y}c\gamma - a\dot{\gamma} \\ 0 \end{bmatrix} . \tag{6}$$

After linearization of Eq. (6) it follows from Eq. (5)

$$\nu_2 = -\tilde{\gamma} + \frac{\dot{\tilde{y}}}{v_0} - \frac{a\dot{\tilde{\gamma}}}{v_0} . \tag{7}$$

Then, the equations of motion follow from Eq. (3) as

$$\underbrace{\begin{bmatrix} m & -\bar{c}m \\ -\bar{c}m & I_G \end{bmatrix}}_{\boldsymbol{M}} \underbrace{\begin{bmatrix} \ddot{\tilde{y}} \\ \ddot{\tilde{\gamma}} \end{bmatrix}}_{\ddot{\boldsymbol{y}}} + \underbrace{\frac{f_{22}}{v_0}\begin{bmatrix} 1 & -a \\ -a & a^2 \end{bmatrix}}_{\boldsymbol{D}} \underbrace{\begin{bmatrix} \dot{\tilde{y}} \\ \dot{\tilde{\gamma}} \end{bmatrix}}_{\dot{\boldsymbol{y}}} + \underbrace{\begin{bmatrix} k_y & -f_{22} \\ 0 & k_\gamma + af_{22} \end{bmatrix}}_{(\boldsymbol{K}+\boldsymbol{N})} \underbrace{\begin{bmatrix} \tilde{y} \\ \tilde{\gamma} \end{bmatrix}}_{\boldsymbol{y}} = \underbrace{\begin{bmatrix} 0 \\ 0 \end{bmatrix}}_{\boldsymbol{0}} , \tag{8}$$

with

$$\boldsymbol{K} = \begin{bmatrix} k_y & -f_{22}/2 \\ -f_{22}/2 & k_\gamma + af_{22} \end{bmatrix} , \quad \boldsymbol{N} = \begin{bmatrix} 0 & -f_{22}/2 \\ f_{22}/2 & 0 \end{bmatrix} . \tag{9}$$

The equation of motion of a trailer with two degrees of freedom has a similar structure to the equation of motion of a railway wheel set, cp. Eqs. (27) and (28) of Problem 3.5. Especially the skew symmetric matrix $\boldsymbol{N}$ of the nonconservative position forces and the damping matrix $\boldsymbol{D}$, whose elements decrease with increasing velocity v_0, occur in both cases.

For the special case of the laterally fixed trailer joint one degree of freedom ($\tilde{y} \equiv 0$) and therewith the first row of Eq. (8) are lost. The following equation remains:

$$I_G\ddot{\tilde{\gamma}} + \frac{1}{v_0}f_{22}a^2\dot{\tilde{\gamma}} + (k_\gamma + af_{22})\tilde{\gamma} = 0 . \tag{10}$$

It describes a damped oscillation of the trailer around the vertical axis through the coupling joint. Remarkable is that even without a torsional spring ($k_\gamma = 0$) a restoring torque occurs. It is generated by the contact force due to lateral slip. This contact force simultaneously affects a damping torque, which decreases with increasing velocity v_0. Concluding it is noted that the calculation of the contact force is based on the simple brush model. This results in simple contact force slip relations. Refined contact models leads to more complex equations. A comparison of different rolling theories and their impact on the motion behavior of a wheel with a drawbar is presented by Sperling (1977). ■

[illegible]

4

Guideway Models

After the discussion of vehicle models and models for support and guidance systems, guideway models are treated now for ground vehicle systems. Depending on the kind of the vehicle and design aspect, different tasks exist. The vehicle dynamicist dealing with the design of road or rail vehicles regards the guideway as rigid in a first approximation. He is interested in the disturbances, which affect the vehicle from the guideway. The structural dynamicist, who designs bridges, pillared guideways or rails, regards the vehicles as moving loads and is interested in the elastic deformations of the structural elements. In contrast to this partial system consideration, in the case of a strong dynamic interaction between vehicle and guideway, the behavior of the vehicle - guideway system has to be analyzed and optimized as a whole. Such strong dynamic interactions exist e.g. for rapid magnetically levitated vehicles on pillared guideways. The consideration of the guideway into the system optimization has also great importance for economic reasons. A good portion of the entire investment - for magnetically levitated transport systems about 75 % - is accounted for the guideway.

In the following models for elastic guideways as well as disturbance models for rigid guideways are developed. It is shown that in both cases linear mathematical relations arise. Therefore, both models can be superposed, if necessary.

4.1 Models for Elastic Guideways

The first theoretical works on the dynamics of elastic structures under moving loads were inspired by the collapse of the railway bridge in Chester, England, in 1847, see e.g. Lewis and Gagg (2004). Since that time much research has been dedicated to this subject. A summary of the engineering literature, the used methods and obtained results can be found in Fryba (1999), Popp (1981) and Kortuem and Wormley (1981).

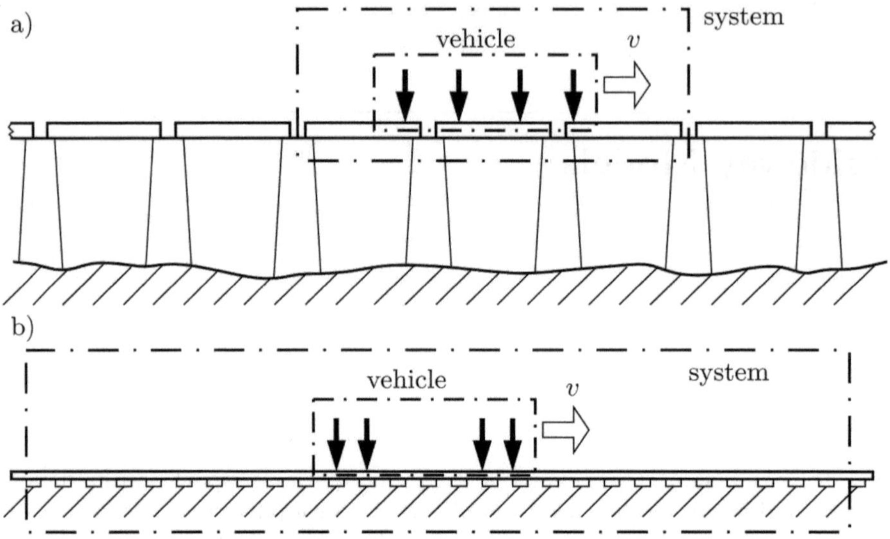

Fig. 4.1. Vehicle - guideway system models for: a) magnetically levitated vehicles; b) railway vehicles

For guided ground vehicles a useful assumption is an infinitely long guideway. Figure 4.1 shows typical designs like pillared guideways for magnetically levitated vehicles and continuously supported tracks for railways. By replacing the loading of the vehicles to the guideway by the free body diagrams resulting in external forces to the elastic structure, in both cases a system of forces moving with the velocity v is achieved. In vertical direction these forces consist of a constant weight and time-dependent magnetic forces or dynamic wheel loads, respectively. It is recognized that the forces in vertical direction are dominating. Therefore, the elastic deformations due to bending have a substantial influence on the vertical dynamics of the total system.

The *guideway for magnetically levitated vehicles*, Fig. 4.1 a), is designed as a periodic structure consisting of identical guideway elements on rigid pillars. Continuous beams have a smaller static displacement as single- and multi-span beams if supported by pillars with the same distance, but the dynamic behavior is less favorable. Therefore, the standard designs consist of single-span beams, double-span beams or two-field structural elements. The total model of the vehicle - guideway system has to contain at least as much guideway elements as coupled by the moving vehicle. For efficiency reasons a model boundary of constant length is used, which contains only the minimum number of coupled guideway elements. The model boundaries are gradually shifted with the vehicle movement if the foremost force reaches the next resting guideway element. There, the initial and transition conditions have to be adapted for the guideway elements.

The *track system for railways*, Fig. 4.1 b), is designed as a continuously bedded, infinitely long structure. The regarded model has to be sufficiently large, so that the elastic deformations at the model boundary disappear. The model boundary may be shifted continuously according to the vehicle movement.

The guideways are modeled and analyzed by simple straight beams. For pillared periodic structures with finite beam length modal methods (Bernoulli's ansatz of standing waves) are used, while continuously bedded beams of infinite length are examined by means of wave methods (D'Alembert's ansatz of progressive waves).

4.1.1 Models for Periodically Pillared Beams

The investigation of the elastic guideway deformations is performed in three stages:

1) mathematical description of a single guideway element,
2) summing up of the guideway elements coupled by the vehicle within the system boundary,
3) calculation of the guideway deformations due to the interactions with the moving vehicle in the context of the total system with consideration of the gradual shift of the system boundary by adaptation of the initial and transfer conditions.

Figure 4.2 shows usual guideway elements. Each element of length L consists of s fields of homogeneous beams of length L_i with constant bending stiffness $(EI)_i$ and distributed mass $(\rho A)_i$, $i = 1(1)s$. Because of the with velocity v moving forces $f_\mu(t) = f_{\mu\, stat} + f_{\mu\, dyn}(t)$, $\mu = 1(1)m$, in each field beam displacements $w_i(\xi_i, t)$, $0 \le \xi_i \le L_i$, occur due to bending. The calculation of these displacements is carried out using the Bernoulli-Euler's beam theory. This theory assumes small deformations, plane cross sections during bending as well as linear elastic material behavior. Rotational inertia and shear deformations are neglected. A comparison with the more accurate Timoshenko beam theory not neglecting both of the last mentioned phenomena, results in a relative error for the first five eigenfrequencies smaller than 5 % if the length to height ratio of the beam is larger than 20. The partial differential equation (PDE) of the beam reads as

$$(EI)_i w_i''''(\xi_i, t) + (\rho A)_i \ddot{w}_i(\xi_i, t) = \sum_\mu f_\mu(t)\delta(\xi_i - \xi_{i\mu}) \, , \tag{4.1}$$

for the field i with $0 \le \xi_i \le L_i$. Here $(\ )'$ and $\dot{(\ )}$ indicate spatial and time derivatives, respectively, $\delta(\ )$ is the Dirac function or distribution and $\xi_{i\mu}$ the point where the force f_μ acts on the field i. The solution of Eq. (4.1) is obtained under consideration of the initial and boundary conditions from the

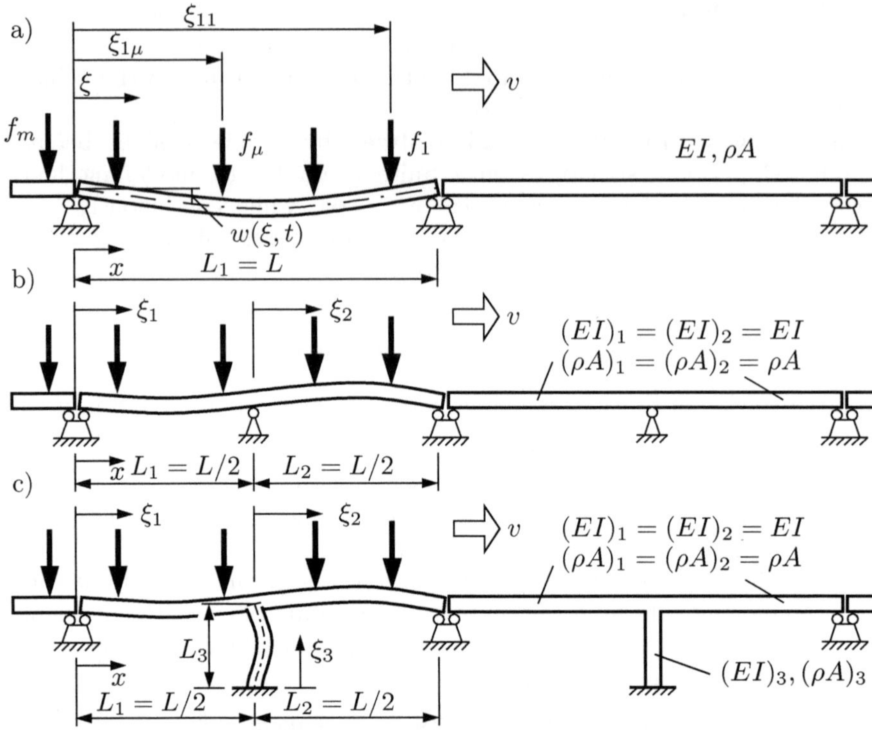

Fig. 4.2. Models for periodically pillared guideway, consisting of: a) single-span beam; b) double-span beam; c) two-field structural element

separation theorem using the spatial-dependent eigenfunctions $\varphi_{ij}(\xi_i)$ and the time-dependent generalized coordinates or modal coordinates $z_j(t)$,

$$w_i(\xi_i, t) = \sum_{j=1}^{\infty} \varphi_{ij}(\xi_i) z_j(t) \ , \quad i = 1(1)s \ . \tag{4.2}$$

The eigenfunction $\varphi_{ij}(\xi_i)$ indicates the j-th eigenmode of the unloaded beam field i and has the following properties:

- The eigenfunction $\varphi_{ij}(\xi_i)$ is a solution of the eigenvalue problem

$$\varphi_i''''(\xi_i) - \left(\frac{\lambda_i}{L_i}\right)^4 \varphi_i(\xi_i) = 0 \ , \tag{4.3}$$

  with the eigenvalue λ_i which is a dimensionless value by introducing the length L_i,

$$\lambda_i^4 = \omega^2 L_i^4 \left(\frac{\rho A}{EI}\right)_i \ . \tag{4.4}$$

- The eigenfunction $\varphi_{ij}(\xi_i)$ fulfills the boundary conditions. The adjustment of the general solution of Eq. (4.3) to the boundary conditions results in a transcendent equation - the frequency equation - for the angular eigenfrequency ω of the beam vibrations. This equation has an infinite number of solutions ω_j with a corresponding number of eigenfunctions $\varphi_{ij}(\xi_i)$, which are to be normalized.
- The eigenfunction $\varphi_{ij}(\xi_i)$ is orthogonal with regard to the whole guideway element. The orthogonality condition is

$$\sum_{i=1}^{s} \int_{0}^{L_i} (\rho A)_i \varphi_{ij}(\xi_i)\varphi_{ik}(\xi_i)\mathrm{d}\xi_i = \begin{cases} 0 & \text{for} \quad j \neq k \\ M_j = \sum\limits_{i=1}^{s} M_{ij} & \qquad\; j = k \end{cases} , \tag{4.5}$$

  where M_j is called modal mass.

With these properties and the fading out characteristic of the Dirac distribution a normalized differential equation follows from Eq. (4.1) for the modal coordinates $z_j(t)$ of a guideway element

$$\ddot{z}_j(t) + \omega_j^2 z_j(t) = \frac{1}{M_j} \sum_{\mu} \varphi_{ij}(\xi_{i\mu}) f_\mu(t) . \tag{4.6}$$

To solve Eq. (4.6) the initial and transition conditions must be considered. For engineering applications the procedure is modified as follows:

- Approximation of the solution of Eq. (4.2) by the first f eigenfunctions $\varphi_{ij}(\xi_i)$, $j = 1(1)f$. As reference value for f can be used $s \leq f \leq 2s$ for the computation of the beam deformation and $f > 3s$ for the evaluation of the bending moments and stresses, whereby s is the number of the fields of a guideway element. A further reference value is the frequency range of interest $0 \leq \omega_j \leq \omega_{max}$ where $\omega_f \leq \omega_{max}$ has to be satisfied.
- A modal damping $2D_j\omega_j\dot{z}_j(t)$ is added on the left side of Eq. (4.6). From measurements arises that the damping ratios D_j vary within a wide range, $0.005 \leq D_j \leq 0.05$, where the smaller ones are for steel structures and the larger ones for reinforced concrete structures. The measurements provide often only the damping ratio D_1 of the first eigenmode. Then, the damping ratios D_j for $j > 1$ must be determined from hypotheses. The most usual assumptions for the damping ratios are viscous damping

$$D_j = D_1 \frac{\omega_j}{\omega_1} , \tag{4.7}$$

  where the higher eigenfunctions according to their higher eigenfrequencies are damped more strongly, and hysteretic damping or structural damping

$$D_j = D_1 , \tag{4.8}$$

where the damping ratios are the same for all eigenmodes.

- Representations of the approximative solution Eq. (4.2) and the differential equations (4.6), extended by a damping term, in vector notation,

$$w_i(\xi_i, t) = \boldsymbol{\varphi}^{\mathrm{T}}(\xi_i)\boldsymbol{z}(t) , \tag{4.9}$$

$$\ddot{\boldsymbol{z}}(t) + \boldsymbol{\Delta}\dot{\boldsymbol{z}}(t) + \boldsymbol{\Omega}\boldsymbol{z}(t) = \boldsymbol{M}^{-1}\sum_{\mu}\boldsymbol{\varphi}(\xi_{i\mu})f_\mu , \tag{4.10}$$

with

$$\boldsymbol{\varphi} = [\varphi_1, ..., \varphi_f]^{\mathrm{T}} , \quad \boldsymbol{z} = [z_1, ..., z_f]^{\mathrm{T}} , \quad \boldsymbol{M} = \mathbf{diag}(M_j) ,$$

$$\boldsymbol{\Delta} = \mathbf{diag}(2D_j\omega_j) , \quad \boldsymbol{\Omega} = \mathbf{diag}(\omega_j^2) , \quad j = 1(1)f . \tag{4.11}$$

It can be recognized that the mathematical description of an individual guideway element (*stage 1*) requires only the determination of the angular eigenfrequencies ω_j and to the associated eigenfunctions $\varphi_{ij}(\xi_i)$. With Eq. (4.5) also the modal masses M_j are known. By adding the modal damping ratios D_j the differential equation (4.10) for the $f \times 1$-vector $\boldsymbol{z}(t)$ of the modal coordinates of a guideway element finally is found. These equations have the same characteristics as the equations of motion (2.99) for ordinary multibody systems. The description of the guideway elements within the system boundaries (*stage 2*), coupled by the vehicle, is therefore straightforward. The final computation of the guideway deformations (*stage 3*) requires models for the interaction forces $f_\mu(t)$ and is performed during the analysis of the complete system. The next section deals with the determination of eigenfrequencies and eigenfunctions for bending vibrations in more detail. This investigation is also called modal analysis.

4.1.2 Modal Analysis of Beam Structures for Bending Vibrations

As a basic example the modal analysis is performed for a single-span beam, see Fig. 4.2 a). Here, the field index i is omitted. The general solution of the eigenvalue problem of Eq. (4.3) for $\lambda \neq 0$ reads as

$$\varphi(\xi) = C_1\cosh(\lambda\xi/L) + C_2\sinh(\lambda\xi/L) + C_3\cos(\lambda\xi/L) + C_4\sin(\lambda\xi/L) \tag{4.12}$$

where for $\lambda = 0$ the solution of the differential equation $\varphi''''(\xi) = 0$ resulting from Eq. (4.3) is given by a polynomial.

Equation (4.12) is now rewritten in vector notation,

$$\varphi(\xi) = \mathbf{c}^{\mathrm{T}}\boldsymbol{a}(\lambda\xi/L) = \boldsymbol{a}^{\mathrm{T}}(\lambda\xi/L)\mathbf{c} , \tag{4.13}$$

with the vector of constants

$$\boldsymbol{c} = [C_1 \,, C_2 \,, C_3 \,, C_4]^{\mathrm{T}} \,, \tag{4.14}$$

and the vector of trigonometric and hyperbolic functions

$$\boldsymbol{a}(\cdot) = [\mathrm{C}(\cdot) \,, \mathrm{S}(\cdot) \,, \mathrm{c}(\cdot) \,, \mathrm{s}(\cdot)]^{\mathrm{T}} \,, \tag{4.15}$$

where

$$\mathrm{C}(\cdot) = \cosh(\cdot) \,, \quad \mathrm{S}(\cdot) = \sinh(\cdot) \,, \quad \mathrm{c}(\cdot) = \cos(\cdot) \,, \quad \mathrm{s}(\cdot) = \sin(\cdot) \,. \tag{4.16}$$

The boundary conditions of the double-sided simply supported beam are

$$\varphi(0) = 0 \,, \quad \varphi''(0) = 0 \,, \quad \varphi(L) = 0 \,, \quad \varphi''(L) = 0 \,. \tag{4.17}$$

The adaptation of the general solution (4.12) to the boundary conditions (4.17) results in the constants $C_1 = C_2 = C_3 = 0$ and the transcendental frequency equation

$$C_4 \sin \lambda = 0 \,. \tag{4.18}$$

The eigenvalues λ_j and the angular eigenfrequencies ω_j follow from the non-trivial solutions of Eq. (4.18),

$$\lambda_j = j\pi \,, \quad \omega_j^2 = \left(\frac{j\pi}{L}\right)^4 \frac{EI}{\rho A} \,, \quad j = 1(1)\infty \,. \tag{4.19}$$

If the free parameter is set to $C_4 = \sqrt{2}$ the eigenfunctions φ_j and the modal masses M_j are

$$\varphi_j = \sqrt{2} \sin j\pi \frac{\xi}{L} \,, \quad M_j = \int_0^L \rho A \varphi_j^2(\xi) \mathrm{d}\xi = \rho A L \,, \quad j = 1(1)\infty \,. \tag{4.20}$$

Table 4.1 shows a listing of the modal values of simple beams with different boundary conditions. For complicated beam structures an analytical modal analysis is not possible any more and numerical methods are applied. Two of the numerous calculation methods, cp. Knothe (1971), are

- the Deformation Method (DEM) or dynamic stiffness method, cp. Kolousek et al. (1973) and
- the Finite Element Method (FEM), cp. e. g. Knothe and Wessels (2008).

Here the deformation method is described. After linearization of the basic relations, this method yields the stiffness and mass matrix found by the finite element method, too. The approach of the modal analysis is equal for both methods. In analogy to the formalism of multibody systems, cp. Sect. 2.5, the procedure can be divided into five steps.

Step 1: Specification of the system and input data.

First the elastic guideway structure element is considered as a whole and is divided into

Table 4.1. Configuration of the modal variables for beams with common boundary conditions. The eigenfunctions φ_j are normalized by the condition $M_j = \rho A \int_0^L \varphi_j^2(x)\mathrm{d}x = \rho AL$.

Boundary conditions $x=0$	$x=L$	Frequency equation	j	Eigenvalues λ_j, form factors γ_j ($\omega_j = \lambda_j^2\sqrt{EI/(\rho AL^4)}$): λ_j	γ_j	Eigenfunctions $\varphi_j = \boldsymbol{c}_j^{\mathrm{T}}\boldsymbol{a}(\lambda_j x/L)$; $\boldsymbol{c}^{\mathrm{T}} = [\,C_1\ C_2\ C_3\ C_4\,]$, $\boldsymbol{a}^{\mathrm{T}} = [\,\mathrm{C}(\cdot)\ \mathrm{S}(\cdot)\ \mathrm{c}(\cdot)\ \mathrm{s}(\cdot)\,]$; $\mathrm{C} \mathrel{\hat{=}} \cosh, \mathrm{S} \mathrel{\hat{=}} \sinh, \mathrm{c} \mathrel{\hat{=}} \cos, \mathrm{s} \mathrel{\hat{=}} \sin$
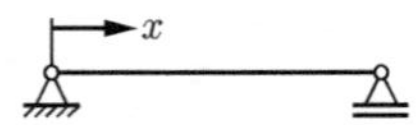 Simply supported	Simply free supported	$\sin\lambda = 0$	1	3.1416		$\boldsymbol{c}_j^{\mathrm{T}} = [\,0\ 0\ 0\ \sqrt{2}\,]$
			2	6.2832		
			3	9.4248		
			4	12.5664		
			$\forall j$	$j\pi$		
 Fixed	Free	$1 + \cos\lambda\cosh\lambda = 0$	1	1.8751	0.7341	$\boldsymbol{c}_j^{\mathrm{T}} = [\,1\ -\gamma_j\ -1\ \gamma_j\,]$
			2	4.6941	1.0185	$\gamma_j = \dfrac{\sinh\lambda_j - \sin\lambda_j}{\cosh\lambda_j + \cos\lambda_j}$
			3	7.8548	0.9992	
			4	10.9955	1.0000	
			$j > 4$	$(2j-1)\pi/2$	1.0000	
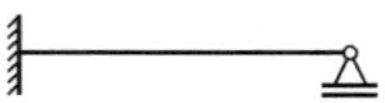 Fixed	Simply free supported	$\tan\lambda - \tanh\lambda = 0$	1	3.9266	1.0008	$\boldsymbol{c}_j^{\mathrm{T}} = [\,1\ -\gamma_j\ -1\ \gamma_j\,]$
			2	7.0686	1.0000	$\gamma_j = \cot\lambda_j = \coth\lambda_j$
			3	10.2102	1.0000	
			4	13.3518	1.0000	
			$j > 4$	$(4j+1)\pi/4$	1.0000	
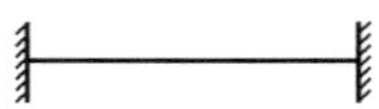 Fixed	Fixed	$1 - \cos\lambda\cosh\lambda = 0$	1	4.7300	0.9825	$\boldsymbol{c}_j^{\mathrm{T}} = [\,1\ -\gamma_j\ -1\ \gamma_j\,]$ (fixed-fixed)
Free	Free		2	7.8532	1.0008	$\boldsymbol{c}_j^{\mathrm{T}} = [\,1\ -\gamma_j\ 1\ -\gamma_j\,]$ (free-free)
			3	10.9956	1.0000	$\gamma_j = \dfrac{\cosh\lambda_j - \cos\lambda_j}{\sinh\lambda_j - \sin\lambda_j}$
			4	14.1372	1.0000	
			$j > 4$	$(2j+1)\pi/2$	1.0000	

$i = 1(1)s$ fields of homogeneous, straight beams with constant profile
and
$\kappa = 1(1)k$ nodes at the boundaries of the fields, with
$\nu = 1(1)n$ possible displacements of the nodes.

Nodes have to be located at the supports and intermediate supports, at locations with variations of cross-sectional area, stiffness or inertia and at points carrying additional springs or masses. In order to improve the computational accuracy of the method, nodes can be located additionally at any position of the structure. Each guideway element has a global, inertially fixed Cartesian frame. Its origin is located preferably at the left support, cp. Fig. 4.2. The node displacements and rotations are described by generalized coordinates $q_\nu, \nu = 1(1)n$. The specific fields i are cut free and characterized by the length of the field L_i, the bending stiffness $(EI)_i$ and the distributed mass $(\rho A)_i$. If there are additional springs or masses at the nodes κ they have to be specified with the mass m_κ, moment of inertia I_κ, spring constant k_κ or torsion spring constant c_κ. The input data are:

$$n \ , \boldsymbol{q} = [q_1, ..., q_\nu, ..., q_n]^{\mathrm{T}} \ , \tag{4.21}$$
$$s, \{L_i, (EI)_i, (\rho A)_i\} \ , \quad i = 1(1)s \ , \tag{4.22}$$
$$k, \{c_\kappa, I_\kappa, k_\kappa, m_\kappa\} \ , \quad \kappa = 1(1)k \ . \tag{4.23}$$

Step 2: Element model, local equations.

A single beam element is regarded and described in a local frame, see Fig. 4.3. For lucidity the index i will be omitted in this section. The equation of motion of an unloaded beam element is given by Eq. (4.1) with $f_\mu \equiv 0$. With the ansatz $w(\xi, t) = \varphi(\xi) \sin \omega t$ the solution is represented by the eigenfunctions $\varphi(\xi)$, cp. Eqs. (4.12)-(4.16). The same applies to the bending moment $M_b(\xi, t) = -EIw''(\xi, t)$ and the shear force $Q(\xi, t) = M_b'(\xi, t)$. After disregarding of the harmonic time function, the internal forces and bending moments, and deformations w and declinations ζ at the left and right boundary of the beam (index l, r), cp. Fig. 4.3, are described as follows:

$$\begin{aligned} M_l &= -EI\varphi''(0) \ , & \zeta_l &= \varphi'(0) \ , \\ Q_l &= EI\varphi'''(0) \ , & w_l &= \varphi(0) \ , \\ M_r &= EI\varphi''(L) \ , & \zeta_r &= \varphi'(L) \ , \\ Q_r &= -EI\varphi'''(L) \ , & w_r &= \varphi(L) \ . \end{aligned} \tag{4.24}$$

The variables in Eq. (4.24) are summarized to a boundary force vector $\boldsymbol{f}$ and a boundary displacement vector $\boldsymbol{v}$

$$\boldsymbol{f} = [M_l \ Q_l \ M_r \ Q_r]^{\mathrm{T}} \ , \quad \boldsymbol{v} = [\zeta_l \ w_l \ \zeta_r \ w_r]^{\mathrm{T}} \ . \tag{4.25}$$

Inserting the eigenfunctions $\varphi(\xi) = \boldsymbol{a}^{\mathrm{T}}(\lambda\xi/L)\boldsymbol{c}$ and their derivatives into Eq. (4.24) yields

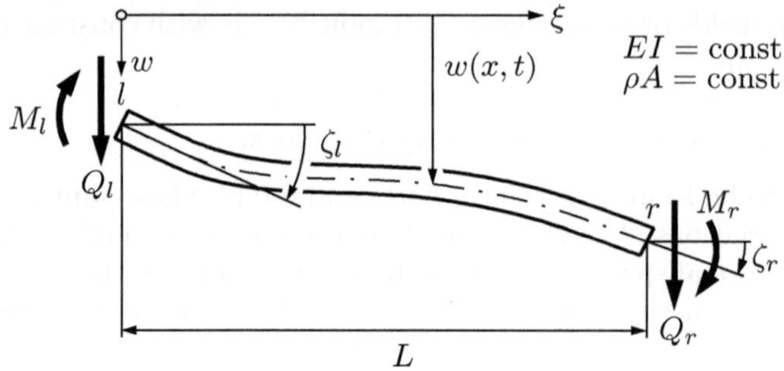

Fig. 4.3. Free body diagram of a beam element with boundary forces and boundary deformations

$$\boldsymbol{f} = \boldsymbol{D}(\lambda)\boldsymbol{c}\,, \quad \boldsymbol{v} = \boldsymbol{E}(\lambda)\boldsymbol{c}\,, \quad \lambda^4 = \omega^2 L^4 \frac{\rho A}{EI}\,, \tag{4.26}$$

with the matrices $\boldsymbol{D}(\lambda), \boldsymbol{E}(\lambda)$, depending on the eigenvalue, and the coefficient vector $\boldsymbol{c}$. The elimination of the vector $\boldsymbol{c}$ results in

$$\boldsymbol{c} = \boldsymbol{E}^{-1}(\lambda)\boldsymbol{v}\,, \quad \boldsymbol{f} = \boldsymbol{D}(\lambda)\boldsymbol{E}^{-1}(\lambda)\boldsymbol{v} = \boldsymbol{F}(\lambda)\boldsymbol{v}\,, \tag{4.27}$$

where

$$\boldsymbol{F}(\lambda) = EI \begin{bmatrix} F_2/L & -F_4/L^2 & F_1/L & -F_3/L^2 \\ -F_4/L^2 & F_6/L^3 & F_3/L^2 & F_5/L^3 \\ F_1/L & F_3/L^2 & F_2/L & F_4/L^2 \\ -F_3/L^2 & F_5/L^3 & F_4/L^2 & F_6/L^3 \end{bmatrix}\,, \tag{4.28}$$

$$\begin{aligned} F_1 &= -\lambda[\mathrm{S}(\lambda) - \mathrm{s}(\lambda)]/N\,, \\ F_2 &= -\lambda[\mathrm{C}(\lambda)\mathrm{s}(\lambda) - \mathrm{S}(\lambda)\mathrm{c}(\lambda)]/N\,, \\ F_3 &= -\lambda^2[\mathrm{C}(\lambda) - \mathrm{c}(\lambda)]/N\,, \\ F_4 &= \lambda^2[\mathrm{S}(\lambda)\mathrm{s}(\lambda)]/N\,, \\ F_5 &= \lambda^3[\mathrm{S}(\lambda) + \mathrm{s}(\lambda)]/N\,, \\ F_6 &= -\lambda^3[\mathrm{C}(\lambda)\mathrm{s}(\lambda) + \mathrm{S}(\lambda)\mathrm{c}(\lambda)]/N\,, \\ N &= \mathrm{C}(\lambda)\mathrm{c}(\lambda) - 1\,. \end{aligned} \tag{4.29}$$

The symmetric field matrix $\boldsymbol{F}(\lambda)$ combines the boundary displacements $\boldsymbol{v}$ and the boundary forces $\boldsymbol{f}$ of one beam element. Due to its eigenvalue dependence the field matrix is also called dynamical stiffness matrix. The elements of the matrix are described by the frequency functions $F_\mu = F_\mu(\lambda)$, $\mu = 1(1)6$, that have been introduced by Kolousek et al. (1973). The frequency functions are also integrated in the matrix

$$E^{-1}(\lambda) = \frac{1}{2\lambda^3} \begin{bmatrix} -L\lambda F_2 & \lambda(\lambda^2 + F_4) & -L\lambda F_1 & \lambda F_3 \\ L(\lambda^2 - F_4) & F_6 & LF_3 & F_5 \\ L\lambda F_2 & \lambda(\lambda^2 - F_4) & L\lambda F_1 & -\lambda F_3 \\ L(\lambda^2 + F_4) & -F_6 & -LF_3 & -F_5 \end{bmatrix} , \tag{4.30}$$

required in the following.

Step 3: Relation between local and global coordinates.

The n nodal degrees of freedom are found considering the boundary and transition conditions and are described by the $n \times 1$-vector $\boldsymbol{q}$, cp. Eq. (4.21). The relation between the boundary displacement vector $\boldsymbol{v}_i$ of the beam element i and the generalized coordinates $\boldsymbol{q}$ is given by the $4 \times n$ Jacobian matrix $\boldsymbol{J}_i$. This matrix describes also the relation between the virtual displacements,

$$\boldsymbol{v}_i = \boldsymbol{J}_i \boldsymbol{q} , \quad \delta \boldsymbol{v}_i = \boldsymbol{J}_i \delta \boldsymbol{q} . \tag{4.31}$$

The global $4s \times n$ Jacobian matrix $\bar{\boldsymbol{J}}$ results from summarizing all boundary displacements into a $4s \times 1$- vector $\bar{\boldsymbol{v}}$,

$$\bar{\boldsymbol{v}} = \bar{\boldsymbol{J}} \boldsymbol{q} , \quad \delta \bar{\boldsymbol{v}} = \bar{\boldsymbol{J}} \delta \boldsymbol{q} , \tag{4.32}$$

$$\bar{\boldsymbol{v}} = [\boldsymbol{v}_1^{\mathrm{T}}, ..., \boldsymbol{v}_i^{\mathrm{T}}, ..., \boldsymbol{v}_s^{\mathrm{T}}]^{\mathrm{T}} , \quad \bar{\boldsymbol{J}} = [\boldsymbol{J}_1^{\mathrm{T}}, ..., \boldsymbol{J}_i^{\mathrm{T}}, ..., \boldsymbol{J}_s^{\mathrm{T}}]^{\mathrm{T}} . \tag{4.33}$$

The Jacobian matrices here are Boolean matrices with the elements 0 and 1.

Step 4: Consideration of the whole system, global equations.

Every beam element i, $i = 1(1)s$, of a guideway element is characterized by the following quantities:

eigenvalue:	$\lambda_i = \lambda_i(\omega) = L_i \sqrt[4]{\omega^2 (\rho A/EI)_i}$,	(4.34)
eigenfunction:	$\varphi_i = \varphi_i(\xi_i) = \boldsymbol{a}_i^{\mathrm{T}} \boldsymbol{c}_i$, $\quad \boldsymbol{a}_i = \boldsymbol{a}(\lambda_i \xi_i / L_i)$,	(4.35)
vector of constants:	$\boldsymbol{c}_i = \boldsymbol{E}_i^{-1}(\lambda_i) \boldsymbol{v}_i$,	(4.36)
dynamic stiffness:	$\boldsymbol{f}_i = \boldsymbol{F}_i(\lambda_i) \boldsymbol{v}_i$.	(4.37)

The field matrices $\boldsymbol{F}_i$ are combined into a global $4s \times 4s$ field matrix $\bar{\bar{\boldsymbol{F}}}$

$$\bar{\bar{\boldsymbol{F}}} = \mathbf{diag}(\boldsymbol{F}_i) . \tag{4.38}$$

The principle of virtual work applied to the inner forces results with Eq. (4.32) in

$$\delta W_{int}^e = \sum_{i=1}^{s} \delta \boldsymbol{v}_i^{\mathrm{T}} \boldsymbol{f}_i = \sum_{i=1}^{s} \delta \boldsymbol{v}_i^{\mathrm{T}} \boldsymbol{F}_i \boldsymbol{v}_i = \delta \bar{\boldsymbol{v}}^{\mathrm{T}} \bar{\bar{\boldsymbol{F}}} \bar{\boldsymbol{v}} = \delta \boldsymbol{q}^{\mathrm{T}} \bar{\boldsymbol{J}}^{\mathrm{T}} \bar{\bar{\boldsymbol{F}}} \bar{\boldsymbol{J}} \boldsymbol{q} = 0 . \tag{4.39}$$

Due to $\delta \boldsymbol{q}^{\mathrm{T}} \neq \boldsymbol{0}$ it follows

$$\bar{\boldsymbol{J}}^{\mathrm{T}} \bar{\bar{\boldsymbol{F}}} \bar{\boldsymbol{J}} \boldsymbol{q} = \bar{\boldsymbol{F}}(\omega) \boldsymbol{q} = \boldsymbol{0} , \quad \bar{\boldsymbol{F}} = \bar{\boldsymbol{F}}^{\mathrm{T}} , \tag{4.40}$$

with the $n \times n$ dynamic stiffness matrix $\bar{\boldsymbol{F}}(\omega)$ of the beam system. Equations (4.40) describe an implicit eigenvalue problem. The influence of the additional

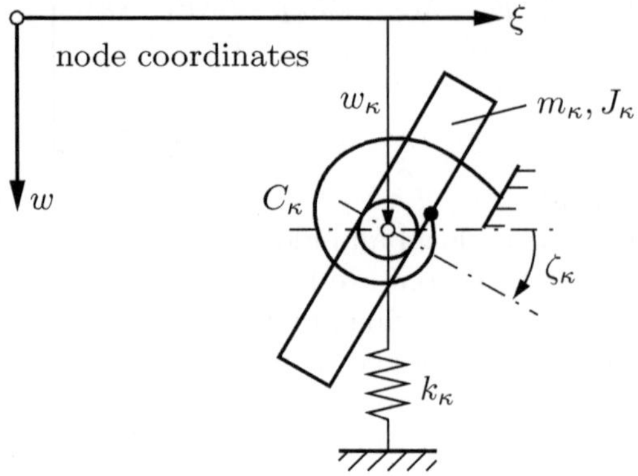

Fig. 4.4. Description of the additional masses and springs at node κ with coordinates w_κ and ζ_κ

masses and springs at node κ, cp. Fig. 4.4, can be considered by the virtual work of the external forces

$$\begin{aligned}\delta W^e_{ext} &= -\sum_{\kappa=1}^{k} \left[\delta\zeta_\kappa(c_\kappa - \omega^2 I_\kappa)\zeta_\kappa + \delta w_\kappa(k_\kappa - \omega^2 m_\kappa)w_\kappa\right] \\ &= -\delta\boldsymbol{q}^{\mathrm{T}}(\boldsymbol{K}^Z - \omega^2\boldsymbol{M}^Z)\boldsymbol{q}\ , \end{aligned} \tag{4.41}$$

with the $n \times n$ stiffness matrix $\boldsymbol{K}^Z$ and the $n \times n$ mass matrix $\boldsymbol{M}^Z$ of the additional elements. The equality of the virtual work of the internal and external forces according to Eqs. (4.39) and (4.41) yields

$$\left[\bar{\boldsymbol{F}}(\omega) + \boldsymbol{K}^Z - \omega^2\boldsymbol{M}^Z\right]\boldsymbol{q} = \hat{\boldsymbol{F}}(\omega)\boldsymbol{q} = \boldsymbol{0}\ , \tag{4.42}$$

with the $n \times n$ dynamic stiffness matrix $\hat{\boldsymbol{F}}(\omega)$ of the whole system including the additional elements.

Step 5: Calculation of the angular eigenfrequencies and eigenfunctions. The angular eigenfrequencies ω_j and the related eigenvectors $\boldsymbol{q}_j$ can be obtained by the numerical solution of the implicit eigenvalue problems (4.40) or (4.42), respectively. The first f values, $j = 1(1)f$, are sufficient to approximate the solution. The eigenfunctions $\varphi_{ij}(\xi_i)$ follow by a reverse calculation from Eqs. (4.32) and Eqs. (4.34)-(4.36) as

$$\lambda_{ij} = \lambda_i(\omega_j) = L_i\sqrt[4]{\omega_j^2(\rho A/EI)_i}\ , \tag{4.43}$$

$$\boldsymbol{v}_{ij} = \boldsymbol{J}_i\boldsymbol{q}_j\ , \tag{4.44}$$

$$\boldsymbol{c}_{ij} = \boldsymbol{E}_{ij}^{-1}\boldsymbol{v}_{ij}\ , \quad \boldsymbol{E}_{ij}^{-1} = \boldsymbol{E}_i^{-1}(\lambda_{ij})\ , \tag{4.45}$$

$$\varphi_{ij}(\xi_i) = \boldsymbol{a}_{ij}^{\mathrm{T}}\boldsymbol{c}_{ij} = \boldsymbol{a}_{ij}^{\mathrm{T}}\boldsymbol{E}_{ij}^{-1}\boldsymbol{v}_{ij}\ , \quad \boldsymbol{a}_{ij} = \boldsymbol{a}_i(\lambda_{ij}\xi_i/L_i)\ . \tag{4.46}$$

The modal analysis applying the deformation method (DEM) is finished after this step. As an example, the results for a two-field structural element (see Fig. 4.2 c)) are shown in Table 4.2.

Table 4.2. Results of the modal analysis of a two-field structural element according to Fig. 4.2 c) consisting of 3 beams

symmetric 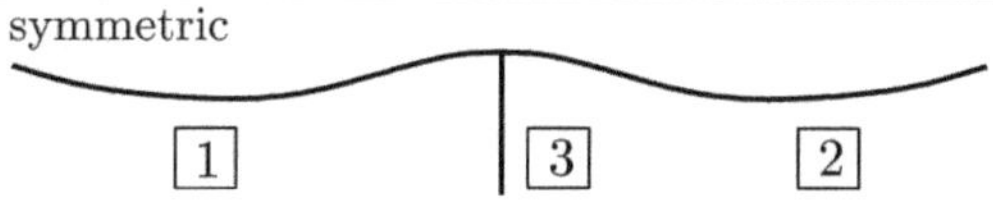	$f = 12.13\,\mathrm{Hz}$ $\omega = 76.22\,\mathrm{s}^{-1}$ $\lambda_1 = 3.926$
	$f = 39.33\,\mathrm{Hz}$ $\omega = 247.1\,\mathrm{s}^{-1}$ $\lambda_1 = 7.068$
antimetric longitudinally fixed 	$f = 9.46\,\mathrm{Hz}$ $\omega = 59.42\,\mathrm{s}^{-1}$ $\lambda_1 = 3.467$ $\lambda_3 = 1.086$
	$f = 33.12\,\mathrm{Hz}$ $\omega = 208.1\,\mathrm{s}^{-1}$ $\lambda_1 = 6.487$ $\lambda_3 = 2.031$
antimetric longitudinally free 	$f = 6.14\,\mathrm{Hz}$ $\omega = 38.55\,\mathrm{s}^{-1}$ $\lambda_1 = 2.792$ $\lambda_3 = 0.874$
	$f = 10.54\,\mathrm{Hz}$ $\omega = 66.25\,\mathrm{s}^{-1}$ $\lambda_1 = 3.660$ $\lambda_3 = 1.146$
	$f = 33.18\,\mathrm{Hz}$ $\omega = 208.5\,\mathrm{s}^{-1}$ $\lambda_1 = 6.494$ $\lambda_3 = 2.034$

$L_{1,2} = 20\,\mathrm{m}$; $(\rho A)_{1,2} = 2.3 \cdot 10^3\,\mathrm{kg/m}$; $(EI)_{1,2} = 0.9 \cdot 10^{10}\,\mathrm{Nm}^2$
$L_3 = 5.66\,\mathrm{m}$; $(\rho A)_3 = 1.725 \cdot 10^3\,\mathrm{kg/m}$; $(EI)_3 = 0.45 \cdot 10^{10}\,\mathrm{Nm}^2$

The finite element method (FEM) requires the same five steps for a solution. If the frequency functions (4.29) are developed in power series of λ and approximated by the first two series terms, the basic relations of the FEM are obtained easily:

$$\begin{aligned} F_1 &\approx 2 + \frac{3}{420}\lambda^4 , & F_4 &\approx -6 + \frac{22}{420}\lambda^4 , \\ F_2 &\approx 4 - \frac{4}{420}\lambda^4 , & F_5 &\approx -12 - \frac{54}{420}\lambda^4 , \\ F_3 &\approx 6 + \frac{13}{420}\lambda^4 , & F_6 &\approx 12 - \frac{156}{420}\lambda^4 . \end{aligned} \tag{4.47}$$

An approximation of the dynamic stiffness matrix for one beam element is obtained by replacing the frequency functions in Eq. (4.28) by Eq. (4.47). Then, this matrix can be divided into two parts,

$$\tilde{\boldsymbol{F}} = \boldsymbol{K} - \tilde{\omega}^2 \boldsymbol{M} , \quad \boldsymbol{K} = \boldsymbol{K}^{\mathrm{T}} , \quad \boldsymbol{M} = \boldsymbol{M}^{\mathrm{T}} , \tag{4.48}$$

$$\begin{aligned} \boldsymbol{K} &= \frac{EI}{L^3} \begin{bmatrix} 4L^2 & 6L & 2L^2 & -6L \\ 6L & 12 & 6L & -12 \\ 2L^2 & 6L & 4L^2 & -6L \\ -6L & -12 & -6L & 12 \end{bmatrix} , \\ \boldsymbol{M} &= \frac{\rho A L}{420} \begin{bmatrix} 4L^2 & 22L & -3L^2 & 13L \\ 22L & 156 & -13L & 54 \\ -3L^2 & -13L & 4L^2 & 22L \\ 13L & 54 & 22L & 156 \end{bmatrix} . \end{aligned} \tag{4.49}$$

The stiffness matrix $\boldsymbol{K}$ and the mass matrix $\boldsymbol{M}$ are identical to the corresponding matrices of the FEM. Executing the modal analysis with the approximation (4.48), instead of Eq. (4.42) an explicit eigenvalue problem is obtained,

$$\left[(\bar{\boldsymbol{K}} + \boldsymbol{K}^Z) - \tilde{\omega}^2 (\bar{\boldsymbol{M}} + \boldsymbol{M}^Z)\right] \tilde{\boldsymbol{q}} = \boldsymbol{0} , \tag{4.50}$$

$$\bar{\boldsymbol{K}} = \bar{\boldsymbol{J}}^{\mathrm{T}} \mathbf{diag}(\boldsymbol{K}_i) \bar{\boldsymbol{J}} , \quad \bar{\boldsymbol{M}} = \bar{\boldsymbol{J}}^{\mathrm{T}} \mathbf{diag}(\boldsymbol{M}_i) \bar{\boldsymbol{J}} , \quad i = 1(1)s . \tag{4.51}$$

From a similar approximation of the eigenfunctions given by Eq. (4.46), the interpolation polynomials of the FEM are found by expanding the product $\boldsymbol{a}_{ij}^{\mathrm{T}}(\lambda_{ij}\xi_i/L_i)\boldsymbol{E}_{ij}^{-1}(\lambda_{ij})$ into a power series of ξ_i and truncating after the third order term,

$$\varphi_{ij}(\xi_i) = \boldsymbol{a}_{ij}^{\mathrm{T}} \boldsymbol{E}_{ij}^{-1} \boldsymbol{v}_{ij} \approx \tilde{\boldsymbol{a}}_{ij}^{\mathrm{T}}(\xi_i) \tilde{\boldsymbol{v}}_{ij} , \tag{4.52}$$

where

$$\tilde{\boldsymbol{a}}_{ij}(\xi_i) = \begin{bmatrix} \xi_i (1 - \xi_i/L_i)^2 \\ 1 - 3(\xi_i/L_i)^2 + 2(\xi_i/L_i)^3 \\ \xi_i^2 (\xi_i/L_i - 1)/L_i \\ 3(\xi_i/L_i)^2 - 2(\xi_i/L_i)^3 \end{bmatrix} . \tag{4.53}$$

Here, the cubical Hermitian polynomials (4.53) were obtained from the series expansion of the exact eigenfunctions. These polynomials are the basis of the

FEM. A comparison of both methods is given in Table 4.3, numerical examples can be found in Popp and Bremer (1983). For comparably good results the FEM requires much more nodes than the DEM. Therefore, the complexity to solve an explicit eigenvalue problem with higher order (FEM) has to be compared with the solution of an implicit eigenvalue problem (DEM). Numerical examples have demonstrated that the computation time of the DEM is less than applying the FEM.

Table 4.3. Comparison of the Deformation Method (DEM) and the Finite Element Method (FEM) for a modal analysis of beam structures

	DEM	FEM
Solution method	exact	approximation
Eigenvalue problem	implicit	explicit
Number of nodes	minimal	large at high solution accuracy

■ **Problem 4.1 Modal analysis of a double-span beam**

The modal analysis has to be performed by the deformation method for a symmetrical double-span beam, cp. Fig. 4.2 b). For a comparison the eigenvalues shall be calculated with the finite element method with minimal number of nodes. The parameters for both fields of the double span beam are: $L_1 = L_2 = L/2$, $(EI)_1 = (EI)_2 = EI$, $(\rho A)_1 = (\rho A)_2 = \rho A$.

Solution

The aforementioned five steps are applied.

Step 1: Specification of the system and input data. Starting from the left hand side, the double-span beam is divided into:

$i = 1(1)s$, $s = 2$ fields with

$\kappa = 1(1)k$, $k = 3$ nodes at the bearings, with

$\nu = 1(1)n$, $n = 3$ possible node rotations ζ_ν.

The 3×1 vector $\boldsymbol{q}$ of the generalized coordinates is

$$\boldsymbol{q} = \left[\zeta_1\ \zeta_2\ \zeta_3\right]^{\mathrm{T}} . \tag{1}$$

Step 2: Element model, local equations. With the element relations (4.25)-(4.29) the local equations are

$$\boldsymbol{f}_i = \boldsymbol{F}_i(\lambda_i)\boldsymbol{v}_i \ , \tag{2}$$

$$\boldsymbol{c}_i = \boldsymbol{E}_i^{-1}(\lambda_i)\boldsymbol{v}_i \ , \quad i = 1, 2 \ , \tag{3}$$

$$\lambda_i^4 = \lambda^4 = \omega^2 \left(\frac{L}{2}\right)^4 \frac{\rho A}{EI} \ . \tag{4}$$

Step 3: Relation between local and global coordinates. The relationship between the local coordinates $\boldsymbol{v}_i, i = 1, 2$, and the global coordinates given in Eq. (1) follows from the boundary and transition conditions

$$w_{l1} = w_{l2} = w_{r1} = w_{r2} = 0\,, \quad \zeta_{l1} = \zeta_1\,, \quad \zeta_{r1} = \zeta_{l2} = \zeta_2\,, \quad \zeta_{r2} = \zeta_3\,, \tag{5}$$

yielding the global 8×3 Jacobian matrix $\bar{\boldsymbol{J}}$,

$$\underbrace{\begin{bmatrix} \zeta_{l1} \\ w_{l1} \\ \zeta_{r1} \\ w_{r1} \\ \zeta_{l2} \\ w_{l2} \\ \zeta_{r2} \\ w_{r2} \end{bmatrix}}_{\bar{\boldsymbol{v}}} = \underbrace{\begin{bmatrix} 1\,0\,0 \\ 0\,0\,0 \\ 0\,1\,0 \\ 0\,0\,0 \\ 0\,1\,0 \\ 0\,0\,0 \\ 0\,0\,1 \\ 0\,0\,0 \end{bmatrix}}_{\bar{\boldsymbol{J}}} \underbrace{\begin{bmatrix} \zeta_1 \\ \zeta_2 \\ \zeta_3 \end{bmatrix}}_{\boldsymbol{q}} . \tag{6}$$

Step 4: Consideration of the whole system, global equation. Combining both of the local Eqs. (2) gives the system description

$$\underbrace{\begin{bmatrix} \boldsymbol{f}_1 \\ \boldsymbol{f}_2 \end{bmatrix}}_{\bar{\boldsymbol{f}}} = \underbrace{\begin{bmatrix} \boldsymbol{F}_1 & 0 \\ 0 & \boldsymbol{F}_2 \end{bmatrix}}_{\bar{\bar{\boldsymbol{F}}}} \underbrace{\begin{bmatrix} \boldsymbol{v}_1 \\ \boldsymbol{v}_2 \end{bmatrix}}_{\bar{\boldsymbol{v}}\,.} , \tag{7}$$

Because of the symmetry $\boldsymbol{F}_1 = \boldsymbol{F}_2 = \boldsymbol{F}$ holds. With the frequency functions F_1, F_2 of Eq. (4.29) and Eq. (4.40), it follows

$$\bar{\boldsymbol{J}}^{\mathrm{T}} \bar{\bar{\boldsymbol{F}}} \bar{\boldsymbol{J}} \boldsymbol{q} = \underbrace{\frac{2EI}{L} \begin{bmatrix} F_2(\lambda) & F_1(\lambda) & 0 \\ F_1(\lambda) & 2F_2(\lambda) & F_1(\lambda) \\ 0 & F_1(\lambda) & F_2(\lambda) \end{bmatrix}}_{\bar{\boldsymbol{F}}[\lambda(\omega)]} \underbrace{\begin{bmatrix} \zeta_1 \\ \zeta_2 \\ \zeta_3 \end{bmatrix}}_{\boldsymbol{q}} = \underbrace{\mathbf{0}}_{\mathbf{0}} . \tag{8}$$

Since the system has no single masses and springs, Eq. (8) is the required global equation. The condition for the solution of the implicit eigenvalue problem given by Eq. (8) reads as

$$\det \bar{\boldsymbol{F}}[\lambda(\omega)] = 2F_2(\lambda)[F_2^2(\lambda) - F_1^2(\lambda)] = 0\,. \tag{9}$$

From $F_2(\lambda) = 0$ the frequency equation follows,

$$\cosh\lambda \sin\lambda - \sinh\lambda \cos\lambda = 0\,. \tag{10}$$

The equation is equivalent to the frequency equation of structure 3 in Table 4.1. The first three eigenvalues are

$$\lambda_1^s = 3.93\,, \quad \lambda_2^s = 7.07\,, \quad \lambda_3^s = 10.21\,. \tag{11}$$

The frequency equation following from $F_2^2(\lambda) - F_1^2(\lambda) = 0$ reads as

$$\sin\lambda = 0 \Rightarrow \lambda_j^a = j\pi\,, \quad j = 1(1)\infty\,. \tag{12}$$

Equation (12) is equal to that of structure 1 in Table 4.1. The corresponding eigenvectors result from insertion of both eigenvalue groups λ_j^s, λ_j^a in Eq. (8),

$$\boldsymbol{q}_j^s = K_j^s \begin{bmatrix} 1 \\ 0 \\ -1 \end{bmatrix}, \quad \boldsymbol{q}_j^a = K_j^a \begin{bmatrix} 1 \\ (-1)^j \\ 1 \end{bmatrix}, \tag{13}$$

where K_j^s and K_j^a are scaling factors. Regarding the algebraic signs of the node rotations, it can be recognized that the first eigenvalue group λ_j^s has symmetrical and the second eigenvalue group λ_j^a asymmetrical eigenmodes.

Step 5: Calculation of the angular eigenfrequencies and eigenfunctions. The angular eigenfrequencies ω_j directly result from Eq. (4) with Eqs. (11) and (12)

$$\omega_j = 4\left(\frac{\lambda_j}{L}\right)^2 \sqrt{\frac{EI}{\rho A}}\,. \tag{14}$$

The eigenfunctions $\varphi_{ij}(\xi_i)$ are obtained by reverse calculation. Firstly, the local displacements $\boldsymbol{v}_{ij}$ are calculated by inserting Eq. (13) into Eq. (6),

$$\begin{aligned} \boldsymbol{v}_{1j}^s &= K_j^s \begin{bmatrix} 1 \\ 0 \\ 0 \\ 0 \end{bmatrix}, \quad \boldsymbol{v}_{1j}^a = K_j^a \begin{bmatrix} 1 \\ 0 \\ (-1)^j \\ 0 \end{bmatrix}, \\ \boldsymbol{v}_{2j}^s &= K_j^s \begin{bmatrix} 0 \\ 0 \\ -1 \\ 0 \end{bmatrix}, \quad \boldsymbol{v}_{2j}^a = K_j^a \begin{bmatrix} (-1)^j \\ 0 \\ 1 \\ 0 \end{bmatrix}. \end{aligned} \tag{15}$$

According to Eq. (4.46) the eigenfunctions $\varphi_{ij}(\xi_i)$ can be expressed by the coefficient vectors $\boldsymbol{c}_{ij} = \boldsymbol{E}_{ij}^{-1}(\lambda_{ij})\boldsymbol{v}_{ij}$. Using Eq. (4.29) and Eq. (4.30) one gets

$$\boldsymbol{c}_{1j}^s = \bar{K}_j^s \begin{bmatrix} 0 \\ 1/\sinh\lambda_j \\ 0 \\ -1/\sin\lambda_j \end{bmatrix} , \quad \boldsymbol{c}_{1j}^a = \sqrt{2}\bar{K}_j^a \begin{bmatrix} 0 \\ 0 \\ 0 \\ 1 \end{bmatrix} ,$$
$$\boldsymbol{c}_{2j}^s = \bar{K}_j^s \begin{bmatrix} 1 \\ -\cot\lambda_1 \\ -1 \\ \cot\lambda_j \end{bmatrix} , \quad \boldsymbol{c}_{1j}^a = \sqrt{2}\bar{K}_j^a \begin{bmatrix} 0 \\ 0 \\ 0 \\ (-1)^j \end{bmatrix} , \tag{16}$$

where $\bar{K}_j^s$ and $\bar{K}_j^a$ are scaling factors. Setting $\bar{K}_j^s = \bar{K}_j^a = 1$, equal generalized masses M_j are obtained for all eigenmodes,

$$M_j = \sum_{i=1}^{2} M_{ij} = \sum_{i=1}^{2} \int_0^{L/2} \rho A \varphi_{ij}^2(\xi_i) d\xi_i = \rho AL \,, \quad j = 1(1)\infty \,. \tag{17}$$

Finally, the eigenvalues are calculated by the finite element method. By replacing the frequency functions F_1 and F_2 approximately by Eq. (4.47), the explicit eigenvalue problem follows from Eq. (8) as

$$\underbrace{\frac{2EI}{L}\left\{\begin{bmatrix} 4 & 2 & 0 \\ 2 & 8 & 2 \\ 0 & 2 & 4 \end{bmatrix} - \frac{\tilde{\lambda}^4}{420}\begin{bmatrix} 4 & -3 & 0 \\ -3 & 8 & -3 \\ 0 & -3 & 4 \end{bmatrix}\right\}}_{\tilde{\boldsymbol{F}}[\tilde{\lambda}(\omega)]} \underbrace{\begin{bmatrix} \tilde{\zeta}_1 \\ \tilde{\zeta}_2 \\ \tilde{\zeta}_3 \end{bmatrix}}_{\tilde{\boldsymbol{q}}} = \underbrace{\boldsymbol{0}}_{\boldsymbol{0}} \,, \quad . \tag{18}$$

The condition for the solution, cp. Eq. (9), is given by

$$\begin{aligned} & F_2(\lambda)\left[F_2(\lambda) - F_1(\lambda)\right]\left[F_2(\lambda) + F_1(\lambda)\right] \\ & \approx \left[4 - \frac{4}{420}\tilde{\lambda}^4\right]\left[2 - \frac{7}{420}\tilde{\lambda}^4\right]\left[6 - \frac{1}{420}\tilde{\lambda}^4\right] = 0 \,. \end{aligned} \tag{19}$$

The approximations $\tilde{\lambda}_j$, $\tilde{\boldsymbol{q}}_j$ of the eigenvalues and corresponding eigenvectors are

$$\begin{aligned} &\tilde{\lambda}_1^s = 4.5270 \,, \quad \tilde{\lambda}_1^a = 3.3098 \,, \quad \tilde{\lambda}_2^a = 7.0872 \,, \\ &(\lambda_1^s = 3.9266 \,, \quad \lambda_1^a = 3.1416 \,, \quad \lambda_2^a = 6.2832) \,, \end{aligned} \tag{20}$$

$$\tilde{\boldsymbol{q}}_1^s = \tilde{K}_1^s \begin{bmatrix} 1 \\ 0 \\ -1 \end{bmatrix} , \quad \tilde{\boldsymbol{q}}_1^a = \tilde{K}_1^a \begin{bmatrix} 1 \\ -1 \\ 1 \end{bmatrix} , \quad \tilde{\boldsymbol{q}}_2^a = \tilde{K}_2^a \begin{bmatrix} 1 \\ 1 \\ 1 \end{bmatrix} , \tag{21}$$

where in Eq. (20) the exact values are given in brackets for comparison. It can be seen that the eigenvalues are overestimated by the FEM approximation. While there are relative errors for the eigenvalues up to 15.3 %, the eigenvectors agree well with the exact values. Applying the FEM, only approximations for the first three eigenvalues and eigenvectors can be computed using the chosen minimal node number. The DEM provides all values exactly. Increasing the number of nodes, a FEM approximation for higher eigenvalues and eigenvectors is possible. The eigenvalues are always overestimated and the error decreases with an increasing number of nodes. ■

4.1.3 Models for Continuously Bedded Beams

Contrary to magnetically levitated vehicles on pillared guideways the dynamic interaction is generally weak for railway vehicles on bedded tracks. Thus, for the investigations of the dynamics in the interesting frequency range up to about 50 Hz even for high-speed trains the dynamic interactions of vehicle and elastic guideway are neglected and rigid tracks are assumed. On the other hand, to design the track and to investigate the vibrations in the foundation, in particular for the analysis of high frequency processes such as structure-borne sound propagation, material wear, rail corrugation, roll and cornering noises the guideway dynamics is of importance. Therefore, numerous models have been developed describing the tracks in different ways, see e.g. Popp and Schiehlen (2003). The track consists of rails, sleepers, pads between rails and sleepers, ballast and foundation. In the following, the classical model of an infinitely long beam on an elastic foundation is presented, cp. Table 4.4 b). This model is treated by Timoshenko (1915) who calculated stresses subject to a static force. The dynamic effects of moving forces are examined by Doerr (1943) for a constant single force and by Mathews (1958) for a harmonic time-variant force. This model has experienced numerous extensions.

Starting point for the modeling is a plane track system as shown in Table 4.4. Both rails were combined to one beam as a result of the assumed symmetry. Distributing the sleeper masses, foundation springs and foundation dampers continuously, see Table 4.4 a), the differential equation can be obtained under assumption of the Bernoulli-Euler beam theory,

$$EIw''''(x,t) + Tw''(x,t) + \mu\ddot{w}(x,t) + \beta\dot{w}(x,t) + \gamma w(x,t) = f(t)\delta(x - vt)\ . \tag{4.54}$$

Here, $(\ )'$ and $(\dot{\ })$ denote the partial derivatives with respect to space and time, respectively, EI is the bending stiffness and $\mu = \rho A$ denotes the distributed mass of the beam. Furthermore, T is the longitudinal pressure force on the beam, β and γ are the viscous damping and stiffness per unit of length of the foundation, respectively, $f\,(t)$ is the force moving with constant velocity v, and δ is the Dirac function. Introducing a frame moving with the force $f(t)$ and applying the transformation $x = vt + \xi$, $v = \text{const}$, Eq. (4.54) yields

$$\begin{aligned} &EI\bar{w}''''(\xi,t) + (T + \mu v^2)\bar{w}''(\xi,t) - \beta v\bar{w}'(\xi,t) - 2\mu v\dot{\bar{w}}'(\xi,t) \\ &\quad + \mu\ddot{\bar{w}}(\xi\ ,t) + \beta\dot{\bar{w}}(\xi\ ,t) + \gamma\bar{w}(\xi\ ,t) = f(t)\delta(\xi)\ . \end{aligned} \tag{4.55}$$

Table 4.4. Track models

Name	Mechanical system	Mathematical model and data
Track system	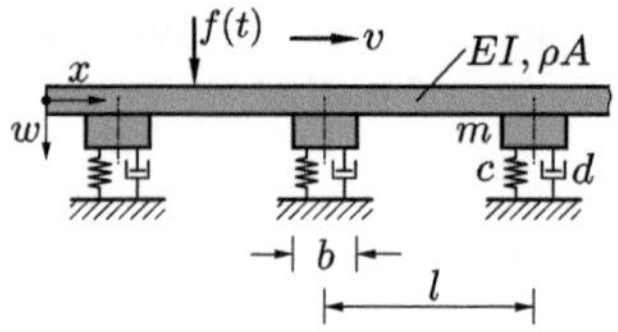	Data: $l = 4b = 0.63\,\mathrm{m}$ $\left.\begin{array}{l} EI = 12.81 \cdot 10^6\,\mathrm{Nm}^2 \\ \rho A = 120\,\mathrm{kg/m} \end{array}\right\}$ 2 rails $m = \begin{cases} 100\,\mathrm{kg} \text{ (wood sleepers)} \\ 300\,\mathrm{kg} \text{ (concrete sleepers)} \end{cases}$ $c = (4 \ldots 10) \cdot 10^7\,\mathrm{N/m}$
a)	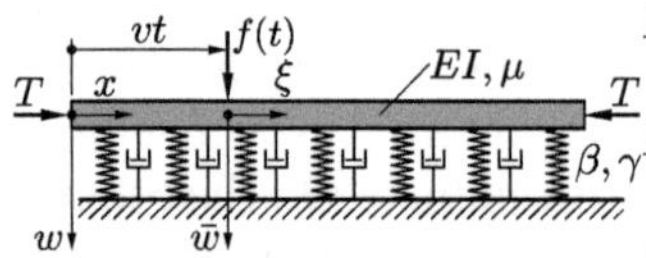	$EIw'''' + Tw'' + \mu\ddot{w} + \beta\dot{w} + \gamma w$ $= f(t)\delta(x - vt)$ $EI\bar{w}'''' + (T + \mu v^2)\bar{w}'' - \beta v\bar{w}' - 2\mu v\dot{\bar{w}}'$ $+\mu\ddot{\bar{w}} + \beta\dot{\bar{w}} + \gamma\bar{w} = f(t)\delta(\xi)$ $\mu = \rho A + m/l\,, \quad \beta = d/l\,, \quad \gamma = c/l$
b)	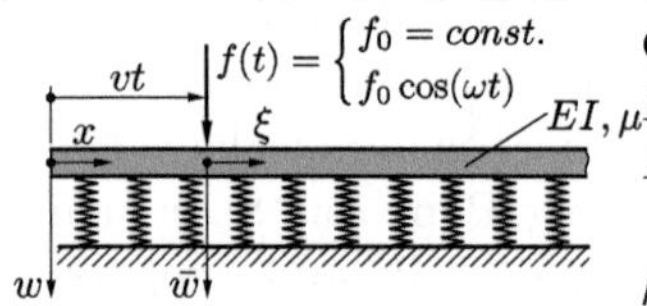	Classical model: $EIw'''' + \mu\ddot{w} + \gamma w = f(t)\delta(x - vt)$ $EI\bar{w}'''' + \mu\ddot{\bar{w}} + \gamma\bar{w}$ $+\mu v^2\bar{w}'' - 2\mu v\dot{\bar{w}}' = f(t)\delta(\xi)$ $\mu = \rho A + m/l\,, \quad \gamma = c/l$
c)	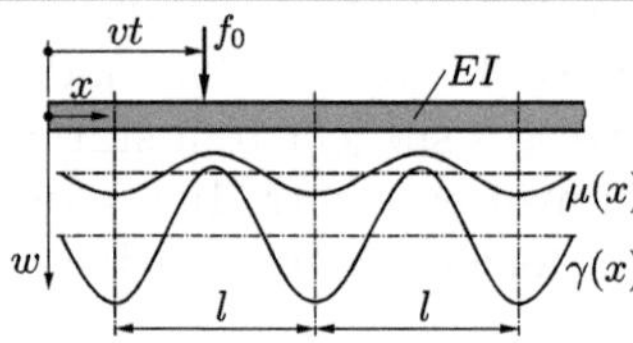	$EIw'''' + \mu(x)\ddot{w} + \gamma(x)w = f_0\delta(x - vt)$ $\mu(x + l) = \mu(x) = \mu_0 + \mu_1\cos(2\pi x/l)$ $\gamma(x + l) = \gamma(x) = \gamma_0 + \gamma_1\cos(2\pi x/l)$ $\mu_0 = \rho A + m/l\,, \quad \mu_1 = 1.8\ m/l$ $\gamma_0 = c/l\,, \quad \gamma_1 = 1.8\ c/l$

The solution of Eq. (4.55) is achieved by methods for propagating waves. The boundary conditions are considered at infinity according to the Sommerfeld radiating condition.

For a travelling constant force $f = f_0 = \text{const}$ the behavior of the solution depends substantially on the velocity ratio ν,

$$\nu = \frac{v}{v_{cr}}\,, \quad v_{cr} = \sqrt{\frac{1}{\mu}\left[\sqrt{4\gamma EI(1 - D^2)} - T\right]}\,, \quad D = \frac{\beta}{2\sqrt{\mu\gamma}}\,, \tag{4.56}$$

where the critical speed v_{cr} and the damping ratio D of the foundation are introduced. The critical speed is about $v_{cr} \approx 1100 \ldots 2050$ km/h for the realistic rail data given in Table 4.4 with weak damping and without longitudinal pressure forces ($D \ll 1, T = 0$). These values are much higher than the velocities of high-speed trains operating nowadays. The critical speed is reduced by larger longitudinal pressure forces and larger foundation damping ratios as well as by the inertia of the ballast.

In the engineering practice with subcritical velocities $\nu < 1$ a stationary wave remains after transient effects vanish for $t \to \infty$. The stationary wave moves with the constant travelling force f_0, and its amplitude $\bar{w}_0(\xi)$ decreases exponentially with the distance ξ from the location where the force is applied. For $D = 0$, $T = 0$, a wave results which is symmetrical to this location,

$$\bar{w}_0(\xi) = \frac{f_0 \mathrm{e}^{-\alpha|\xi|}}{4\alpha\sqrt{\gamma EI}} \left(\cos\delta\xi + \frac{\alpha}{\delta}\sin\delta\,|\xi|\right) ,$$

$$\alpha = \sqrt{1-\nu^2}\,\sqrt[4]{\gamma/(4EI)}\,, \quad \delta = \sqrt{1+\nu^2}\,\sqrt[4]{\gamma/(4EI)}\,. \tag{4.57}$$

The static beam displacements $\bar{w}_s(\xi)$ follow from Eq. (4.57) for $\nu = 0$. The influence of the velocity v on the beam displacements is found by examining the dynamic amplification function φ, i. e. the relation between dynamic and static beam displacement at the location where the force is applied,

$$\varphi = \frac{\bar{w}_0(\xi = 0)}{\bar{w}_s(\xi = 0)} = \frac{1}{\sqrt{1-\nu^2}}\,, \quad (D = 0\,, T = 0)\,. \tag{4.58}$$

With the track system data of Table 4.4, values of $\varphi \leq 1.071$ result from Eq. (4.58) for $v = 400\,\mathrm{km/h}$, i. e. in the worst case a dynamic amplification of only about 7 % follows related to the static displacement. For $D > 0$ the symmetry of the wave disappears and the maximum beam displacement occurs at $\xi < 0$.

For overcritical velocities with $\nu > 1$ waves with constant amplitude propagate to both sides from the location where the force is applied. The amplitudes and wavelengths in front of this location are smaller than behind it. The dynamic amplification functions φ are shown in Fig. 4.5 depending on the velocity ratio ν and the foundation damping ratio D. They were calculated by Fryba (1999). The quality of the model based on a continuous foundation can be checked by assuming the sleeper mass and foundation to be homogeneously distributed over the width b. Expanding them into a Fourier series, considering only the constant term and the fundamental harmonic and neglecting the damping influence, a sine-shaped distributed mass and foundation stiffness is obtained, cp. Table 4.4 c). This model can be treated in the steady state by applying a perturbation method with the ansatz $\bar{w}(\xi) = \bar{w}_0(\xi) + \bar{w}_1(\xi)$, $w_1(\xi) \ll \bar{w}_0(\xi)$, where the basic solution $\bar{w}_0(\xi)$ is used according to Eq. (4.57), cp. Popp and Mueller (1982). It follows that the maximum amplitude $\bar{w}_1(\xi = 0)$ of the first approximation is only about 1/1000 of the corresponding amplitude of the basic solution $\bar{w}_0(\xi = 0)$. The amplitude $\bar{w}_1(\xi = 0)$ also represents the periodic oscillations at the location where the moving force is applied and it is decisive for dynamic interactions.

The case of a harmonic time-dependent travelling force $f(t) = f_0 \cos\omega t$, cp. Table 4.4 b), has been examined first by Matthews (1958) and Matthews (1959). The behavior of the solution depends on the velocity ratio ν and on the frequency ratio η, with

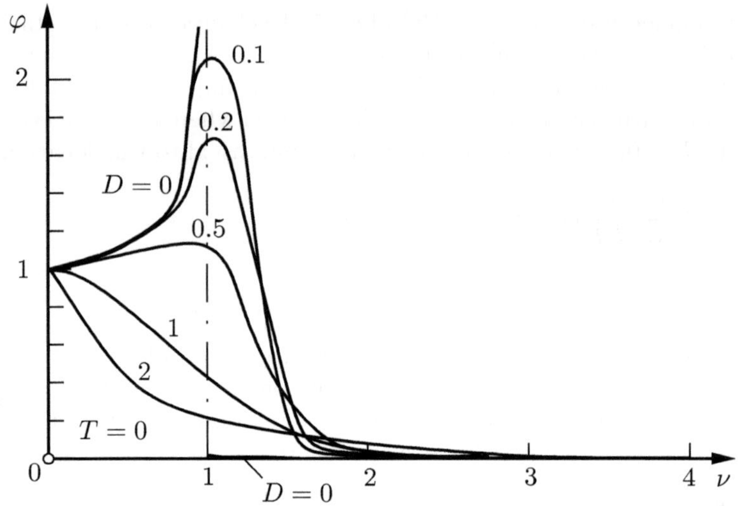

Fig. 4.5. Dynamic amplification functions φ dependent on the velocity ratio ν and the foundation damping ratio D ($T = 0$)

$$\eta = \frac{\omega}{\omega_{cr}}, \quad \omega_{cr} = \sqrt{\frac{\gamma}{\mu}}, \tag{4.59}$$

where resonance occurs for the critical angular frequency ω_{cr} and $v = 0$, $D = 0$. For the track system data given in Table 4.4, a resonance frequency $f_{cr} = \omega_{cr}/2\pi \geq 52\,\text{Hz}$ is found. For velocity ratios between $0 < \nu < 1$ the resonance already occurs at lower frequency ratios $1 > \eta > 0$. In the case without damping and steady state condition the solution can be described by waves of the form

$$\bar{w}_0(\xi, t) = \bar{y}_1(\xi)\cos\omega t + \bar{y}_2(\xi)\sin\omega t\,. \tag{4.60}$$

In the subcritical region $0 < \nu < 1$ which is limited according to Matthews (1958) by a declining curve in the η, ν-space, cp. Fig. 4.6, the amplitudes of the wave decrease exponentially at both sides of the location where the force is applied, as with the model of a constant moving force. In the supercritical region the waves propagate with constant amplitude.

The results of Mathews have been generalized for a Timoshenko beam, cp. Bogacz et al. (1989). Then, the relations are much more complex. For example, two critical speeds appear, one for the shear waves and one for the longitudinal waves. Instead of the one solution area, examined by Mathews, there are 22 areas and 8 border lines in the η, ν-space, with qualitatively different behavior of the solution. The solution for the Bernoulli-Euler beam is included in the solution variety for the Timoshenko beam as a special case. In general, the

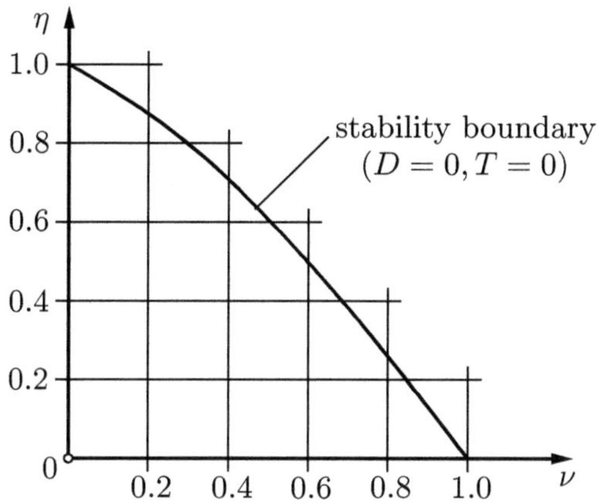

Fig. 4.6. Subcritical region in dependence of the speed ratio ν and the frequency ratio η, $(D = 0,\ T = 0)$

Bernoulli-Euler beam model provides precise results in the subcritical region, while it is insufficient in the supercritical region.

The solution for a moving harmonic time-dependent single force is the basis for the investigation of coupled vehicle - guideway systems with several points of force application. In general, every contact point is a transmitter and receiver for propagating waves at the same time. The resulting much more complex relations are not discussed here.

4.2 Perturbation Models for Rigid Guideways

Vehicles on rigid guideways are subject to perturbations due to various reasons. Generally, individual obstacles and the guideway unevenness can be distinguished. The obstacles comprise steps (rail joints, curbstone edges) and cambers (potholes and bumps) that can be expressed mathematically by step and impulse functions, primarily resulting in free vibrations of the vehicle. The guideway unevenness causes forced vibrations of the vehicle. The unevenness is a permanent and, therefore, very important source of excitation. In the past, it has been assumed sometimes to be sinusoidal and, thus, it was treated deterministically as well. Nowadays, the unevenness is regarded as random and statistical methods are applied. In principle, the guideway unevenness could be measured locally with sufficient accuracy and reproducibility, and be treated deterministically. Nevertheless, the main focus in vehicle dynamics is the consideration of the variety of local unevenness representing the globally roads or tracks. For this purpose, stochastic processes provide suitable mathematical models discussed in the following. At first, the guideway unevenness is described by an unevenness profile, depending on a position coordinate,

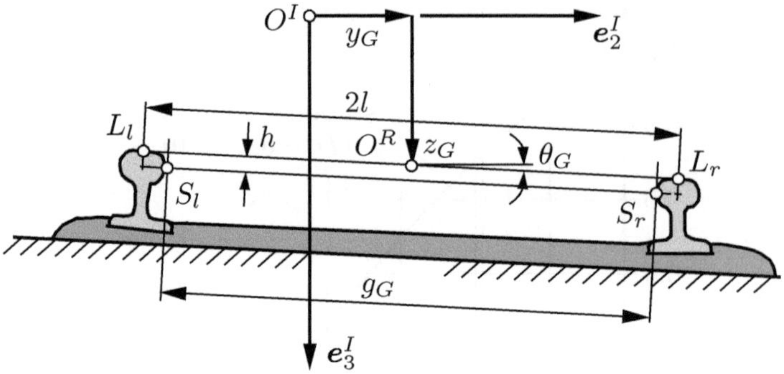

Fig. 4.7. Rail track coordinates for perturbation profile definition

which is later transformed into the time domain by means of the vehicle velocity. The unevenness is represented by a stochastic excitation process acting on the vehicle causing random vibrations. The resulting system response is a stochastic process, too, discussed later in this section.

In the case of railway tracks or pillared guideways for magnetically levitated vehicles, four unevenness profiles occur while in the simpler case of roads the unevenness is characterized by only two perturbation profiles. Figure 4.7 shows the cross-section of a railway track with its center defined by the coordinates y_G and z_G, measured in the inertial system I. The bank angle is indicated by $\theta_G \ll 1$, and the gauge of the track by g_G which is measured between the points S_l and S_r in a depth h below the top of the rail head. The coordinates y_l, y_r and z_l, z_r of the points S_l, S_r (gauge) and the points L_l, L_r (head of the rail) on the left and right rail, respectively, are given by

$$y_l = y_G - g_G/2 \,, \quad y_r = y_G + g_G/2 \,, \tag{4.61}$$

$$z_l = z_G - 2l\theta_G/2 \,, \quad z_r = z_G + 2l\theta_G/2 \,, \tag{4.62}$$

where the length $2l$ is denoted as span. The deviations $\Delta(\bullet)$ of the individual coordinates from their nominal values are introduced as perturbation or unevenness profiles $\zeta_{(\bullet)}(x)$ which depend on the longitudinal coordinate x. Usual perturbation profiles for rails are

$$\text{lateral alignment:} \quad \zeta_y(x) = \Delta y_G(x) \,, \tag{4.63}$$

$$\text{vertical alignment:} \quad \zeta_z(x) = \Delta z_G(x) \,, \tag{4.64}$$

$$\text{bank variation:} \quad \zeta_\theta(x) = 2l\Delta\theta_G(x) \,, \tag{4.65}$$

$$\text{gauge variation:} \quad \zeta_g(x) = \Delta g_G(x) \,. \tag{4.66}$$

These perturbations are uncorrelated and defined consistently as lengths. From here, the deviations $\Delta y_{l,r}$, $\Delta z_{l,r}$ can be calculated by means of Eqs. (4.61) and (4.62).

Perturbation profiles of roads are related only to the vertical unevenness. They can be determined by either Eqs. (4.64) and (4.65) or according to Eq. (4.62). In contrast to rails, both resulting perturbation profiles are correlated in the case of parallel lanes.

The investigation of perturbation profiles has been a subject of research activities for a long time. Numerous measurements of road profiles (Mitschke and Wallentowitz (2004), Braun (1969), Voy (1977), Bormann (1978)), railway tracks (ORE (1971), Helms and Strothmann (1977)) and magnetic guideways (Sussmann (1974), Snyder III and Wormley (1977), Kropac and Mucka (2005)) show that the different perturbation profiles may be characterized by

- stationary, normally distributed, ergodic random processes.

Prior to modeling the individual perturbation profiles, general stochastic processes with the properties mentioned above are discussed, cp. Crandall and Mark (1963), Newland (1975), Heinrich and Henning (1978).

4.2.1 Mathematical Description of Stochastic Processes

Considering a particular type of road, e. g. a freeway, the unevenness may be characterized by a large number of measurements $\zeta^{(r)}(x)$, $r = 1, 2, ...$, with $\zeta^{(r)}(x) \neq \zeta^{(s)}(x)$ for $r \neq s$, each representing a small section of the road surface, see Fig. 4.8 a). The independent variable x denotes the distance to an arbitrarily chosen starting point of each measurement. At large, these measurements are assumed to represent the freeway with sufficient accuracy. However, all individual profiles are different from each other $\zeta^{(r)}(x) \neq \zeta^{(s)}(x)$, $r \neq s$. The whole of the profiles $\zeta^{(r)}(x)$ is representing the stochastic process $\zeta(x)$ of the freeway unevenness. Each measurement $\zeta^{(r)}(x)$ is called a realization or

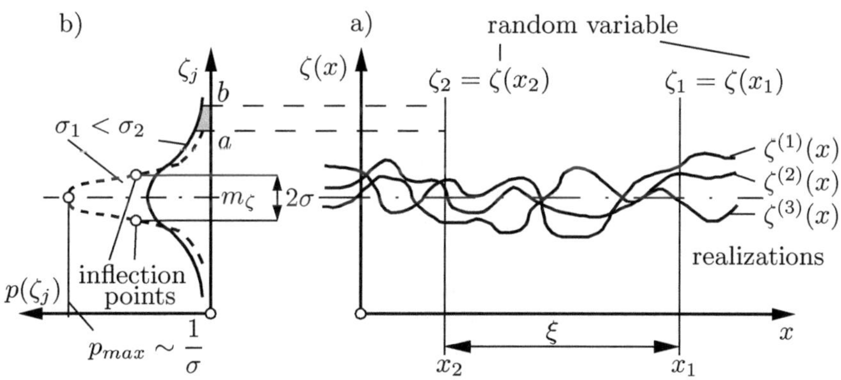

Fig. 4.8. Guideway unevenness described as a stochastic process: a) realizations $\zeta^{(r)}(x)$ and random variable ζ_j; b) probability density function $p(\zeta_j)$

pattern function of the process. The unevenness values $\zeta_j = \zeta(x_j)$ at certain distances $x = x_j, j = 1, 2, ...$, are random variables that can be analyzed statistically. The probability density function $p(\zeta_j)$, see Fig. 4.8 b), can be used to evaluate the probability Pr for the unevenness value included in a given interval $[a, b]$

$$Pr(a \le \zeta_j \le b) = \int_a^b p(\zeta_j)\mathrm{d}\zeta_j \ , \quad 0 \le Pr \le 1 \ . \tag{4.67}$$

Replacing the integration limits by infinite values, all unevenness values ζ_j are included, i. e. the probability $Pr = 1$ is found,

$$\int_{-\infty}^{\infty} p(\zeta_j)\mathrm{d}\zeta_j = 1 \ . \tag{4.68}$$

The so-called moments of the probability density are referred to as expected values E{•}, characterizing the random variables. The most important are the mean value $m_\zeta(x_j)$, a moment of first order, and the variance $P_\zeta(x_j)$ or square of the standard deviation $\sigma_\zeta^2(x_j)$, a central moment of second order. For $j = 1$ it holds

$$m_\zeta(x_1) = \mathrm{E}\{\zeta(x_1)\} \equiv \int_{-\infty}^{\infty} \zeta_1 p(\zeta_1)\mathrm{d}\zeta_1 \ , \tag{4.69}$$

$$m_{\zeta^2}(x_1) = \mathrm{E}\{\zeta^2(x_1)\} \equiv \int_{-\infty}^{\infty} \zeta_1^2 p(\zeta_1)\mathrm{d}\zeta_1 \ , \tag{4.70}$$

$$P_\zeta(x_1) \equiv \sigma_\zeta^2(x_1) = \mathrm{E}\{[\zeta(x_1) - m_\zeta(x_1)]^2\} = \mathrm{E}\{\zeta^2(x_1)\} - m_\zeta^2(x_1) \ . \tag{4.71}$$

The averaging is based on the ensemble of all realizations at a fixed location $x = x_j$. The calculation of the mean or expected value $\mathrm{E}\{\cdot\}$ is a linear operation where $\mathrm{E}\{c\} = c$ holds if $c =$ const. These properties have been used in Eq. (4.71). Moments of first and second order are well-known from the mechanical properties of the center of mass and the moment of inertia. For the reader not familiar with mean values and variances, this mechanical analogy may be helpful to illustrate the stochastic relationships of Eqs. (4.69) - (4.71): For an infinitely long rod with a mass density of $p(\zeta)$ where, according to Eq. (4.68), the total mass is normalized to 1, the mean value of Eq. (4.69) is equivalent to the coordinate of the center of mass. The mean square value given in Eq. (4.70) corresponds to the rod's moment of inertia related to the point of origin. The variance of Eq. (4.71) is comparable to the moment of inertia related to the center of mass. Thus, the relationship $\sigma_\zeta^2 = \mathrm{E}\{\zeta^2\} - m_\zeta^2$ in Eq. (4.71) can be interpreted as a mechanical analogy to the Huygens-Steiner theorem. The moments of first and second order provide essential information about the distribution of a certain variable, here mass and probability, respectively.

The statistical relations of two random variables $\zeta_1 = \zeta(x_1)$ and $\zeta_2 = \zeta(x_2)$ of the same stochastic process ζ give a further insight into the process. For this purpose, the (auto)correlation $R_\zeta(x_1, x_2)$ or, by using the central moments of second order, the (auto)covariance $P_\zeta(x_1, x_2)$ are introduced by

$$R_\zeta(x_1, x_2) = \mathrm{E}\{[\zeta(x_1)\zeta(x_2)]\} = \int_{-\infty}^{\infty}\int_{-\infty}^{\infty} \zeta_1\zeta_2 p(\zeta_1, \zeta_2)\mathrm{d}\zeta_1\mathrm{d}\zeta_2 \, , \tag{4.72}$$

$$\begin{aligned} P_\zeta(x_1, x_2) &= \mathrm{E}\{[\zeta(x_1) - m_\zeta(x_1)][\zeta(x_2) - m_\zeta(x_2)]\} \\ &= \mathrm{E}\{\zeta(x_1)\zeta(x_2)\} - m_\zeta(x_1)m_\zeta(x_2) \, , \end{aligned} \tag{4.73}$$

where $p(\zeta_1, \zeta_2)$ denotes the joint probability density. The prefix 'auto' indicates that both random variables considered belong to the same stochastic process. If they belong to different processes, the prefix 'cross' is used instead. In the case of identical correlation positions $x_1 = x_2$, the correlation of Eq. (4.72) turns into the quadratic mean value given in Eq. (4.70) and the covariance of Eq. (4.73) into the variance of Eq. (4.71).

Now the properties of stationarity, normal (Gaussian) distribution and ergodicity are explained and described mathematically.

A stochastic process is called *stationary* (or homogeneous) if its statistical properties are invariant with respect to a change of the origin of the x-axis. Thus, all probability densities are equal, $p(\zeta_1) = p(\zeta_2) = p(\zeta)$, and the joint probability density only depends on the correlation distance $\xi = x_1 - x_2$. Consequently, all random variables and the overall stochastic process as well are characterized by the same mean and square mean value, respectively. A stationary stochastic process can always be centralized such that the mean value vanishes. Hence, for centralized stationary processes from Eqs. (4.69)-(4.73) it yields

$$m_\zeta(x) = \text{const} = 0 \, , \tag{4.74}$$

$$R_\zeta(x_1, x_2) = P_\zeta(x_1, x_2) = R_\zeta(\xi = x_1 - x_2) \, , \tag{4.75}$$

$$m_{\zeta^2}(x) = \sigma_\zeta^2(x) = R_\zeta(\xi = 0) = \text{const} \, . \tag{4.76}$$

For centralized processes, correlation and covariance are identical. According to Eq. (4.75), the so-called correlation function, characterizing the entire stochastic process, depends only on the correlation distance ξ.

A stochastic process is called *normally distributed* or *Gaussian* if the random variable $\zeta_j = \zeta(x_j)$ for each x_j shows a probability density function of the form

$$p(\zeta_j) = \frac{1}{\sigma_j\sqrt{2\pi}} \exp\left[-\frac{(\zeta_j - m_j)^2}{2\sigma_j^2}\right] \tag{4.77}$$

with the mean value $m_j = m_\zeta(x_j)$ and the standard deviation or dispersion $\sigma_j = \sigma_\zeta(x_j)$, respectively. It is obvious that a normally distributed process can be characterized completely by its first and second moment, i. e. mean value and variance. Higher moments can be reduced to these two quantities. The density function given in Eq. (4.77) shows the well-known Gaussian bell-shaped curve, see Fig. 4.8 b). The curve is symmetric with respect to $\zeta_j = m_j$ and shows the maximum value $p_{max} = 1/(\sigma_j\sqrt{2\pi})$ at $\zeta_j = m_j$ and inflection points at $\zeta_j = m_j \pm \sigma_j = \zeta_w$ with $p(\zeta_w) = p_{max}/\sqrt{\mathrm{e}}$. Small dispersions σ_j indicate that the random variable is concentrated at the vicinity of the mean value m_j. The probability that the random variable ζ_j is included in an interval of the width $2k\sigma_j$, $k = 1, 2, ...$, around the mean value m_j can be calculated by means of Eq. (4.67). These so-called reliability intervals are also referred to as $k\sigma$-limits with $k = 1, 2, ...$:

$$\begin{aligned} k = 1: \; & Pr(m - \sigma \le \zeta \le m + \sigma) = Pr(|\zeta - m| \le \sigma) = 0.6827 \,, \\ k = 2: \; & Pr(|\zeta - m| \le 2\sigma) = 0.9545 \,, \\ k = 3: \; & Pr(|\zeta - m| \le 3\sigma) = 0.9973 \,, \\ k = 4: \; & Pr(|\zeta - m| \le 4\sigma) = 0.99994 \,, \end{aligned} \tag{4.78}$$

where the index j has been neglected due to reasons of clarity. Thus, the probability that a Gaussian random variable is outside the 2σ-limit is only $Pr = 0.0455 \mathrel{\widehat{=}} 4.55\ \%$. For stationary, normally distributed processes the index j in Eq. (4.77) may be completely omitted according to Eqs. (4.74) and (4.76).

A stationary stochastic process is called *ergodic* if the expected values of the random variables $\zeta_j = \zeta(x_j)$ can be obtained from the mean values of a pattern function or realization $\zeta^{(r)}(x)$ of sufficient length over the coordinate x (or time t, respectively). Thus,

- the ensemble average is equivalent to the mean value over the space (or time, respectively).

Therefore, for an ergodic process it follows that

$$m_\zeta = \lim_{X\to\infty} \frac{1}{2X} \int_{-X}^{X} \zeta^{(r)}(x)\mathrm{d}x \,, \tag{4.79}$$

$$R_\zeta(\xi) = \lim_{X\to\infty} \frac{1}{2X} \int_{-X}^{X} \zeta^{(r)}(x)\zeta^{(r)}(x-\xi)\mathrm{d}x \,, \tag{4.80}$$

$$m_{\zeta^2} = R_\zeta(\xi = 0) = \lim_{X\to\infty} \frac{1}{2X} \int_{-X}^{X} [\zeta^{(r)}(x)]^2\mathrm{d}x \,, \tag{4.81}$$

$$P_\zeta(\xi) = \lim_{X\to\infty} \frac{1}{2X} \int_{-X}^{X} [\zeta^{(r)}(x) - m_\zeta][\zeta^{(r)}(x-\xi) - m_\zeta]\mathrm{d}x \,, \tag{4.82}$$

$$\sigma_\zeta^2 = P_\zeta(\xi = 0) = \lim_{X\to\infty} \frac{1}{2X} \int\limits_{-X}^{X} [\zeta^{(r)}(x) - m_\zeta]^2 \mathrm{d}x \ . \tag{4.83}$$

In many cases, it is not easy to decide if a process is ergodic. In the present case of the road unevenness, ergodicity is plausible, considering the background of Fig. 4.8. Regardless of the measurement method - analyzing one single, very long measurement of the perturbation profile or using several shorter measurements whose ensemble is analyzed - the statistical variables remain the same. However, the experimental costs to determine a space (or time) mean are lower than to average an ensemble. In practice, neither an infinitely large ensemble nor an infinitely long realization can be obtained, always resulting in approximated expected values.

In the following, centralized processes with a mean value $m_\zeta = 0$ are assumed. Thus, the quadratic mean value is equivalent to the variance and the correlation function is equivalent to the covariance function, respectively.

In engineering applications, stationary processes are usually characterized by their spectral power density or, briefly, spectral density $S_\zeta(\Omega)$ that can be determined by the Fourier transform of the correlation function $R_\zeta(\xi)$,

$$S_\zeta(\Omega) = \frac{1}{2\pi} \int\limits_{-\infty}^{\infty} R_\zeta(\xi) \mathrm{e}^{-i\Omega\xi} \mathrm{d}\xi \ , \tag{4.84}$$

$$R_\zeta(\xi) = \int\limits_{-\infty}^{\infty} S_\zeta(\Omega) \mathrm{e}^{i\Omega\xi} \mathrm{d}\Omega \ , \tag{4.85}$$

$$m_\zeta = 0: \quad P_\zeta = \sigma_\zeta^2 = R_\zeta(\xi = 0) = \int\limits_{-\infty}^{\infty} S_\zeta(\Omega) \mathrm{d}\Omega \ . \tag{4.86}$$

Here, Ω is the spatial angular frequency (wave number) given in rad/m with $\Omega = 2\pi/\lambda = 2\pi F$ where the wave length λ and the spatial frequency F occur. Equations (4.84) and (4.85) are usually referred to as Wiener Chintschin relation, where in the literature the factor $1/2\pi$ is sometimes assigned to Eq. (4.85), or the modified factor $1/\sqrt{2\pi}$ is assigned to both equations. The advantage of the definition of the spectral density chosen here becomes evident in Eq. (4.86), where the variance is directly related to the spectral power. Therefore, the stochastic process of the guideway unevenness can be characterized equivalently by the correlation function $R_\zeta(\xi)$ in the spatial domain as well as by the spectral density $S_\zeta(\Omega)$ in the frequency domain. The variance follows from Eq. (4.86) with $R_\zeta(\xi = 0)$ or as an integral over the contributions $S_\zeta(\Omega)\mathrm{d}\Omega$ to the variance, respectively. A contribution $S_\zeta(\Omega)\mathrm{d}\Omega$ can also be interpreted as the variance of the process after passing a narrow-band filter of the bandwidth $\mathrm{d}\Omega$. Both functions are even,

$$R_\zeta(\xi) = R_\zeta(-\xi)\,, \quad S_\zeta(\Omega) = S_\zeta(-\Omega)\,, \tag{4.87}$$

allowing a simpler analysis of the integrals given in Eqs. (4.84)-(4.86) by using the so-called one-sided spectral density

$$\Phi_\zeta(\Omega) = \begin{cases} 2S_\zeta(\Omega) & \Omega \geq 0 \\ & \text{for} \\ 0 & \Omega < 0 \end{cases} \tag{4.88}$$

without negative angular frequencies. Accordingly, $S_\zeta(\Omega)$ is denoted as two-sided spectral density. Table 4.5 gives twelve typical correlation functions $R_\zeta(\xi)$ and their corresponding one-sided spectral densities $\Phi_\zeta(\Omega)$.

So far, just one scalar stochastic process has been considered. Taking another scalar process into account, e. g. the perturbation profiles ζ_l and ζ_r of the left and right lane of a road, it has to be distinguished between auto- and crosscorrelation functions and their spectra accordingly. This can be achieved by introducing a double subscript for the cross quantities, their pair of Fourier transforms is written analogously to Eqs. (4.84) and (4.85) as

$$S_{lr}(\Omega) = \frac{1}{2\pi}\int_{-\infty}^{\infty} R_{lr}(\xi)\mathrm{e}^{-i\Omega\xi}\mathrm{d}\xi\,, \tag{4.89}$$

$$R_{lr}(\xi) = \int_{-\infty}^{\infty} S_{lr}(\Omega)\mathrm{e}^{i\Omega\xi}\mathrm{d}\Omega\,. \tag{4.90}$$

In general, the cross spectral density is a complex quantity. The following special kind of symmetry conditions hold:

$$R_{lr}(\xi) = R_{rl}(-\xi)\,, \quad S_{lr}(\Omega) = S_{rl}(-\Omega)\,. \tag{4.91}$$

The coherence function $\gamma(\Omega)$ with

$$\gamma^2(\Omega) = \frac{|S_{lr}(\Omega)|^2}{S_l(\Omega)S_r(\Omega)}\,, \quad 0 \leq \gamma \leq 1\,, \tag{4.92}$$

is used to characterize the correlation between two processes. A value of $\gamma \equiv 0$ indicates that two processes, e. g. the perturbation profile of the left and right lane, are completely uncorrelated, whereas they are completely correlated for $\gamma \equiv 1$. Generally, the coherence depends on the frequency, see Eq. (4.92).

In the case of n scalar processes $\zeta_\nu, \nu = 1(1)n$, they can be merged into an $n \times 1$ vector process $\boldsymbol{\zeta}$. The mean value m_ζ is then replaced by the $n \times 1$ mean value vector $\boldsymbol{m}_\zeta$ and the variance P_ζ by the $n \times n$ covariance matrix $\boldsymbol{P}_\zeta$, see Mueller and Schiehlen (1985). Corresponding to Eq. (4.71) it follows that

$$\begin{aligned} \boldsymbol{m}_\zeta(x) &= \mathrm{E}\{\boldsymbol{\zeta}(x)\}\,, \\ \boldsymbol{P}_\zeta(x) &= \mathrm{E}\{[\boldsymbol{\zeta}(x) - \boldsymbol{m}_\zeta(x)][\boldsymbol{\zeta}(x) - \boldsymbol{m}_\zeta(x)]^{\mathrm{T}}\} = \boldsymbol{P}_\zeta^{\mathrm{T}}(x)\,. \end{aligned} \tag{4.93}$$

Table 4.5. Correlation functions $R(\xi)$ and according one-sided spectral densities $\Phi(\Omega)$ (all constants are positive)

Correlation function $R(\xi) = \int_0^\infty \Phi(\Omega)\cos\Omega\xi\,\mathrm{d}\Omega$	One-sided spectral density $\Phi(\Omega) = \frac{2}{\pi}\int_0^\infty R(\xi)\cos\Omega\xi\,\mathrm{d}\xi$	
Colored noise:		
$\sigma^2 \mathrm{e}^{-\alpha\lvert\xi\rvert}$	$\frac{2\alpha\sigma^2}{\pi}\frac{1}{\Omega^2+\alpha^2}$	(I)
$\sigma^2 \mathrm{e}^{-\alpha\lvert\xi\rvert}\cos\beta\xi$	$\frac{2\alpha\sigma^2}{\pi}\frac{\Omega^2+\alpha^2+\beta^2}{(\Omega^2-\alpha^2-\beta^2)^2+4\alpha^2\Omega^2}$	(II)
$\sigma^2 \mathrm{e}^{-\alpha\lvert\xi\rvert}\left(\cos\beta\xi - \frac{\alpha}{\beta}\sin\beta\lvert\xi\rvert\right)$	$\frac{4\alpha\sigma^2}{\pi}\frac{\Omega^2}{(\Omega^2-\alpha^2-\beta^2)^2+4\alpha^2\Omega^2}$	(III)
$\sigma^2 \mathrm{e}^{-\alpha\lvert\xi\rvert}\left(\cos\beta\xi + \frac{\alpha}{\beta}\sin\beta\lvert\xi\rvert\right)$	$\frac{4\alpha\sigma^2}{\pi}\frac{\alpha^2+\beta^2}{(\Omega^2-\alpha^2-\beta^2)^2+4\alpha^2\Omega^2}$	(IV)
$\sigma^2 \mathrm{e}^{-D\Omega_0\lvert\xi\rvert}\left(\cos\Omega_d\xi + \frac{D\Omega_0}{\Omega_d}\sin\Omega_d\lvert\xi\rvert\right)$ with $\Omega_d = \sqrt{1-D^2}\Omega_0$	$\frac{4D\Omega_0\sigma^2}{\pi}\frac{\Omega_0^2}{(\Omega^2-\Omega_0^2)^2+4D^2\Omega_0^2\Omega^2}$	(V)
$\frac{\sigma^2}{\beta-\alpha}\left[\beta\mathrm{e}^{-\alpha\lvert\xi\rvert}-\alpha\mathrm{e}^{-\beta\lvert\xi\rvert}\right]$	$\frac{2\alpha\beta(\alpha+\beta)\sigma^2}{\pi}\frac{1}{(\Omega^2+\alpha^2)(\Omega^2+\beta^2)}$	(VI)
$\sigma^2\mathrm{e}^{-\alpha\xi^2}$	$\frac{\sigma^2}{\sqrt{\pi\alpha}}\mathrm{e}^{-\frac{\Omega^2}{4\alpha}}$	(VII)
$\sigma^2\mathrm{e}^{-\alpha\xi^2}\cos\beta\xi$	$\frac{\sigma^2}{\sqrt{4\pi\alpha}}\left[\mathrm{e}^{-\frac{(\Omega+\beta)^2}{4\alpha}}+\mathrm{e}^{-\frac{(\Omega-\beta)^2}{4\alpha}}\right]$	(VIII)
Harmonic function:		
$\sigma^2\cos\beta\xi$	$\sigma^2\delta(\Omega-\beta)$	(IX)
White noise:		
$q_w\delta(\xi)$	$\frac{1}{\pi}q_w$	(X)
Band-limited white noise:		
$q_w\frac{\sin\alpha\xi}{\pi\xi}$	$\begin{cases}\frac{1}{\pi}q_w & \text{if } 0\le\Omega\le\alpha\\ 0 & \text{otherwise}\end{cases}$	(XI)
Ideal band filter:		
$2q_w\frac{\sin\alpha\xi/2\cos\beta\xi}{\pi\xi}$	$\begin{cases}\frac{1}{\pi}q_w & \text{if } 0\le\beta-\frac{\alpha}{2}\le\Omega\le\beta+\frac{\alpha}{2}\\ 0 & \text{otherwise}\end{cases}$	(XII)

The variances of the individual processes occur here as the diagonal elements of the matrix $\boldsymbol{P}_\zeta$. Obviously, the variances have significant importance in stationary ergodic Gaussian processes. They can be interpreted as:

- central moment of second order of random variables, see Eqs. (4.71), (4.93),
- mean square value of centralized realizations, see Eq. (4.83),
- spectral power of a centralized process, see Eq. (4.86),
- square of the dispersion of the probability density and, thus, determining the Gaussian distribution, see Eq. (4.77), and the width of the confidence intervals, see Eq. (4.78).

■ Problem 4.2 White and colored noise

The spectral densities have to be derived and interpreted for the correlation functions given in Table 4.5 X, I and IV.

Solution

Firstly, the Fourier transform of the one-sided spectral density $\Phi(\Omega)$ is calculated. Using Euler's identity, $e^{\pm i\Omega\xi} = \cos\Omega\xi \pm i\sin\Omega\xi$, and dividing the integral given in Eq. (4.84) into two integration intervals yields

$$S(\Omega)=\frac{1}{2\pi}\left[\int_{-\infty}^{0} R(\xi)(\cos\Omega\xi - i\sin\Omega\xi)\mathrm{d}\xi+\int_{0}^{\infty} R(\xi)(\cos\Omega\xi - i\sin\Omega\xi)\mathrm{d}\xi\right]. \tag{1}$$

Applying the coordinate transformation $\xi = -\bar{\xi}$ to the first integral and considering $R(\xi) = R(-\xi) = R(\bar{\xi})$ according to Eq. (4.87) gives

$$2S(\Omega) = \frac{1}{\pi}\left[\int_{0}^{\infty} R(\bar{\xi})(\cos\Omega\bar{\xi} + i\sin\Omega\bar{\xi})\mathrm{d}\xi + \int_{0}^{\infty} R(\xi)(\cos\Omega\xi - i\sin\Omega\xi)\mathrm{d}\xi\right]. \tag{2}$$

With $2S(\Omega) = \Phi(\Omega)$ it follows that

$$\Phi(\Omega) = \frac{2}{\pi}\int_{0}^{\infty} R(\xi)\cos\Omega\xi\mathrm{d}\xi\,. \tag{3}$$

Analogously, the backward transformation is derived,

$$R(\xi) = \int_{0}^{\infty} \Phi(\Omega)\cos\Omega\xi\mathrm{d}\Omega\,. \tag{4}$$

Now, inserting the correlation functions of Table 4.5 in Eq. (3) yields the following results that are depicted in Table 4.6 additionally.

Table 4.6. Comparison of the correlation function $R(\xi)$ and the spectral density $\Phi(\Omega)$ of white and colored noise

name of noise process	correlation function $R(\xi)$	spectral density $\Phi(\Omega)$ (double log. scale)
white noise (X)	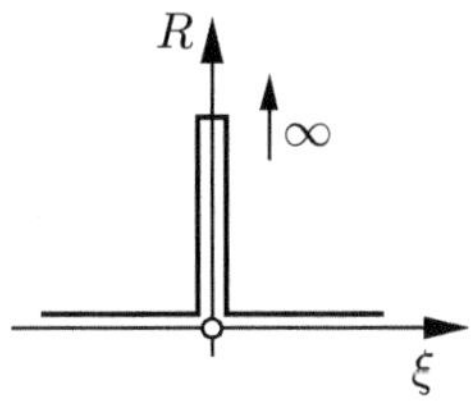 $R = q_w \cdot \delta(\xi)$	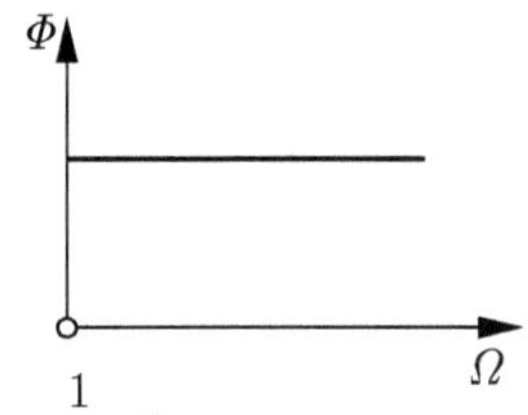 $\Phi = \frac{1}{\pi} \cdot q_w$
colored noise (I)	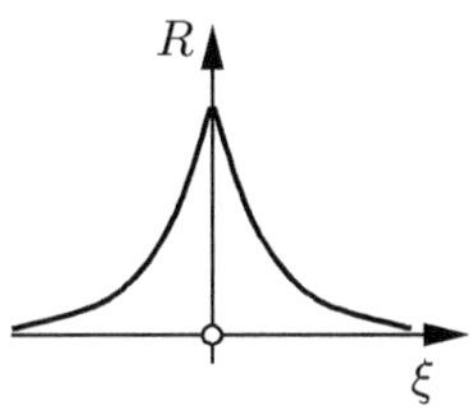 $R = \sigma^2 e^{-\alpha\|\xi\|}$	 $\Phi = \frac{2\alpha\sigma^2}{\pi} \cdot \frac{1}{\Omega^2+\alpha^2}$
colored noise (IV)	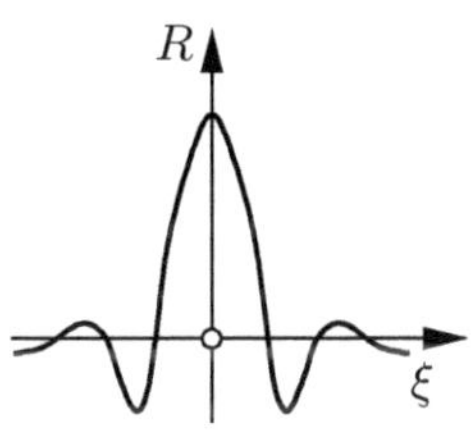 $R = \sigma^2 e^{-\alpha\|\xi\|} \cdot (\cos\beta\xi + \frac{\alpha}{\beta}\sin\beta \mid \xi \mid)$	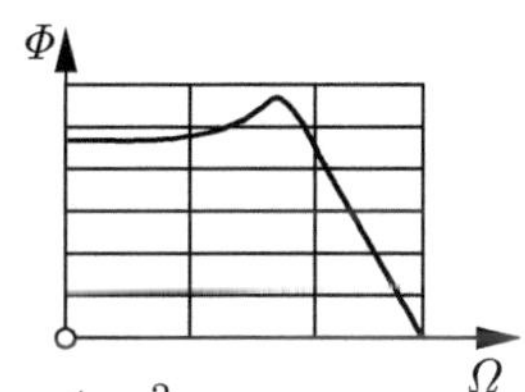 $\Phi = \frac{4\alpha\sigma^2}{\pi} \cdot \frac{\alpha^2+\beta^2}{(\Omega^2-\alpha^2-\beta^2)^2+4\alpha^2\Omega^2}$

1) $R(\xi) = q_w \delta(\xi)$: White noise (X)
Evaluating the integral given in Eq. (3) yields

$$\Phi(\Omega) = \frac{2}{\pi} \int_0^\infty q_w \delta(\xi) \cos \Omega\xi \mathrm{d}\xi = \frac{1}{\pi} q_w = \text{const} . \tag{5}$$

Evaluating the integral of Eq. (5), attention should be paid to the fact that the discontinuity of the Dirac function occurs at the lower limit of integration. Thus, at $\xi = 0$ the fading out by the Dirac function is limited to only half of the integrand. The same result can be obtained by the transformation given in Eq. (4.84) with $S(\Omega) = \Phi(\Omega)/2$. A stochastic process with the property given in Eq. (X) is completely uncorrelated for $\xi \neq 0$. It is characterized by a uniform spectral density at all frequencies. According to the visible white light that contains spectral contributions of all frequencies, such a random process is called white noise $w(x)$. White noise is characterized by an infinite variance,

$$P_w = \sigma_w^2 = R_w(\xi = 0) = \int_0^\infty \Phi(\Omega) \mathrm{d}\Omega = \infty . \tag{6}$$

It becomes evident that such a process is just a mathematical idealization that can not occur in reality. Nevertheless, it is helpful and frequently applied as a limit case of real processes. Stationary, normally distributed white noise can be characterized by the mean value $m_w = 0$ and the intensity q_w, abbreviated by

$$w(x) \sim N(0, q_w) . \tag{7}$$

2) $R(\xi) = \sigma^2 \mathrm{e}^{-\alpha|\xi|}$, $\alpha > 0$: Coloured noise (I)
The solution is given by Eq. (3):

$$\begin{aligned} \Phi(\Omega) &= \frac{2\sigma^2}{\pi} \int_0^\infty \mathrm{e}^{-\alpha\xi} \cos \Omega\xi \mathrm{d}\xi \\ &= \frac{2\sigma^2}{\pi} \left[\frac{\mathrm{e}^{-\alpha\xi}}{\Omega^2 + \alpha^2} (-\alpha \cos \Omega\xi + \Omega \sin \Omega\xi) \right]_{\xi=0}^{\xi=\infty} \\ &= \frac{2\sigma^2}{\pi} \frac{\alpha}{\Omega^2 + \alpha^2} . \end{aligned} \tag{8}$$

The evaluation is quite simple in this case because $|\xi| = \xi$ holds for $\xi \geq 0$ and the term in square brackets vanishes at the upper integration limit. The resulting so-called colored noise shows a maximum spectral density of $\Phi_{max} = 2\sigma^2/\pi\alpha$ at the angular frequency $\Omega = 0$ and an asymptotic behavior

$\Phi \sim 1/\Omega^2$ for $\Omega \to \infty$. The variance $\sigma^2 = R(\xi = 0)$ is finite and the correlation decreases monotonously with increasing correlation distance ξ. From Table 4.6, the difference between colored and white noise is obvious.

3) $R(\xi) = \sigma^2 \mathrm{e}^{-\alpha|\xi|}(\cos\beta\xi + \frac{\alpha}{\beta}\sin\beta\,|\xi|)$, $\alpha > 0$: Coloured noise (IV)

The solution is performed analogously to 2) by means of the trigonometric identities for products of harmonic functions,

$$
\begin{aligned}
\Phi(\Omega) &= \frac{2\sigma^2}{\pi}\int_0^\infty \mathrm{e}^{-\alpha\xi}(\cos\beta\xi + \frac{\alpha}{\beta}\sin\beta\xi)\cos\Omega\xi \mathrm{d}\xi \\
&= \frac{2\sigma^2}{\pi}\int_0^\infty \{\mathrm{e}^{-\alpha\xi}\frac{1}{2}[\cos(\beta+\Omega)\xi + \cos(\beta-\Omega)\xi] \\
&\quad + \mathrm{e}^{-\alpha\xi}\frac{\alpha}{2\beta}[\sin(\beta+\Omega)\xi + \sin(\beta-\Omega)\xi]\}\mathrm{d}\xi \\
&= \frac{\sigma^2}{\pi}\Bigg\{\mathrm{e}^{-\alpha\xi}\Bigg[\frac{1}{\alpha^2+(\beta+\Omega)^2}(-\alpha\cos(\beta+\Omega)\xi \\
&\quad + (\beta+\Omega)\sin(\beta+\Omega)\xi) \\
&\quad + \frac{1}{\alpha^2+(\beta-\Omega)^2}(-\alpha\cos(\beta-\Omega)\xi + (\beta-\Omega)\sin(\beta-\Omega)\xi) \\
&\quad + \frac{\alpha}{\beta(\alpha^2+(\beta+\Omega)^2)}(-\alpha\sin(\beta+\Omega)\xi - (\beta+\Omega)\cos(\beta+\Omega)\xi) \\
&\quad + \frac{\alpha}{\beta(\alpha^2+(\beta-\Omega)^2)}(-\alpha\sin(\beta-\Omega)\xi \\
&\quad - (\beta-\Omega)\cos(\beta-\Omega)\xi)\Bigg]\Bigg\}_{\xi=0}^{\xi=\infty} \\
&= \frac{\sigma^2}{\pi}\Bigg[\frac{\alpha}{\alpha^2+(\beta+\Omega)^2} + \frac{\alpha}{\alpha^2+(\beta-\Omega)^2} \\
&\quad + \frac{\alpha}{\beta}\left(\frac{\beta+\Omega}{\alpha^2+(\beta+\Omega)^2} + \frac{\beta-\Omega}{\alpha^2+(\beta-\Omega)^2}\right)\Bigg] \\
&= \frac{4\alpha\sigma^2}{\pi}\frac{\alpha^2+\beta^2}{N} \,.
\end{aligned} \tag{9}
$$

Herein, the denominator can be expressed in different ways,

$$
\begin{aligned}
N &= [\alpha^2 + (\beta+\Omega)^2][\alpha^2 + (\beta-\Omega)^2] \\
&= (\Omega^2 - \alpha^2 - \beta^2)^2 + 4\alpha^2\Omega^2 \\
&= (\Omega^2 + \alpha^2 - \beta^2)^2 + 4\alpha^2\beta^2 \,.
\end{aligned} \tag{10}
$$

Again, the result is a colored noise process with the maximum spectral density

$$\Phi_{max} = \begin{cases} \dfrac{\alpha\sigma^2}{\pi}\dfrac{\alpha^2+\beta^2}{\alpha^2\beta^2} & \Omega = \sqrt{\beta^2-\alpha^2} & \beta \geq \alpha\,, \\ & \text{at} \qquad\qquad \text{for} & \\ \dfrac{\alpha\sigma^2}{\pi}\dfrac{4}{\alpha^2+\beta^2} & \Omega = 0 & \beta \leq \alpha \end{cases} \tag{11}$$

and the asymptotic behavior $\Phi \sim 1/\Omega^4$ for $\Omega \to \infty$. The variance $\sigma^2 = R(\xi = 0)$ is finite again. With increasing correlation distance ξ, the correlation oscillatory approaches a value of zero. ■

4.2.2 Models for Unevenness Profiles

From the analysis of the numerous measured data of guideway unevenness it emerges a tendency to standardization. A simple and frequently used model for road unevenness reads as

$$\Phi_\zeta(\Omega) = \Phi_0 \left(\frac{\Omega_0}{\Omega}\right)^w, \quad 0 < \Omega_\mathrm{I} \leq \Omega \leq \Omega_\mathrm{II} < \infty\,, \tag{4.94}$$

cp. Mitschke and Wallentowitz (2004), Voy (1977), where the frequency range is limited. Here, Ω_0 [rad/m] denotes a reference spatial angular frequency and $\Phi_0 = \Phi_\zeta(\Omega_0)$ [m^2/(rad/m)] serves as degree of unevenness and specifies whether the road is good or bad. The wavelength is given by $\lambda = 2\pi/\Omega$. The waviness w is a measure if the road contains mainly long waves (high w) or as well short waves (small w) with significant spectral densities. For roads the waviness ranges in an interval $1.75 \leq w \leq 2.25$ with a mean value of $w \approx 2$. Applying the model (4.94) to the rail unevenness a mean value of $w \approx 4$ is obtained in the higher frequency range. Equation (4.94) is characterized by decreasing straight lines (slope $-w$) if plotted in an Ω-Φ_ζ-diagram with a double logarithmic scale, cp. Fig. 4.9 a).

A similar, slightly more complicated model can be used for both road and rail unevenness, cp. Dodds and Robson (1973), Garg and Dukkipati (1984),

$$\Phi_\zeta(\Omega) = \begin{cases} \Phi_0 \left(\dfrac{\Omega_0}{\Omega}\right)^{w_1} & 0 < \Omega_\mathrm{I} \leq \Omega \leq \Omega_0\,, \\ & \text{for} \\ \Phi_0 \left(\dfrac{\Omega_0}{\Omega}\right)^{w_2} & \Omega_0 \leq \Omega \leq \Omega_\mathrm{II} < \infty\,, \end{cases} \tag{4.95}$$

where different values of the waviness w_1, w_2 with $w_1 < w_2$ occur in the two frequency ranges, cp. Fig. 4.9 b). In the case of tracks, Eq. (4.95) is applied for all four perturbation profiles.

The models (4.94) and (4.95) represent approximations of measured spectral densities in the bounded frequency range $0 < \Omega_\mathrm{I} \leq \Omega \leq \Omega_\mathrm{II} < \infty$. Within both models the limit $\Omega \to 0$ would result in infinite spectral densities

$\Phi_\zeta \to \infty$, hence, leading to infinite variances $P_\zeta \to \infty$ as well. To avoid this unrealistic fact, improved unevenness models have been proposed which are valid in the entire frequency range,

$$\text{I: } \Phi_{\zeta_{\text{I}}}(\Omega) = \frac{2\alpha\sigma^2}{\pi}\frac{1}{\Omega^2+\alpha^2}, \quad 0 \le \Omega < \infty, \tag{4.96}$$

$$\text{II: } \Phi_{\zeta_{\text{II}}}(\Omega) = \frac{2\alpha\sigma^2}{\pi}\frac{\Omega^2+\alpha^2+\beta^2}{(\Omega^2-\alpha^2-\beta^2)^2+4\alpha^2\Omega^2}, \quad 0 \le \Omega < \infty, \tag{4.97}$$

with the positive constants α, β and σ^2. For these spectral densities the corresponding correlation functions are now calculated. Adapting Eq. (4.85) to the one-sided spectral density $\Phi_\zeta(\Omega)$ yields

$$R_\zeta(\xi) = \int_0^\infty \Phi_\zeta(\Omega)\cos\Omega\xi \, \mathrm{d}\Omega . \tag{4.98}$$

Inserting Eqs. (4.96), (4.97) into Eq. (4.98) yields, cp. Table 4.5,

$$\text{I: } R_{\zeta_{\text{I}}}(\xi) = \sigma^2 \mathrm{e}^{-\alpha|\xi|}, \tag{4.99}$$

$$\text{II: } R_{\zeta_{\text{II}}}(\xi) = \sigma^2 \mathrm{e}^{-\alpha|\xi|}\cos\beta\xi, \tag{4.100}$$

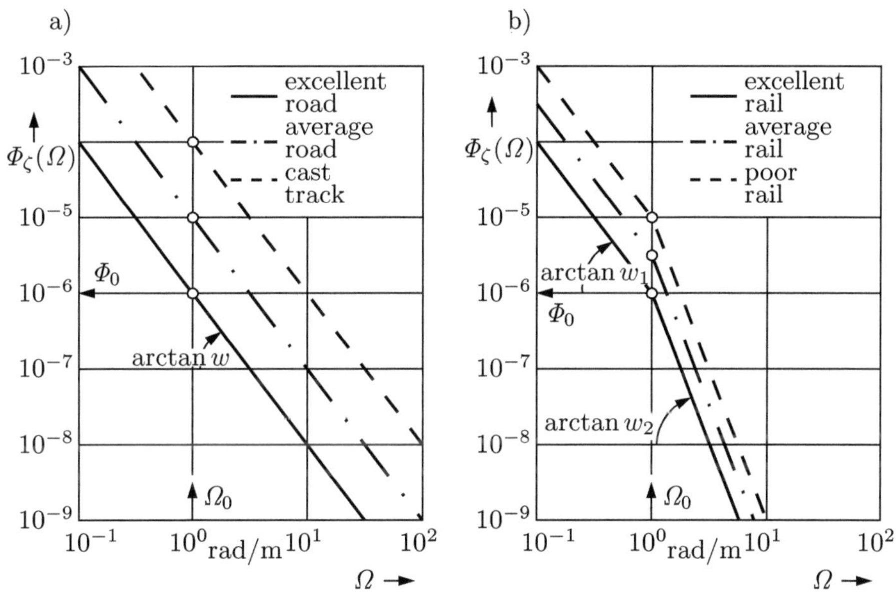

Fig. 4.9. Simple guideway models with coarse values for the degree of unevenness $\Phi_0 = \Phi_\zeta(\Omega_0)$ in $\text{m}^2/(\text{rad/m})$, $\Omega_0 = 1$ rad/m: a) Model (4.94) for road unevenness with $w = 2$; b) Model (4.95) for lateral and vertical rail alignment with $w_1 = 2$, $w_2 = 4$

where $\sigma^2 = R_\zeta\,(\xi = 0)$ denotes the finite variance of the guideway unevenness. For $\Omega \to \infty$ the spectral densities of Eqs. (4.96) and (4.97) follow the relation $\Phi \sim 1/\Omega^2$ which leads to a good approximation of measured road spectra. Models IV to VI in Table 4.5 are suitable as rail spectra. Data for corresponding parameters can be found e. g. in Garg and Dukkipati (1984).

4.2.3 Models for Vehicle Excitation Processes

From the presented models for the guideway unevenness $\zeta(x)$ corresponding models for the vehicle excitation $\zeta(t)$ depending on the time t can be found where permanent guideway contact is assumed and only one contact point is regarded. The transformation of the spatial into the time domain $\zeta(x) \to \zeta(t)$ is carried out via the velocity $v(t)$ of the vehicle,

$$\mathrm{d}x(t) = v(t)\mathrm{d}t\;, \quad x(t) = x(t_0) + \int_{t_0}^{t} v(\tau)\mathrm{d}\tau\;. \tag{4.101}$$

In the case of time-variant velocities $v(t)$ one gets non-stationary vehicle excitation processes in the time domain even from stationary stochastic unevenness processes in spatial domain, cp. Rill (1983), Czerny (1987). This topic is not treated here. In the following, a constant velocity, $v = \text{const}$, is assumed resulting in stationary vehicle excitation processes. Setting $t_0 = 0$ and $x(t_0) = 0$, it follows from Eq. (4.101)

$$x(t) = vt\;, \quad \tau = \frac{\xi}{v}\;, \quad \omega = v\Omega\;, \quad v = \text{const}\;. \tag{4.102}$$

Here, τ denotes the correlation time according to the correlation distance ξ and ω [rad/s] is the angular frequency with respect to time. Since the variances $R_\zeta(0)$ of the guideway unevenness $\zeta(x)$ and of the vehicle excitation $\zeta(t)$ must be equal, from (4.98) the relation $\Phi_\zeta(\omega)\mathrm{d}\omega = \Phi_\zeta(\Omega)\mathrm{d}\Omega$ proves. Hence, with Eq. (4.102) the one-sided spectral density $\Phi_\zeta(\omega)$ [m^2/(rad/s)] of the vehicle excitation follows as

$$\Phi_\zeta(\omega) = \frac{1}{v}\Phi_\zeta\left(\Omega = \frac{\omega}{v}\right)\;. \tag{4.103}$$

From the unevenness model given in Eq. (4.94), one gets the spectral density in the time domain for an average waviness $w = 2$ of roads as

$$\tilde{\Phi}_\zeta(\omega) = \frac{1}{v}\tilde{\Phi}_0\left(\frac{v\Omega_0}{\omega}\right)^2 = v\tilde{\Phi}_0\left(\frac{\Omega_0}{\omega}\right)^2\;. \tag{4.104}$$

Up to here exclusively the stochastic process $\zeta(t)$ of the vehicle displacement has been regarded. However, additionally the processes $\dot{\zeta}(t)$ and $\ddot{\zeta}(t)$ of the excitation velocity and the excitation acceleration have to be considered. The respective spectral densities then yield, cp. e. g. Newland (1975),

$$\Phi_{\dot{\zeta}}(\omega) = \omega^2 \Phi_\zeta(\omega) \,, \quad \Phi_{\ddot{\zeta}}(\omega) = \omega^4 \Phi_\zeta(\omega) \,. \tag{4.105}$$

Applying Eq. (4.105) to (4.104), a white noise process with the spectral density

$$\tilde{\Phi}_{\dot{\zeta}}(\omega) = \omega^2 v \tilde{\Phi}_0 \left(\frac{\Omega_0}{\omega}\right)^2 = v \tilde{\Phi}_0 \Omega_0^2 = \text{const} \tag{4.106}$$

is obtained for the excitation velocity $\dot{\zeta}(t)$. The corresponding correlation function reads

$$R_{\dot{\zeta}}(\tau) = q_{\dot{\zeta}} \delta(\tau) \,, \quad q_{\dot{\zeta}} = \pi v \tilde{\Phi}_0 \Omega_0^2 \,, \tag{4.107}$$

where $q_{\dot{\zeta}}$ denotes the noise intensity and $\delta(\tau)$ is the Dirac distribution. In analogy, from model (4.94) it follows a white noise process for the excitation acceleration $\ddot{\zeta}(t)$ using the characteristic waviness of $w = 4$ for rails as

$$\tilde{\tilde{\Phi}}_{\ddot{\zeta}}(\omega) = \omega^4 v \tilde{\tilde{\Phi}}_0 \left(\frac{\Omega_0}{\omega}\right)^4 = v \tilde{\tilde{\Phi}}_0 \Omega_0^4 = \text{const} \,, \tag{4.108}$$

$$R_{\ddot{\zeta}}(\omega) = q_{\ddot{\zeta}} \delta(\tau) \,, \quad q_{\ddot{\zeta}} = \pi v \tilde{\tilde{\Phi}}_0 \Omega_0^4 \,. \tag{4.109}$$

The white noise processes show an infinite variance and are, consequently, unrealistic. On the other hand it is possible to reduce the effort to calculate stochastic vehicle vibrations for vehicle excitations according to Eqs. (4.106)-(4.109). Therefore, the velocity white noise for road vehicles and the acceleration white noise for rail vehicles are important engineering approximations.

Improved models for the vehicle excitation $\zeta(t)$ are based on Eqs. (4.96) and (4.97). These models represent normally distributed, stationary colored noise processes. They can be obtained by solving a linear time-invariant system of differential equations excited by a white noise process $w(t)$. Interpretive-ly explained, such a system of differential equations changes the shape of the correlation function of excitation, therefore, it is termed shape filter, cp. Fig. 4.10. In general it holds that stationary, Gaussian and ergodic processes affecting a linear time-invariant system of differential equations do not change

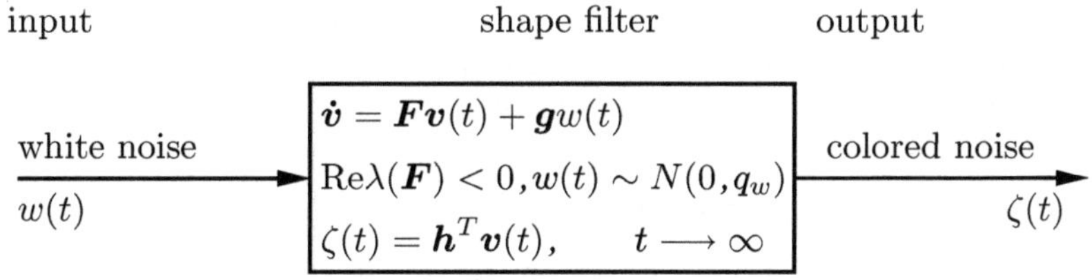

Fig. 4.10. Illustration of the mode of operation of a shape filter

their properties. Stationarity, normal distribution and ergodicity are transferred from the input process to the output process. The equations for the shape filter read

$$\zeta(t) = \boldsymbol{h}^{\mathrm{T}}\boldsymbol{v}(t) \ , \tag{4.110}$$

$$\dot{\boldsymbol{v}}(t) = \boldsymbol{F}\boldsymbol{v}(t) + \boldsymbol{g}w(t) \ , \quad \mathrm{Re}\lambda(\boldsymbol{F}) < 0 \ , \quad w(t) \sim N(0, q_w) \ . \tag{4.111}$$

At steady-state conditions the colored noise process $\zeta(t)$ is obtained by the superposition of the state variables $v_i(t), i = 1(1)m$, which are merged into an $m \times 1$ state vector $\boldsymbol{v}(t)$, cp. Eq. (4.110). The state vector $\boldsymbol{v}$ satisfies Eq. (4.111) for an asymptotically stable system matrix $\boldsymbol{F}$, where the excitation is due to a white noise $w(t)$ with mean value zero and intensity q_w. The quantities $\boldsymbol{F}$, $\boldsymbol{g}$ and $\boldsymbol{h}$ characterize completely the shape filter. For the processes given by Eqs. (4.96), (4.99) and Eqs. (4.97), (4.100), respectively, these quantities read as

$$\mathrm{I}: \boldsymbol{F} = -\alpha v \ , \quad \boldsymbol{g} = g \ , \quad \boldsymbol{h} = 1 \ , \quad (g^2 q_w = 2\alpha v \sigma^2) \ , \tag{4.112}$$

$$\mathrm{II}: \boldsymbol{F} = \begin{bmatrix} 0 & 1 \\ -(\alpha^2+\beta^2)v^2 & -2\alpha v \end{bmatrix} , \quad \boldsymbol{g} = g \begin{bmatrix} 0 \\ 1 \end{bmatrix} , \quad \boldsymbol{h} = \begin{bmatrix} v\sqrt{\alpha^2+\beta^2} \\ 1 \end{bmatrix} ,$$
$$(g^2 q_w = 2\alpha v \sigma^2) \ . \tag{4.113}$$

Since some of the model parameters are multiplicatively connected there are some options to define them. By setting e. g. $g = 1$, it follows for both shape filters an exciting noise intensity of $q_w = 2\alpha v \sigma^2$. It turns out that the intensity of the vehicle excitation increases proportionally with the velocity v.

In general the shape filter parameters are not found by means of analytical approximations but directly from measured values of the spectral density $\Phi_\zeta(\Omega)$ where the order m of the filter has to be preset. For $m = 2$ the procedure is described in detail by Mueller et al. (1980).

In all models for vehicle excitations mentioned above only one single contact point with the guideway is assumed, leading always to scalar stochastic excitation processes. However, real multi-wheeled vehicles show multiple contacts with the guideway. For $i = 1(1)q$ axles and $k = 1(1)s$ lanes there are $q \cdot s$ contact points with an identical number of scalar excitation processes. They may be summarized in a $(q \cdot s) \times 1$ vector process $\boldsymbol{\zeta}$ of the excitations.

If the contact points are located in one lane it is possible to get the corresponding exciting processes $\zeta_i(t), i = 1(1)q$, from the process $\zeta_1(1) \equiv \zeta(t)$ of the foremost contact point by a time delay t_i,

$$\zeta_i(t) = \zeta(t - t_i) \ , \quad t_i = \frac{l_i}{v} \ , \quad 0 = t_1 < t_2 < ... < t_q \ , \quad i = 1(1)q \ , \tag{4.114}$$

where l_i denotes the distance between the foremost and the i-th axle and v describes the constant velocity. Further details are given by Mueller and Popp (1979).

For parallel driving lanes their statistical dependence has to be considered when required. For road unevenness, correlation measurements can be found in Bormann (1978), which Rill (1983) applied to a shape filter and considered them in the analysis. A typical vector process for the excitation of a four wheeled vehicle reads ($v \mathrel{\widehat{=}}$ front, $h \mathrel{\widehat{=}}$ rear, $l \mathrel{\widehat{=}}$ left, $r \mathrel{\widehat{=}}$ right),

$$\boldsymbol{\zeta}(t,T) = \begin{bmatrix} \zeta_{lv}(t) \\ \zeta_{rv}(t) \\ \zeta_{lh}(t) \\ \zeta_{rh}(t) \end{bmatrix} = \begin{bmatrix} \zeta_l(t) \\ \zeta_r(t) \\ \zeta_l(t-T) \\ \zeta_r(t-T) \end{bmatrix}, \quad T \equiv t_2 = \frac{l_2}{v}, \tag{4.115}$$

where the delay T results from the axle-center distance l_2 and the constant velocity v.

5

Models for Vehicle-Guideway-Systems

For modeling purposes the ground vehicle systems have been decomposed into the subsystems vehicle, guidance and suspension system, and guideway. These subsystems have been considered extensively in Chap. 2 to Chap. 4. Now they will be assembled to the mathematical model of the complete system. This means the final step of the modeling of a vehicle system.

5.1 State Equations of the Subsystems

The dynamics of all the subsystems can be described consistently by state equations. The state space representation is widely used in control and system theory, and it proves to be most appropriate for vehicle dynamics, too. References of the state representation for control problems are found in Foellinger (2008) and for vibration engineering in Mueller and Schiehlen (1985). The equations of motion of a *vehicle* read according to (2.96)

$$\boldsymbol{M}(\boldsymbol{y},t)\ddot{\boldsymbol{y}}(t) + \boldsymbol{k}(\boldsymbol{y},\dot{\boldsymbol{y}},t) = \boldsymbol{q}(\boldsymbol{y},\dot{\boldsymbol{y}},t) \tag{5.1}$$

or in linearized form according to (2.99)

$$\boldsymbol{M}\ddot{\boldsymbol{y}}(t) + (\boldsymbol{D}+\boldsymbol{G})\dot{\boldsymbol{y}}(t) + (\boldsymbol{K}+\boldsymbol{N})\boldsymbol{y}(t) = \boldsymbol{h}(t) \ , \tag{5.2}$$

where the vector $\boldsymbol{h}(t)$ includes the forces acting on the vehicle generated by the suspension and guidance system.

The equations of motion (5.1) and (5.2) are systems of differential equations of second order, they have to be completed by the initial conditions for position and velocity

$$\boldsymbol{y}(0) = \boldsymbol{y}_0 \ , \quad \dot{\boldsymbol{y}}(0) = \dot{\boldsymbol{y}}_0 \ . \tag{5.3}$$

The state vector of the vehicle is composed correspondingly by the state variables of position and velocity as

$$\boldsymbol{x}_F(t) = \begin{bmatrix} \boldsymbol{y}(t) \\ \dot{\boldsymbol{y}}(t) \end{bmatrix} . \tag{5.4}$$

Thus, $\boldsymbol{x}_F$ is the $n_F \times 1$-state vector of all state variables. For ordinary multibody systems, and therefore also for vehicles, it yields $n_F = 2f$ where f is the number of degrees of freedom again.

Using the state vector (5.4) the equations of motion are easily transferred in the corresponding state equations. In the nonlinear case one gets

$$\begin{aligned} \dot{\boldsymbol{y}}(t) &= \dot{\boldsymbol{y}}(t) \\ \underbrace{\ddot{\boldsymbol{y}}(t)}_{\dot{\boldsymbol{x}}_F(t)} &= \underbrace{\boldsymbol{M}^{-1}(\boldsymbol{y},t)[\boldsymbol{q}(\boldsymbol{y},\dot{\boldsymbol{y}},t) - \boldsymbol{k}(\boldsymbol{y},\dot{\boldsymbol{y}},t)]}_{\boldsymbol{a}_F(\boldsymbol{x}_F,t)} , \end{aligned} \tag{5.5}$$

where $\boldsymbol{a}_F$ is a $n_F \times 1$-vector function depending on the state variables and the time t. For vehicles performing small linear motions it remains

$$\dot{\boldsymbol{x}}_F(t) = \boldsymbol{A}_F \boldsymbol{x}_F(t) + \boldsymbol{B}_F \boldsymbol{u}_F(t) , \tag{5.6}$$

where

$$\boldsymbol{A}_F = \begin{bmatrix} \boldsymbol{0} & \boldsymbol{E} \\ -\boldsymbol{M}^{-1}(\boldsymbol{K}+\boldsymbol{N}) & -\boldsymbol{M}^{-1}(\boldsymbol{D}+\boldsymbol{G}) \end{bmatrix} \tag{5.7}$$

is the $n_F \times n_F$-system matrix, $\boldsymbol{B}_F$ the $n_F \times r_F$-input matrix and $\boldsymbol{u}_F$ the $r_F \times 1$-input vector of the excitations acting on the vehicle.

The state equations are completed according to (5.3) by the initial conditions reading as

$$\boldsymbol{x}_F(0) = \boldsymbol{x}_{F0} . \tag{5.8}$$

Since the state equations are systems of differential equations of the first order, the initial conditions are represented by one $n_F \times 1$-vector the dimension of which is doubled compared to (5.3).

The *suspension systems* of a vehicle are described in the most general case according to (3.38) by linear differential equations of the first order,

$$c_{\dot{f}} \dot{f}(t) + c_f f(t) = c_s s(t) + c_{\dot{s}} \dot{s}(t) + c_u u(t) , \quad f(0) = f_0 . \tag{5.9}$$

Considering that a vehicle has several suspension devices, one gets from (5.9) state equations of the form

$$\dot{\boldsymbol{x}}_T(t) = \boldsymbol{A}_T \boldsymbol{x}_T(t) + \boldsymbol{B}_T \boldsymbol{u}_T(t) , \quad \boldsymbol{x}_T(0) = \boldsymbol{x}_{T0} , \tag{5.10}$$

where $\boldsymbol{x}_T$ is a $n_T \times 1$-state vector of force variables, $\boldsymbol{A}_T$ is the corresponding $n_T \times n_T$-system matrix and $\boldsymbol{u}_T$ is the $r_T \times 1$-input vector of the relative motions of the suspension elements and the control variables of the control device.

The *guidance systems* of wheeled vehicles are described only statically in Chap. 3. For the contact forces it yields according to (3.86) or (3.143), respectively, the relation

$$\boldsymbol{f}^K = -\boldsymbol{F}\boldsymbol{v}^K \; . \tag{5.11}$$

Therefore, the state equations are degenerated and only algebraic equations occur reading as

$$\boldsymbol{A}_L\boldsymbol{x}_L(t) + \boldsymbol{B}_L\boldsymbol{u}_L(t) = \boldsymbol{0} \; . \tag{5.12}$$

The $n_L \times 1$-state vector $\boldsymbol{x}_L$ summarizes the force variables whereas $\boldsymbol{u}_L$ is the $r_L \times 1$-input vector of the relative or slip motions in the contact point. The matrix $\boldsymbol{A}_F$ is according to the assumptions on Chap. 3 a unit matrix while the $n_L \times r_L$-input matrix $\boldsymbol{B}_L$ includes, e.g., the Kalker coefficients or the corresponding characteristics of the tire.

It has to be mentioned that guidance systems show instationary behavior with more complex models, too. Then, the algebraic equations (5.12) mutate to real state equations with time derivatives. This may be of engineering relevance for road vehicles with rubber tires. More details are presented in the comprehensive report by Lugner and Ploechl (2005).

Rigid guideways are characterized by their special profile functions which are transformed into pure temporal functions $\boldsymbol{x}_s(t)$ using the vehicle speed. In the deterministic case the excitation functions due to the guideway are summarized in the $n_W \times 1$-vector

$$\boldsymbol{x}_W(t) = [\boldsymbol{x}_s^{\mathrm{T}}(t) \;\; \dot{\boldsymbol{x}}_s^{\mathrm{T}}(t)]^{\mathrm{T}} \; , \tag{5.13}$$

which has to be supplemented by the first derivative if damping is considered as shown in (5.13). The deterministic functions are mainly used to describe individual obstacles. Randomly uneven guideways require the modeling by stochastic processes. Starting with a white noise excitation it follows a colored noise process via a shape filter as described in (4.110) and (4.111) resulting in

$$\boldsymbol{x}_W(t) = \boldsymbol{C}_{WR}\boldsymbol{x}_R(t) \; , \tag{5.14}$$

$$\dot{\boldsymbol{x}}_R(t) = \boldsymbol{A}_R\boldsymbol{x}_R(t) + \boldsymbol{B}_R\boldsymbol{w}(t) \; ,$$

$$\boldsymbol{x}_R(0) = \boldsymbol{0} \; , \quad \boldsymbol{w}(t) \sim (0, \boldsymbol{Q}_W) \; , \tag{5.15}$$

where the shape filter is characterized by the time-invariant matrices $\boldsymbol{A}_R$, $\boldsymbol{B}_R$ and $\boldsymbol{C}_{WR}$. The $n_R \times 1$-vector $\boldsymbol{x}_R$ represents the random process generated by the $r_R \times 1$-vector process of white noise $\boldsymbol{w}(t)$ with the $r_R \times r_R$-intensity matrix $\boldsymbol{Q}_w$.

Elastic guideways consisting of beam structures are mechanical systems described by equations of motion, again. According to (4.10) it yields for a finite number of eigenvalues and the assumption of modal damping

$$\ddot{\boldsymbol{z}}(t) + \Delta\dot{\boldsymbol{z}}(t) + \boldsymbol{\Omega}\boldsymbol{z}(t) = \boldsymbol{M}^{-1}\sum_k \boldsymbol{\varphi}[x_k(t)]f_k(t) \ . \tag{5.16}$$

Based on the vector $\boldsymbol{z}$ of the modal beam coordinates a $n_E \times 1$-state vector can be defined for the elastic guideway as

$$\boldsymbol{x}_E(t) = \begin{bmatrix} \boldsymbol{z}(t) \\ \dot{\boldsymbol{z}}(t) \end{bmatrix} \ . \tag{5.17}$$

Then, the equations of motion (5.16) are transferred once again into state equations as

$$\dot{\boldsymbol{x}}_E(t) = \boldsymbol{A}_E\boldsymbol{x}_E(t) + \boldsymbol{B}_E(t)\boldsymbol{u}_E(t) \ , \quad \boldsymbol{x}_E(0) = \boldsymbol{0} \ . \tag{5.18}$$

In contrary to the system matrix $\boldsymbol{A}_E$, the input matrix $\boldsymbol{B}_E(t)$ is now time-variant. Furthermore, the $r_E \times 1$-input vector $\boldsymbol{u}_E$ includes the dynamic and static forces acting via the primary suspension from the vehicle on the guideway. The static force result due to the beam deflection in a forced excitation of the complete system.

The vector $\boldsymbol{x}_W$ of the effective deflection of the beam can be found from Eq. (4.9) by the eigenmodes considering the vehicle speed as a pure temporal function

$$\boldsymbol{x}_W(t) = \boldsymbol{C}_{WE}(t)\boldsymbol{x}_E(t) \ . \tag{5.19}$$

The matrix $\boldsymbol{C}_{WE}$ summarizes the eigenmodes of the beam related to the vector $\boldsymbol{z}$. In contrary to (5.14) the vector $\boldsymbol{x}_E$ depends dynamically on the guideway acting forces as shown in (5.18). Therefore, there is a feedback between vehicle and guideway in the case of elastic structures what does not occur for rigid guideways.

5.2 State Equations of the Complete System

The state equations of the subsystems or vehicle system components, respectively, have been consistently introduced in Sect. 5.1.

These equations can now be assembled resulting in the state equations of the complete system as required by the engineering tasks. There will be only linear systems considered sufficient for the investigation of small vibrations and for a stability analysis. As a basic principle the procedure can be transferred directly to nonlinear systems.

For the assembly of the complete system the input and output variable of the subsystems have to be connected properly to each other. The generalized forces acting on the vehicle are summarized in the $r_F \times 1$-vector $\boldsymbol{u}_F$. Further, forces are generated by the suspension and guidance system and they serve as state variables featuring the $n_F \times 1$-vector $\boldsymbol{x}_T$ of the suspension devices and the $n_L \times 1$-vector $\boldsymbol{x}_L$ of the guidance systems. Therefore, it yields for the suspension system

$$\boldsymbol{u}_F(t) = \boldsymbol{C}_{FT}\boldsymbol{x}_T(t) \,, \tag{5.20}$$

and for the guidance system

$$\boldsymbol{u}_F(t) = \boldsymbol{C}_{FL}\boldsymbol{x}_L(t) \,, \tag{5.21}$$

with the $r_F \times n_{T,L}$-coupling matrices $\boldsymbol{C}_{FT,L}$.

On the other hand, the suspension system is controlled by the $r_T \times 1$-input vector $\boldsymbol{u}_T$. This input depends on the relative motions of the vehicle and the guideway as well as the control variables of the control device. In particular, the controller feeds back the measured motions of the vehicle and the guideway. With the assumption of a simple PD-controller, it yields

$$\boldsymbol{u}_T(t) = \boldsymbol{C}_{TF}\boldsymbol{x}_F(t) + \boldsymbol{C}_{TW}\boldsymbol{x}_W(t) \,, \tag{5.22}$$

where the $r_T \times n_{F,W}$-coupling matrices $\boldsymbol{C}_{TF,W}$ occur. In analogous manner it follows for the guidance system

$$\boldsymbol{u}_L(t) = \boldsymbol{C}_{LF}\boldsymbol{x}_F(t) + \boldsymbol{C}_{LW}\boldsymbol{x}_W(t) \,, \tag{5.23}$$

where the slips $\boldsymbol{u}_L$ in the wheel contact points are obtained via the $r_L \times n_{F,W}$-coupling matrices $\boldsymbol{C}_{LF,W}$ from the relative velocities with respect to the guideway as indicated, e.g., in (3.111).

An elastic guideway is loaded by the dynamic forces of the primary suspension collected in the $r_E \times 1$-vector $\boldsymbol{u}_E$. Furthermore, the weight of the vehicle distributed via the different axles results in static loads summarized in the $r_P \times 1$-vector $\boldsymbol{w}_P$. Then, it yields

$$\boldsymbol{u}_E(t) = \boldsymbol{C}_{ET}\boldsymbol{x}_T(t) + \boldsymbol{C}_{EP}\boldsymbol{w}_P(t) \tag{5.24}$$

with the corresponding $r_E \times n_{T,P}$-coupling matrices $\boldsymbol{C}_{ET,P}$.

That is the end of the mathematical description of the complete vehicle system. Due to the large number of engineering tasks there is a great variety of different systems which may represent a complete system. As examples one complete system for the lateral motion and two complete systems for the vertical motion are presented in more detail.

For small lateral motions the state equations of the complete system follow from (5.6), (5.21), (5.12) and (5.23) as

$$\begin{aligned} \dot{\boldsymbol{x}}_F(t) &= (\boldsymbol{A}_F - \boldsymbol{B}_F\boldsymbol{C}_{FL}\boldsymbol{A}_L^{-1}\boldsymbol{B}_L\boldsymbol{C}_{LF})\boldsymbol{x}_F(t) - \boldsymbol{B}_F\boldsymbol{C}_{FL}\boldsymbol{A}_L^{-1}\boldsymbol{B}_L\boldsymbol{C}_{LW}\boldsymbol{x}_W(t) \\ \boldsymbol{x}_F(0) &= \boldsymbol{x}_{F0} \,. \end{aligned} \tag{5.25}$$

For an even track, $\boldsymbol{x}_W = \boldsymbol{0}$, a homogeneous system of differential equations is found which is the basis for an analysis of the driving stability.

For small vertical vibrations on an elastic guideway the state equations of the complete systems follow from (5.6), (5.10), (5.18), (5.19), (5.20), (5.22) and (5.24) as

$$\underbrace{\begin{bmatrix} \dot{\boldsymbol{x}}_F \\ \dot{\boldsymbol{x}}_T \\ \dot{\boldsymbol{x}}_E \end{bmatrix}}_{\dot{\boldsymbol{x}}(t)} = \underbrace{\begin{bmatrix} \boldsymbol{A}_F & \boldsymbol{B}_F\boldsymbol{C}_{FT} & \boldsymbol{0} \\ \boldsymbol{B}_T\boldsymbol{C}_{TF} & \boldsymbol{A}_T & \boldsymbol{B}_T\boldsymbol{C}_{TW}\boldsymbol{C}_{WE} \\ \boldsymbol{0} & \boldsymbol{B}_E\boldsymbol{C}_{ET} & \boldsymbol{A}_E \end{bmatrix}}_{\boldsymbol{A}(t)} \underbrace{\begin{bmatrix} \boldsymbol{x}_F \\ \boldsymbol{x}_T \\ \boldsymbol{x}_E \end{bmatrix}}_{\boldsymbol{x}(t)} + \underbrace{\begin{bmatrix} \boldsymbol{0} \\ \boldsymbol{0} \\ \boldsymbol{B}_E C_{EP} \end{bmatrix}}_{\boldsymbol{B}(t)} \underbrace{w_P}_{\boldsymbol{w}(t)} . \tag{5.26}$$

On the other hand, vertical vibrations on a randomly uneven rigid road result in state equations of the complete system based on (5.6), (5.10), (5.14), (5.15), (5.20), and (5.22),

$$\underbrace{\begin{bmatrix} \dot{\boldsymbol{x}}_F \\ \dot{\boldsymbol{x}}_T \\ \dot{\boldsymbol{x}}_R \end{bmatrix}}_{\dot{\boldsymbol{x}}(t)} = \underbrace{\begin{bmatrix} \boldsymbol{A}_F & \boldsymbol{B}_F\boldsymbol{C}_{FT} & \boldsymbol{0} \\ \boldsymbol{B}_T\boldsymbol{C}_{TF} & \boldsymbol{A}_T & \boldsymbol{B}_T\boldsymbol{C}_{TW}\boldsymbol{C}_{WR} \\ \boldsymbol{0} & \boldsymbol{0} & \boldsymbol{A}_R \end{bmatrix}}_{\boldsymbol{A}(t)} \underbrace{\begin{bmatrix} \boldsymbol{x}_F \\ \boldsymbol{x}_T \\ \boldsymbol{x}_R \end{bmatrix}}_{\boldsymbol{x}(t)} + \underbrace{\begin{bmatrix} \boldsymbol{0} \\ \boldsymbol{0} \\ \boldsymbol{B}_R \end{bmatrix}}_{\boldsymbol{B}(t)} \underbrace{\boldsymbol{w}(t)}_{\boldsymbol{w}(t)} . \tag{5.27}$$

By comparison of (5.26) and (5.27) it turns out that an elastic guideway experiences an additional coupling via the primary suspension. Both complete systems (5.26) and (5.27) are truly inhomogeneous systems of differential equations which are the basis for an analysis of the driving comfort.

In the global system matrix $\boldsymbol{A}(t)$ matrix product like $\boldsymbol{B}_F\boldsymbol{C}_{FT}$, $\boldsymbol{B}_T\boldsymbol{C}_{TF}$ and $\boldsymbol{B}_E\boldsymbol{C}_{ET}$ appear. Thus, the definition of the matrices $\boldsymbol{B}$ and $\boldsymbol{C}$ is not unique what facilitates the modeling in many cases. The dynamicist is free to choose the most suitable vectors and matrices.

The general form of the state equations of a linear model of the complete vehicle-guideway system reads as

$$\dot{\boldsymbol{x}}(t) = \boldsymbol{A}(t)\boldsymbol{x}(t) + \boldsymbol{B}(t)\boldsymbol{w}(t) \tag{5.28}$$

where $\boldsymbol{x}$ means the $n \times 1$-state vector, $\boldsymbol{w}$ the $r \times 1$-excitation vector, $\boldsymbol{A}$ the $n \times n$-system matrix and $\boldsymbol{B}$ the $n \times r$-input matrix. In many cases the matrices $\boldsymbol{A}$ and $\boldsymbol{B}$ are time-invariant so that the computations are more simple. The general form of the nonlinear state equations is given by

$$\dot{\boldsymbol{x}}(t) = \boldsymbol{a}(\boldsymbol{x}, \boldsymbol{w}, t) \tag{5.29}$$

where the $n \times 1$-vector function $\boldsymbol{a}$ can be found from linear and nonlinear subsystems correspondingly.

■ **Problem 5.1 State equations of the vertical motion of an actively controlled automobile**
The car model shown in Fig. 5.1 a) has an active suspension on the front and rear axle. This means an extension of the automobile considered in Problem 2.12. Now the subsystems vehicle, suspension devices and road, including a PD-controller, are described and assembled to a complete vehicle system.

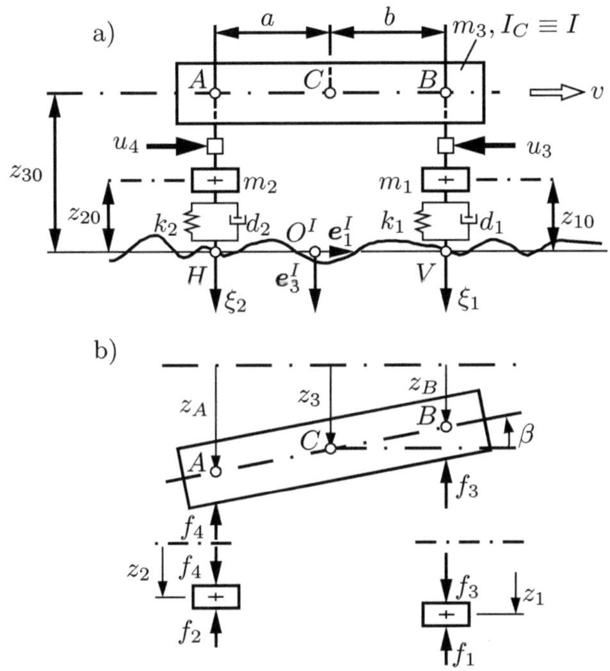

Fig. 5.1. Automobile with active suspension a) equilibrium position b) free body diagram

Solution
The subsystem vehicle consists of the car body and the wheel and axles masses which are subject to the applied forces shown in Fig. 5.1 b). For the given system parameters the state equations of the vehicle read as

$$
\underbrace{\begin{bmatrix} \dot{z}_1 \\ \dot{z}_2 \\ \dot{z}_3 \\ \dot{\beta} \\ \cdots \\ \ddot{z}_1 \\ \ddot{z}_2 \\ \ddot{z}_3 \\ \ddot{\beta} \end{bmatrix}}_{\dot{\boldsymbol{x}}_F} = \underbrace{\begin{bmatrix} \mathbf{0} & \vdots & \boldsymbol{E} \\ \cdots & & \cdots \\ \mathbf{0} & \vdots & \mathbf{0} \end{bmatrix}}_{\boldsymbol{A}_F} \underbrace{\begin{bmatrix} z_1 \\ z_2 \\ z_3 \\ \beta \\ \cdots \\ \dot{z}_1 \\ \dot{z}_2 \\ \dot{z}_3 \\ \dot{\beta} \end{bmatrix}}_{\boldsymbol{x}_F} + \underbrace{\begin{bmatrix} & & \mathbf{0} & \\ \cdots & \cdots & \cdots & \cdots \\ -\frac{1}{m_1} & 0 & \frac{1}{m_1} & 0 \\ 0 & -\frac{1}{m_2} & 0 & \frac{1}{m_2} \\ 0 & 0 & -\frac{1}{m_3} & -\frac{1}{m_3} \\ 0 & 0 & \frac{b}{I} & -\frac{a}{I} \end{bmatrix}}_{\boldsymbol{B}_F} \underbrace{\begin{bmatrix} f_1 \\ f_2 \\ f_3 \\ f_4 \end{bmatrix}}_{\boldsymbol{u}_F} . \tag{1}
$$

The suspension system is composed of the passive primary suspension with relative motions u_1, u_2, $\dot{u}_1$, $\dot{u}_2$ and the active secondary suspension with the control variables u_3 and u_4 compare (5.22). Its degenerated state equations read according to (5.10) as

$$
\underbrace{[\boldsymbol{E}]}_{\boldsymbol{A}_T} \underbrace{\begin{bmatrix} f_1 \\ f_2 \\ f_3 \\ f_4 \end{bmatrix}}_{\boldsymbol{x}_T} = \underbrace{\begin{bmatrix} k_1 & 0 & d_1 & 0 & \vdots & 0 & 0 \\ 0 & k_2 & 0 & d_2 & \vdots & 0 & 0 \\ 0 & 0 & 0 & 0 & \vdots & 1 & 0 \\ 0 & 0 & 0 & 0 & \vdots & 0 & 1 \end{bmatrix}}_{-\boldsymbol{B}_T} \underbrace{\begin{bmatrix} u_1 \\ u_2 \\ \dot{u}_1 \\ \dot{u}_2 \\ \cdots \\ u_3 \\ u_4 \end{bmatrix}}_{\boldsymbol{u}_T} . \tag{2}
$$

The automobile is running on a rigid road with a given profile $\zeta(t)$. Then, the vector of the excitation function reads as

$$
\boldsymbol{x}_W = \begin{bmatrix} \zeta_1 & \zeta_2 & \dot{\zeta}_1 & \dot{\zeta}_2 \end{bmatrix}^{\mathrm{T}} . \tag{3}
$$

The subsystems (1), (2) and (3) are assembled by the coupling relations to the complete system. To begin with, the 4×4-coupling matrix between vehicle and suspension device is

$$
\boldsymbol{C}_{FT} = \boldsymbol{E} , \tag{4}
$$

what means that the generalized forces are acting immediately from the suspension devices on the vehicle. The kinematical relations of the primary suspension and the control gains $k_5, ..., k_8, d_5, ..., d_8$ of the PD-controller are integrated in the coupling matrices between the suspension device and the vehicle as well as between the vehicle and the guideway, respectively,

$$
\boldsymbol{C}_{TF} = \left[\begin{array}{cccc:cccc}
1 & 0 & 0 & 0 & 0 & 0 & 0 & 0 \\
0 & 1 & 0 & 0 & 0 & 0 & 0 & 0 \\
0 & 0 & 0 & 0 & 1 & 0 & 0 & 0 \\
0 & 0 & 0 & 0 & 0 & 1 & 0 & 0 \\
\hdashline
-k_6 & 0 & k_5 & -bk_5 & -d_6 & 0 & d_5 & -bd_5 \\
0 & -k_8 & k_7 & ak_7 & 0 & -d_8 & d_7 & ad_7
\end{array}\right] , \tag{5}
$$

$$
\boldsymbol{C}_{TW} = \left[\begin{array}{cccc}
-1 & 0 & 0 & 0 \\
0 & -1 & 0 & 0 \\
0 & 0 & -1 & 0 \\
0 & 0 & 0 & -1 \\
\hdashline
0 & 0 & 0 & 0 \\
0 & 0 & 0 & 0
\end{array}\right] . \tag{6}
$$

Now the complete state equations of the automobile can be compiled. With (1), (5.20), (2), (5.22) and (3) it follows as discussed in Sect. 5.2 for the complete vehicle system

$$
\begin{aligned}
&\dot{\boldsymbol{x}}_F = (\boldsymbol{A}_F - \boldsymbol{B}_F\boldsymbol{B}_T\boldsymbol{C}_{TF})\boldsymbol{x}_F - \boldsymbol{B}_F\boldsymbol{B}_T\boldsymbol{C}_{TW}\boldsymbol{x}_w , \\
&\boldsymbol{x}_F(0) = \boldsymbol{x}_{F0} .
\end{aligned} \tag{7}
$$

In particular, it is considered that the matrices $\boldsymbol{A}_T$ and $\boldsymbol{C}_{FT}$ are unit matrices. For the special choice of the control gains

$$
k_5 = k_6 = k_3 , \quad k_7 = k_8 = k_4 , \quad d_5 = d_6 = d_3 , \quad d_7 = d_8 = d_4 \tag{8}
$$

one gets again the passively suspensed automobile treated on Problem 2.12. Then, the corresponding state equations follow from (7), too. ■

6

Assessment Criteria

The assessment of a vehicle's dynamical properties is related to three essential motions:

- longitudinal motion (driving and braking),
- lateral motion (guidance and steering),
- vertical motion (suspension and damping).

These three motions of a vehicle are more or less decoupled, so that the assessment criteria can be formulated separately for each motion. Of course, coupled criteria can be developed if necessary. But the more sophisticated criteria have to meet also the vehicle's rating by test drivers. Thus, the question of a mathematical formulation of assessment criteria is more difficult and can often be answered empirically only. Sometimes an interdisciplinary approach is helpful taking into account ergonomical knowledge.

The criteria to assess the longitudinal motion are known as vehicle performance comprising maximum speed, gradeability and vehicle acceleration. These criteria are self-explanatory and do not need any further consideration. However, it should be mentioned that the exact measurement of the vehicle performance in driving tests is a nontrivial problem.

The essential criterium for the lateral motion is the driving stability. The vehicle should follow safely a trajectory prescribed by the track or steering system, respectively. Thus, a stability problem is given, that can be solved applying the well developed methods of stability analysis in system dynamics.

For the vertical motion the criteria ride comfort and driving safety are essential. The wheel suspension should absorb the disturbances due to the road irregularities so that a comfortable ride is achieved. At the same time the weight of the vehicle should be transferred to the road with minor wheel vibrations so that the vertical load variations remain small avoiding lateral force reductions and driving safety deterioration.

Besides the vehicle motions the static and dynamic forces acting on the vehicle parts are important criteria, because the durability of the vehicle components depends strongly on the resulting stresses. These loads can be calculated in advance by model based numerical simulations. In the following sections the criteria driving stability, ride comfort and safety as well as some aspects of durability are considered.

6.1 Driving Stability

The lateral motion is generally characterized by linear or nonlinear state equations, respectively, see e. g. (5.25)

$$\dot{\boldsymbol{x}}(t) = \boldsymbol{A}\boldsymbol{x}(t) \tag{6.1}$$

$$\dot{\boldsymbol{x}}(t) = \boldsymbol{a}(\boldsymbol{x}, t)\ , \quad \boldsymbol{x}(t_0) = \boldsymbol{x}_0\ . \tag{6.2}$$

Therefore, the driving stability is defined as stability in the sense of Lyapunov. Then it yields: the equilibrium position $\boldsymbol{x}(t) = \boldsymbol{0}$ of the dynamical system (6.2) subject to $\boldsymbol{a}(\boldsymbol{0}, t) = \boldsymbol{0}$ is called stable in the sense of Lyapunov if for any initial time t_0 and each $\varepsilon > 0$ a positive number $\delta = \delta(\varepsilon, t_0) > 0$ exists such that for all initial conditions bounded by

$$\|\boldsymbol{x}_0\| < \delta \tag{6.3}$$

the corresponding trajectories $\boldsymbol{x}(t)$ remain bounded for all $t \geq 0$ as

$$\|\boldsymbol{x}(t)\| < \varepsilon\ , \quad t \geq t_0\ . \tag{6.4}$$

Here $\|\boldsymbol{x}(t)\|$ means the norm of the vector $\boldsymbol{x}$. An often used norm is, e.g., the Euclidian norm $\|\boldsymbol{x}(t)\| = \sqrt{\boldsymbol{x}^{\mathrm{T}}\boldsymbol{x}}$. If, in addition,

$$\lim_{t\to\infty} \boldsymbol{x}(t) = \boldsymbol{0}\ , \tag{6.5}$$

then the system is called asymptotically stable. Further, the equilibrium position $\boldsymbol{x}(t) = \boldsymbol{0}$ is called instable if it is not stable.

■ **Problem 6.1 Stability of a system of second order**
Present for a system of second order with the state vector $\boldsymbol{x} = [x_1\ x_2]^{\mathrm{T}}$ the solutions of an asymptotic stable, a stable and an instable system graphically in the state plane.

Solution
The trajectories of three different systems of second order are shown in the state plane, all of them starting from the same initial condition, Fig. 6.1. At first an ε-neighborhood of the equilibrium position located in the origin of the frame is chosen. Then, a δ-neighborhood is specified in such a way that the stable and asymptotic stable trajectories remain completed within the ε-neighborhood. This means that the δ-neighborhood may be possibly very small. The Euclidian norm serves as vector norm

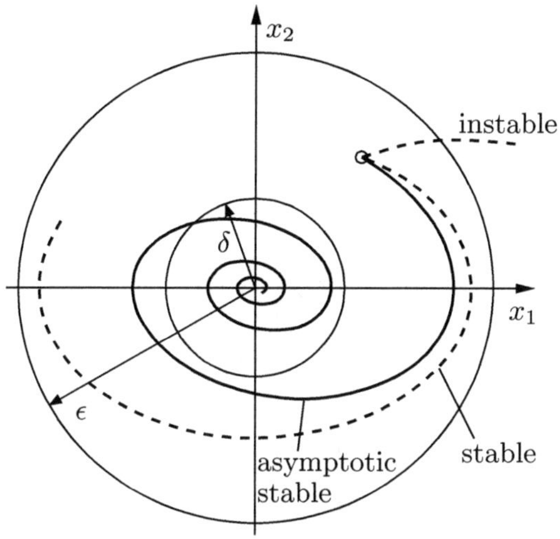

Fig. 6.1. Trajectories of a system of second order shown in the state space

$$\|\boldsymbol{x}\| = \sqrt{x_1^2 + x_2^2}\,, \tag{1}$$

where the neighborhoods are characterized by circles with the center at the origin. ∎

6.2 Ride Comfort

Ride comfort is based on the subjective human perception. Numerous experimental investigations in ergonomics have shown that the human perception of vibrations depends on the acceleration,

$$K = K(a)\,, \tag{6.6}$$

where K is a nondimensional perception measure and a is the absolute value of the acceleration in horizontal (x,y) or vertical (z) direction, respectively, see Fig. 6.2. Furthermore, the position of the human body (sitting, standing or lieing) is of importance. In ergonomics the vibration tests have been performed using deterministic, especially harmonic, excitations. In vehicles, however, the human body usually is exposed to random vibrations. Therefore, in the following the different types of excitation are considered separately.

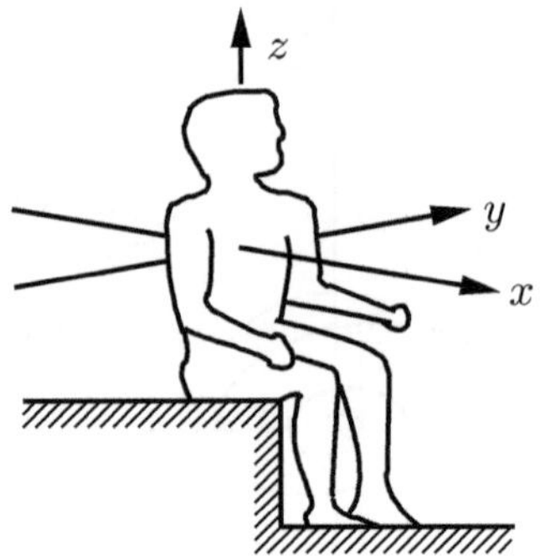

Fig. 6.2. Direction of exposure to vibration of a seated human

6.2.1 Deterministic Excitation

For a rough qualitative assessment of ride comfort during deterministic excitation it yields

$$K \sim a_{max} \ , \tag{6.7}$$

i.e. the maximum acceleration value is a first perception measure. As a thumb rule $a_{max} \leq 0.5$ m/s^2 results in a good ride comfort. More refined relations between K and a are given in international or national standards that are based on extensive ergonomical investigations. The ISO International Standard 2631 (1974) or VDI - Richtlinie 2057 (1979), respectively, are better qualified for the understanding of the assessment problem and will be used firstly. Then, the revised standards published more recently are discussed.

For harmonic excitations the aforementioned standards give a precise relation that suits well for vehicle dynamics. Starting from the harmonic acceleration

$$a(t) = A \sin \omega t \ , \quad \omega = 2\pi f \ , \tag{6.8}$$

with amplitude A[m/s^2] and frequency f [Hz] the root mean square (rms) value $a_{\rm rms}$ of the acceleration can be determined,

$$a_{rms} = \sqrt{\frac{1}{T}\int_0^T a^2(t)\,\mathrm{d}t} = \frac{A}{\sqrt{2}}\,[\mathrm{m/s^2}] \ , \tag{6.9}$$

where the unit m/s^2 has to be used. Then the perception measure K reads for vertical excitation

$$\begin{aligned} K &= 10 a_{rms}\sqrt{f} \ , & 1 \leq f \leq 4 \ , \\ K &= 20 a_{rms} \ , & 4 \leq f \leq 8 \ , \\ K &= 160 a_{rms} f^{-1} \ , & 8 \leq f \leq 80 \ , \end{aligned} \tag{6.10}$$

and for horizontal excitation

$$\begin{aligned} K &= 28a_{rms} \,, & 1 \le f \le 2 \,, \\ K &= 56a_{rms} f^{-1} \,, & 2 \le f \le 80 \,. \end{aligned} \tag{6.11}$$

Equations (6.10) and (6.11) can be interpreted as experimentally determined frequency response functions of the human perception, they are depicted in Fig. 6.3.

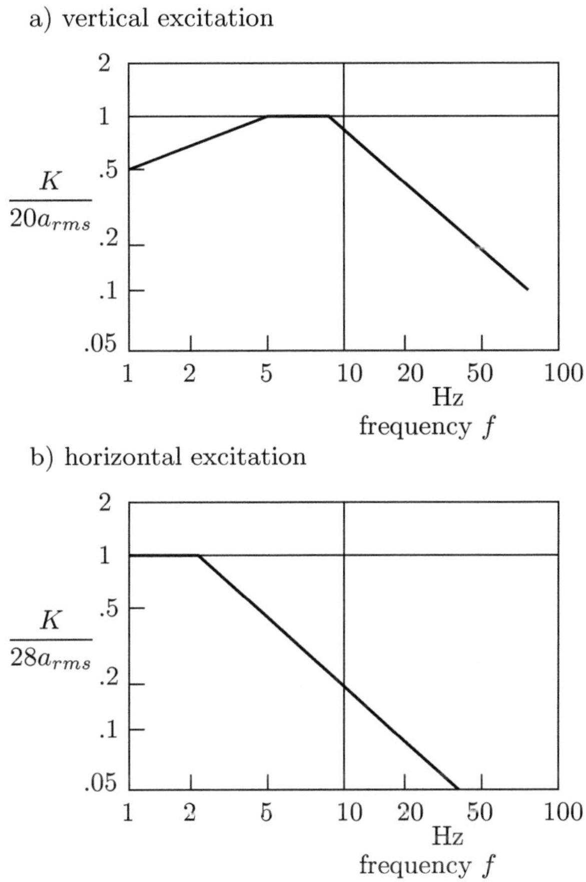

Fig. 6.3. Frequency responses of the perception measure: a) vertical exposure; b) horizontal exposure for original standards (1974 or 1975-1979)

From Figure 6.3 a) it can be seen that a vertical excitation between 4 and 8 Hz is perceived as very unpleasant, because in this frequency range the resonance of the human stomach occurs. The perception K is a nondimensional measure characterizing the subjective impression of test persons related to the criteria well-being and comfort, efficiency of labor and impairment of health, respectively. Here, the exposure time plays an essential role. The original standards ISO 2631 and VDI 2057, respectively, describe the relation between exposure time and the mentioned criteria, see Fig. 6.4. The perception measure required for road vehicles is about

$$2 < K < 10 \,. \tag{6.12}$$

Subclasses of the perception can be defined and the corresponding vibration levels are described as:

$$\begin{aligned} &C1\,/\,C2 \text{ noticeable} \,, \\ &D1\,/\,D2 \text{ strongly noticeable} \,, \\ &E1\,/\,E4 \text{ very strongly noticeable} \,. \end{aligned} \tag{6.13}$$

Obviously, the description of perception shows some uncertainties, due to the subjective human nature.

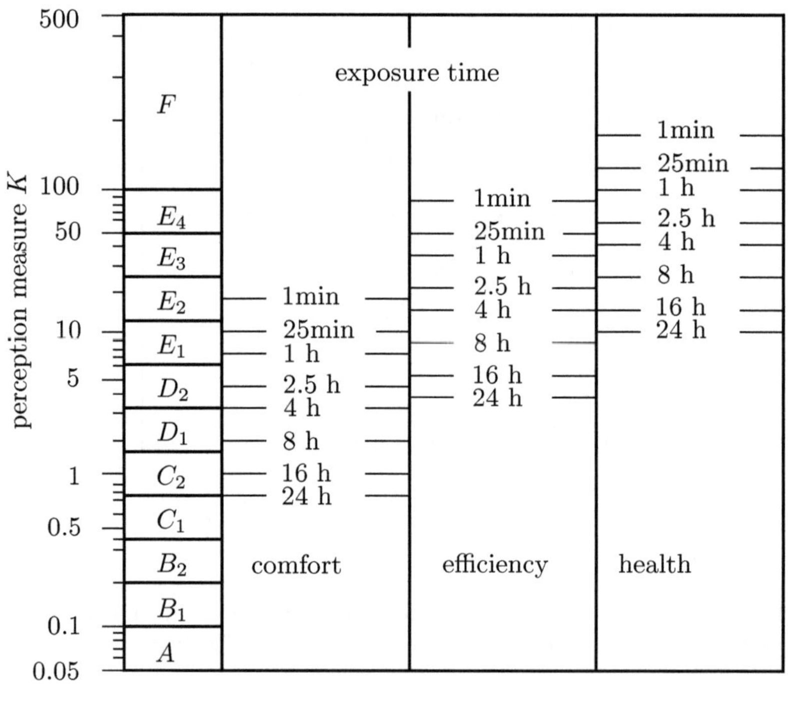

Fig. 6.4. Perception measure and exposure time of vibrations

6.2.2 Stochastic Excitation

Experimental results for a stochastic excitation of test persons are limited. However, they confirm the interpretation of the subjective human perception as response of a vibration system to stochastic excitation, too, and the possibility to describe the perception using methods of linear system theory. The rms-value of a Gaussian process $a(t)$ is equivalent to its standard deviation σ_a,

$$a_{\mathrm{rms}} = \sigma_a = \sqrt{\int_0^\infty \Phi_a(\omega)\,\mathrm{d}\omega}\ , \tag{6.14}$$

where $\Phi_a(\omega)$ is the single sided power spectral density (PSD) of the process under consideration. The standard deviation is a scalar parameter characterizing globally the stochastic process, however, it is frequency independent. Thus, the weighting of frequencies and the calculation of the rms-value have to be interchanged what is allowed for linear stochastic systems. This results in the variance $\sigma_{\overline{a}}^2$ of the weighted process $\overline{a}(t)$,

$$\sigma_{\overline{a}}^2 = \int_0^\infty \alpha^2 |F(\omega)|^2 \Phi_a(\omega)\,\mathrm{d}\omega\ . \tag{6.15}$$

Here, α is a dimensional factor, $F(\omega)$ is the frequency response function of a weighting filter and $\Phi_a(\omega)$ is the single sided PSD of the mechanical acceleration process $a(t)$. The still unknown quantities of the frequency weighting in (6.15) are found by comparison with the results for deterministic excitation. For vertical excitation the dimensional factor α reads

$$\alpha = 20\ \mathrm{s}^2/\mathrm{m} \tag{6.16}$$

and the frequency response function is given by

$$|F(\omega)| = \frac{1}{20} K(a_{rms}, f)\ , \quad f = \omega/2\pi\ , \tag{6.17}$$

where the perception measure K according to (6.10) is used.

The frequency response function (6.17) can be well approximated by a linear weighting filter or shape filter. This yields in the frequency domain

$$\begin{aligned} F(\omega) &= \overline{\boldsymbol{h}}^{\mathrm{T}} (i\omega \boldsymbol{E} - \overline{\boldsymbol{F}})^{-1} \overline{\boldsymbol{g}} \\ &= \frac{b_0 + b_1(i\omega) + \ldots + b_r(i\omega)^r}{a_0 + a_1(i\omega) + \ldots + a_{s-1}(i\omega)^{s-1} + (i\omega)^s}\ . \end{aligned} \tag{6.18}$$

Here, $\overline{\boldsymbol{h}}$ and $\overline{\boldsymbol{g}}$ are $s \times 1$-vectors and $\overline{\boldsymbol{F}}$ is a $s \times s$-matrix characterizing the shape filter. Furthermore, $a_i, i = 0(1)s-1$, and $b_j, j = 0(1)r \leq s$ are coefficient of the shape-filter response function. The following relation holds,

$$\overline{\boldsymbol{F}} = \begin{bmatrix} 0 & 1 & 0 & \cdots & 0 \\ 0 & 0 & 1 & \cdots & 0 \\ \vdots & \vdots & \vdots & \ddots & \vdots \\ 0 & 0 & 0 & & 1 \\ -a_0 & -a_1 & -a_2 & \cdots & -a_{s-1} \end{bmatrix}, \quad \overline{\boldsymbol{g}} = \begin{bmatrix} 0 \\ 0 \\ \vdots \\ 0 \\ 1 \end{bmatrix}, \quad \overline{\boldsymbol{h}}^{\mathrm{T}} = [b_0 b_1 \ldots b_r \ldots 0] , \tag{6.19}$$

i.e. the system matrix $\overline{\boldsymbol{F}}$ is chosen as Frobenius matrix. In time domain the shape filter can be described equivalently by differential equations,

$$\begin{aligned} \overline{a}(t) &= \alpha \overline{\boldsymbol{h}}^{\mathrm{T}} \overline{\boldsymbol{v}}(t) , \\ \dot{\overline{\boldsymbol{v}}}(t) &= \overline{\boldsymbol{F}} \overline{\boldsymbol{v}}(t) + \boldsymbol{g} a(t) . \end{aligned} \tag{6.20}$$

Here, $\overline{\boldsymbol{v}}$ is the $s \times 1$-state vector of the shape filter that is excited by the acceleration $a(t)$. The scalar product of the $s \times 1$-vectors $\overline{\boldsymbol{h}}$ and $\overline{\boldsymbol{v}}(t)$ results in the frequency weighted scalar acceleration $\overline{a}(t)$. The coefficients of the shape filter have to be determined in such a way that the given frequency response function (6.17) is approximated sufficiently well. This can always be achieved by choosing the order s of the filter sufficiently large.

6.2.3 Shape Filter for the Human Perception

Now a second order shape filter is determined so that the frequency response function (6.10) for the human perception of vibration is approximated well.

The shape filter of second order, $s = 2$, has the following form according to (6.19) and (6.20):

$$\overline{a} = \alpha [b_0 \ b_1] \begin{bmatrix} \overline{v}_1 \\ \overline{v}_2 \end{bmatrix} , \tag{6.21}$$

$$\begin{bmatrix} \dot{\overline{v}}_1 \\ \dot{\overline{v}}_2 \end{bmatrix} = \begin{bmatrix} 0 & 1 \\ -a_0 & -a_1 \end{bmatrix} \begin{bmatrix} \overline{v}_1 \\ \overline{v}_2 \end{bmatrix} + \begin{bmatrix} 0 \\ 1 \end{bmatrix} a(t) . \tag{6.22}$$

Choosing the numbers

$$\begin{aligned} a_0 &= 1200 \ \mathrm{s}^{-2}, & b_0 &= 500 \ \mathrm{s}^{-2}, \\ a_1 &= 50 \ \mathrm{s}^{-1}, & b_1 &= 50 \ \mathrm{s}^{-1}, \\ \alpha &= 20 \ \mathrm{s}^2\mathrm{m}^{-1}. \end{aligned}$$

in the case of vertical excitation results in the frequency response function depicted in Fig. 6.5. It turns out, that already a second order shape filter approximates the standard ISO 2631 (1974) surprisingly well. The deviations remain completely within the tolerances allowed by the standard.

Similarly, a second order shape filter regarding the frequency weighting (6.11) for the horizontal excitation can be determined using the numbers

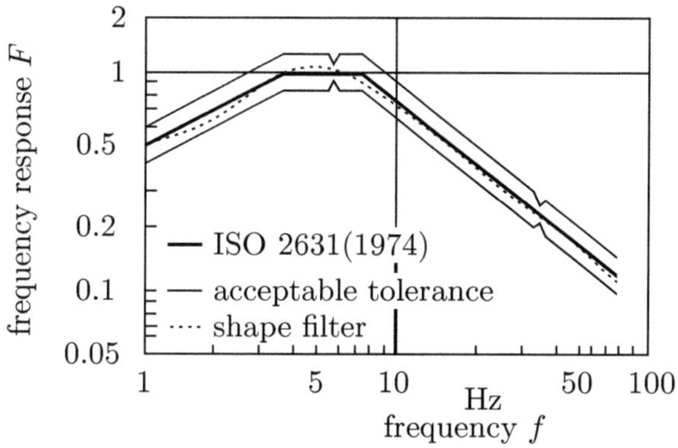

Fig. 6.5. Shape filter frequency response of perception in vertical direction

$$a_0 = 75 \text{ s}^{-2}, \quad b_0 = 31.25 \text{ s}^{-2},$$
$$a_1 = 12.5 \text{ s}^{-1}, \quad b_1 = 12.5 \text{ s}^{-1},$$
$$\alpha = 28 \text{ s}^2\text{m}^{-1}.$$

Thus, the computation of the perception measure K results simply in

$$K = \sigma_{\overline{a}} , \tag{6.23}$$

that includes the frequency weighting automatically. Using the K-value from (6.23), the exposure time admissible for the criteria comfort, efficiency and health follows from Fig. 6.4.

6.2.4 Revised Standards for Human Exposure to Whole-body Vibration

The revised standards, ISO 2631 (2004) and VDI 2057 (2002) use a slightly changed perception measure, see Fig. 6.6. Moreover, the exposure time is no longer included in the standards. Instead, it is assumed that the perception is proportional to the energy involved. Then, it yields for the variances of the frequency-weighted acceleration $\sigma^2_{\overline{a}}$ and the energy-equivalent acceleration $\sigma^2_{\overline{a}e}$ during exposure time T_e the relation

$$\sigma^2_{\overline{a}} T_0 = \sigma^2_{\overline{a}e} T_e \tag{6.24}$$

where the assessment period T_0 is usually 8 hours and $T_e < T_0$ is the exposure time which may be composed by several segments. The perception is described immediately with respect to the root mean square of the frequency weighted accelerations $\sigma_{\overline{a}}$ as shown in Table 6.1.

The shape filter of the human perception can be easily adapted to the revised perception measure. Then, the coefficients of (6.18) read as

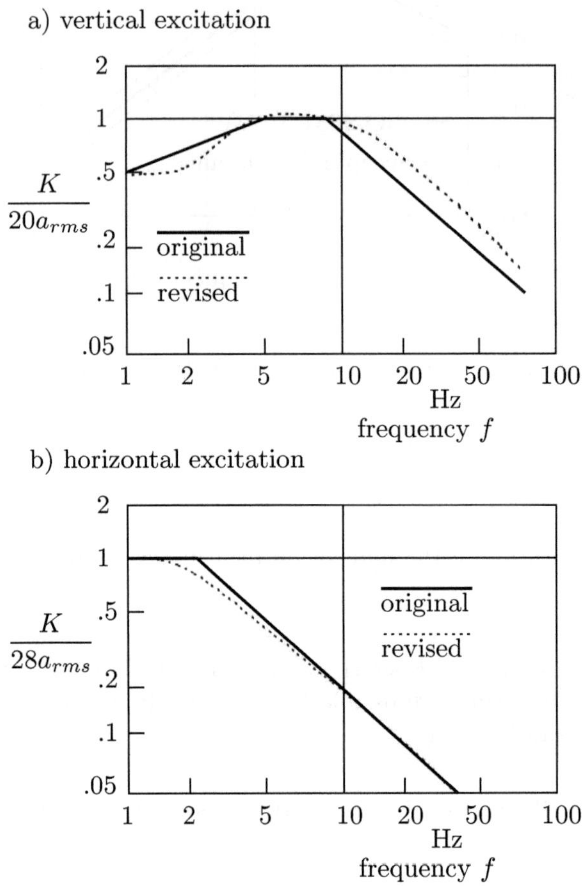

Fig. 6.6. Frequency response of perception measure K: original and revised standards.

$$a_0 = 2170\,\mathrm{s}^{-2}\,, \quad a_1 = 77.5\,\mathrm{s}^{-1}\,, \quad b_0 = 875\,\mathrm{s}^{-2}\,, \quad b_1 = 82\,\mathrm{s}^{-1}$$

as shown by Rill (2007).

6.3 Ride Safety

Ride safety is related to the longitudinal braking forces and the lateral guidance forces of a vehicle. The tangential contact forces between tire and road

Table 6.1. Subjective perception and frequency-weighted acceleration

Root-mean-square $\sigma_{\overline{a}e}$ of frequency-weighted acceleration $\overline{a}(t)$ [m/s^2]	Description of perception
< 0.01	not perceptible
$0.01 < \sigma_{\overline{a}e} < 0.02$	barely perceptible
$0.02 < \sigma_{\overline{a}e} < 0.08$	easily perceptible
$0.08 < \sigma_{\overline{a}e} < 0.32$	strongly perceptible
$0.32 < \sigma_{\overline{a}e}$	extremely perceptible

are depending directly on the normal forces in the contact area. Hence, the remaining minimum wheel load is a criterion for ride safety.

The total wheel load f_{Wheel} can be divided into a static load f_{Stat} due to the vehicle weight and a critical dynamic load $f_{Dyn,crit}$ generated by vehicle vibrations,

$$f_{Wheel}(t) = f_{Stat} + f_{Dyn,crit}(t) \,, \tag{6.25}$$

and visualized in Fig. 6.7.

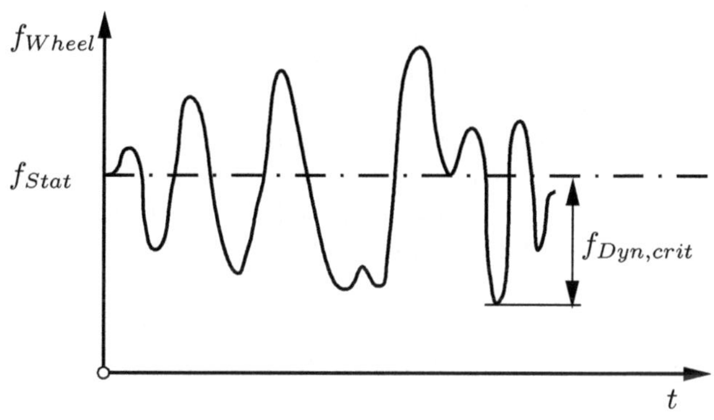

Fig. 6.7. Statical and dynamical wheel load

The remaining minimum total wheel load reads

$$f_{Wheel,min} = f_{Stat} - f_{Dyn,crit} \,. \tag{6.26}$$

This defines the safety margin R,

$$R = \frac{f_{Stat} - f_{Dyn,crit}}{f_{Stat}} \,, \tag{6.27}$$

a scalar quantity varying in the range of $0 \leq R \leq 1$. The safety margin for a vehicle in standstill is $R = 1$, it can drop for a fast moving vehicle on a rough road down to $R = 0$.

For roads with a randomly disturbed surface the dynamic wheel load variations can be determined from numerical simulations. For an assessment of the ride safety the standard deviation σ_f of the vertical wheel load variations may be used. This yields

$$R^* = \frac{f_{Stat} - \sigma_f}{f_{Stat}}, \tag{6.28}$$

where R^* has to be considered as a statistical quantity.

The computation of the dynamical wheel load variations, i.e. $f_{Dyn,crit}$ or σ_f, respectively, is based on the state and input quantities of the complete vehicle system. Depending on the tire model also the relative displacements and velocities have to be known. Hence, the following general expressions can be given,

$$f_{Dyn} = \boldsymbol{b}^{\mathrm{T}}\boldsymbol{x} \quad \text{or} \quad \sigma_f^2 = \boldsymbol{b}^{\mathrm{T}}\boldsymbol{P}_x\boldsymbol{b} \tag{6.29}$$

where $\boldsymbol{b}$ denotes a $n \times 1$-vector of coefficients, $\boldsymbol{x}(t)$ the $n \times 1$-state vector and $\boldsymbol{P}_x$ the corresponding $n \times n$-covariance matrix.

For railway vehicles the driving safety is closely related to the derailment safety. According to the many reasons for derailments different safety boundaries have been defined, see Krugmann (1982) and Brabie (2007). A primary reason for derailment is the climbing up of the wheel flange during curving. This risk can be avoided if the ratio of the acting lateral and vertical forces in the contact point does not exceed certain boundaries. In this case the dynamical wheel loads play also a central role.

For trucks the driving safety may be also affected by the motion of sloshing cargo. In particular, granular and fluid materials are excited by the vehicle's manoeuvres as shown in Fig. 6.8.

Fleissner, Lehnart and Eberhard (2009) introduced a new method for the dynamic simulation of tank trucks. The approach couples Multibody Systems methods (MBS) and Lagrangian particle methods such as the Discrete Element Method (DEM) and Smoothed Particles (SPH) using co-simulation. For the simulation of both, sloshing cargo with free surfaces and granular materials in tank trucks, the co-simulation approach couples a Lagrangian framework for the 3D simulation of granular materials and fluid models implemented in Pasimodo (2009) with the commercial multibody system simulation software Simpack (2009). Simulations can be used to investigate the impact of different tank designs on the stability of the silo vehicle system. Comparisons between two different tank designs showed the positive effect of a subdivision of the tank into compartments in terms of braking stability. Moreover, simulations show that the lateral motion of a sloshing cargo can be beneficial in terms of rolling stability in lane change maneuvers.

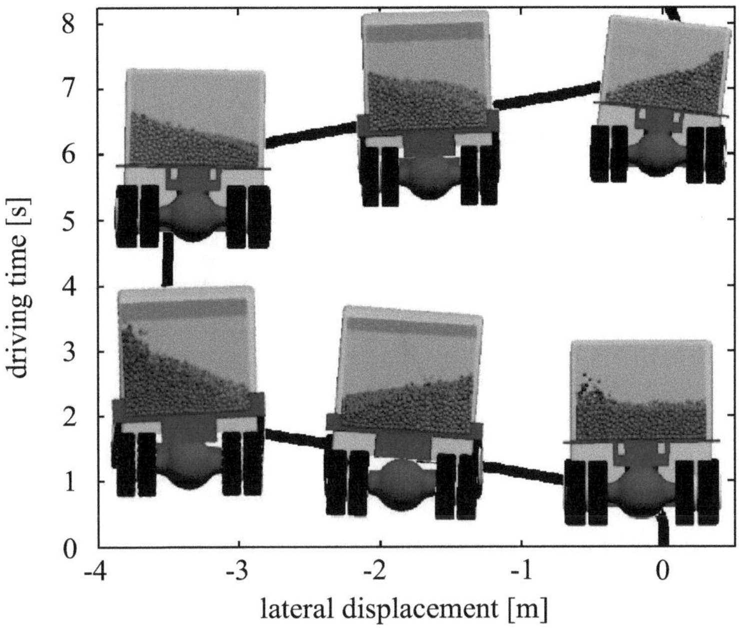

Fig. 6.8. Sloshing motion (lateral displacement) of granular cargo during a double lane change manuever with a speed of 20 m/s^2, see Fleissner et al. (2009)

6.4 Durability of Components

The assessment criteria discussed have been related to the motion of the vehicle, they are concerned with the passengers of the vehicle itself. For the design of a vehicle additional information on the loads acting on the components for proper dimensioning is required. Since vehicles running on roads or tracks, respectively, are subject to high dynamical loads, for the design of the components the dynamical stresses are most important. Thus dimensioning is only possible with respect to the durability. The dynamical analysis of vehicles may provide data for the durability design of vehicle components, too.

For the dimensioning of a suspension control arm, Fig. 6.9 the forces $f_1(t)$, $f_2(t)$ and $f_3(t)$ and the torque $l_3(t)$ acting on the joints are essential. These quantities are reaction forces within the vehicle and do not occur in the equations of motion. Therefore it is necessary to use the Newton-Euler equations (2.119) which still include the reactions. An alternative is offered by the evaluation of all reactions of the system according to (2.123). However, the computational cost is high. A computationally less expensive method for the evaluation of the reaction forces was presented by Schramm (1986) by treating the joints independently from each other.

The reaction forces and torques are the basis for the stress evaluation of the component under consideration by finite element approaches. A railway

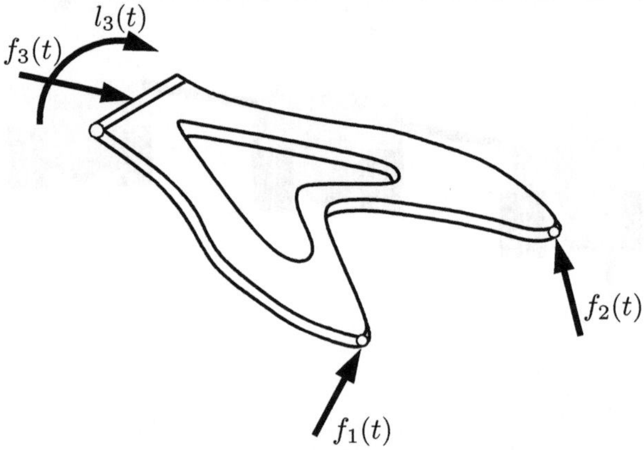

Fig. 6.9. Dynamical loads acting on a suspension control arm

bogie stress analysis was presented by Claus (2004) the theory of which is described in Claus and Schiehlen (2002) Finally, with the stress dynamics characterizing the cycles of the stress, the durability can be estimated see, e.g., Melzer (1994).

In addition to the durability the wear of railway wheelset is an important problem where short-term dynamics and long-term wear have to be considered in a feedback loop, see Meinders and Meinke (2003).

7

Computational Methods

The global vehicle system is completely described by the state equations (5.28) or (5.29), respectively. The state equations represent immediately the basis for the assessment of the driving stability, too, see (6.1) and (6.2). For the assessment of the driving comfort, in contrary, additional dynamical effects have to be considered as described by the shape filter (6.20). Thus, the state equations of the vehicle system have to be supplemented by the shape filter prior to the computational analysis.

The input of the shape filter is the acceleration $a(t)$ at some seat position, a quantity composed of the states and excitation variables of the entire vehicle system,

$$a(t) = \boldsymbol{c}^{\mathrm{T}}\boldsymbol{x}(t) + \boldsymbol{d}^{\mathrm{T}}\boldsymbol{w}(t)\,, \tag{7.1}$$

where $\boldsymbol{c}$ is a weighting vector of the states and $\boldsymbol{d}$ a weighting vector of the excitations. Thus, the state equations (5.28) extended by the shape filter (6.20) read

$$\begin{aligned}\underbrace{\begin{bmatrix}\dot{\boldsymbol{x}}\\ \dot{\boldsymbol{v}}\end{bmatrix}}_{\dot{\tilde{\boldsymbol{x}}}} &= \underbrace{\begin{bmatrix}\boldsymbol{A} & \boldsymbol{0}\\ g\boldsymbol{c}^{\mathrm{T}} & \boldsymbol{F}\end{bmatrix}}_{\tilde{\boldsymbol{A}}}\underbrace{\begin{bmatrix}\boldsymbol{x}\\ \boldsymbol{v}\end{bmatrix}}_{\tilde{\boldsymbol{x}}} + \underbrace{\begin{bmatrix}\boldsymbol{B}\\ g\boldsymbol{d}^{\mathrm{T}}\end{bmatrix}}_{\tilde{\boldsymbol{B}}}\boldsymbol{w}\,,\end{aligned} \tag{7.2}$$

The extended system (7.2) has exactly the same structure as the vehicle system (5.28), so that system (5.28) is considered subsequently without loss of generality. The required computational methods can be subdivided into numerical and analytical methods that are described in the following sections.

7.1 Numerical Simulation

The given set of linear or nonlinear differential equations (5.28) or (5.29), respectively, can be analyzed by numerical time integration. For random excitations a single realization is sufficient due to ergodicity. In the case of nonlinear systems the random response in general is not normal distributed any more. Even so, in many cases one computes the first two moments only for assessment purposes. However, the effort of numerical simulations is always large due to the complexity of the vehicle system models. Thus, the proper choice of the integration methods is very important. Nevertheless, it is not possible to give general recommendations since on the one hand new integration methods are developed by numerical mathematicians, on the other hand the performance of the computers is increasing continuously. Often the engineer applies an integration methods at hand and performs test runs for comparison, e.g., by using just Matlab.

A theoretical review of time integration methods can be found in Hairer et al. (2008). The numerical time integration for the simulation of automobile motions is considered by Rill (1994).

7.1.1 Simulation of Vertical Motions of Vehicles

Different integration methods are applied to the simulation of the vertical motion of the front and rear axle of a nonlinear 16 degrees of freedom (16-DOF) vehicle model under a ramp excitation. Furthermore, a vehicle with 5 degrees of freedom (5-DOF) in vertical direction is considered as a test example for a random excitation.

The ramp is crossed first by the front axle and then by the rear axle, this results in a time delay in the corresponding vibration responses, see Fig. 7.1. According to Rill (1981), the following time integration codes are compared:

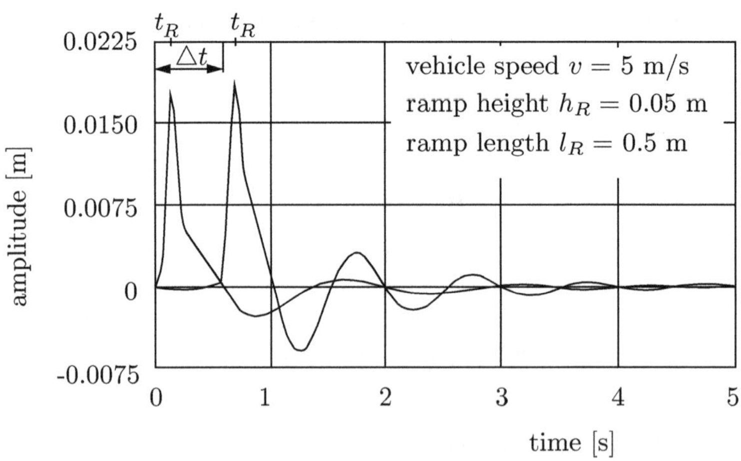

Fig. 7.1. Vertical response of front and rear wheels of a 16-DOF vehicle model after ramp excitation

- Single step methods, Runge-Kutta-Fehlberg procedure of order 5 and 6 (RKF5, RKF6),
- Predictor-corrector multi-step methods, Shampine-Gordon procedure (SGDE),
- Extrapolation methods, Stoer-Bulirsch procedure (SB).

The results of the large vehicle model are found in Fig. 7.2 where the maximal deviation in the state variables from the reference solution is shown. It can be seen that the Shampine-Gordon procedure (Shampine and Gordon, 1984) leads to the best results with respect to computing time and accuracy.

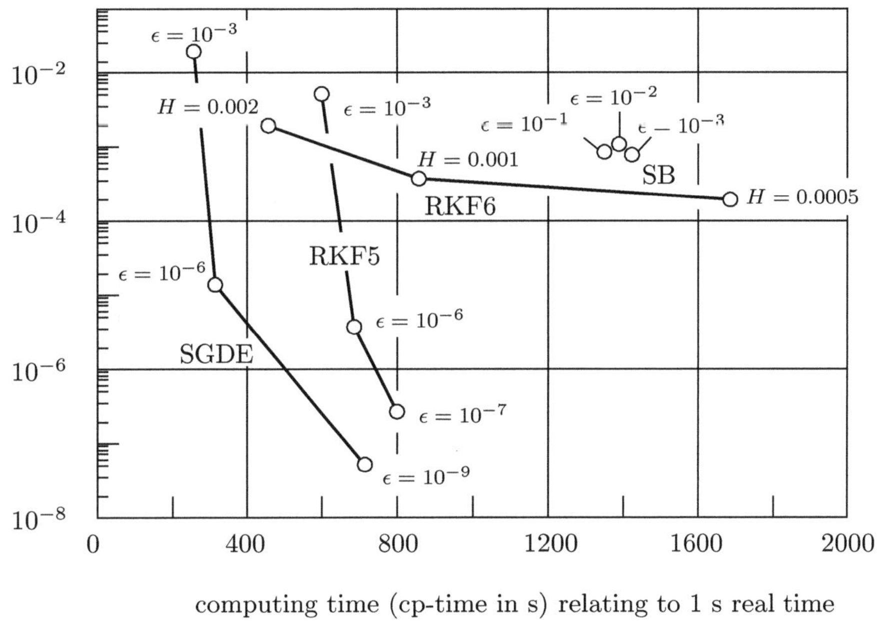

Fig. 7.2. Comparison of time integration codes for a 16-DOF vehicle model after ramp excitation

The random excitation of the small vehicle model is generated by a random superposition of harmonic functions characterizing the road roughness profile, see Fig. 7.3 and Table 7.1. The vertical motion of the front axle, see Fig. 7.4, follows the road profile, whereas the high frequency components of the excitation are no longer visible. Nevertheless, the integration procedure must take into account the high frequencies. Applying the procedures mentioned above, lead to the results presented in Fig. 7.5 where the maximal deviation in the state variables from the reference solution is shown, too. Here, the simple

Runge-Kutta-Fehlberg procedures are superior due to the fast evaluation of the right hand side of the differential equations.

The numerical integration procedures result in time histories that have to be evaluated in order to assess ride comfort and safety. In the case of random excitation the computation of the performance criteria requires time averaging what can be easily executed during simulation by an additional time integration.

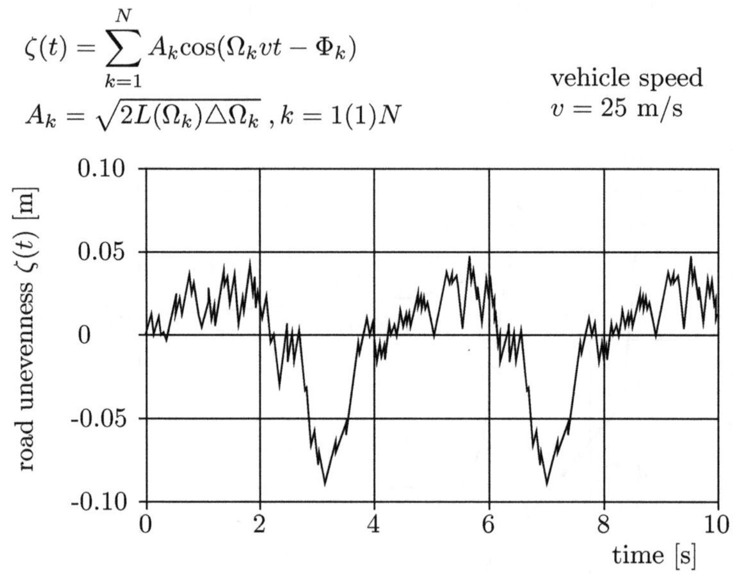

Fig. 7.3. Realization of a random process of road roughness

Table 7.1. Approximation of random excitation $N = 28$

k	$1 - 10$	$11 - 19$	$19 - 28$
$\Delta\Omega_k$ [rad/m]	0.0628	0.628	6.28
Ω_k [rad/m]	$k\Delta\Omega_k$	$(k-9)\,\Delta\Omega_k$	$(k-8)\,\Delta\Omega_k$
Φ_k	Uniformly distributed random number $0 \leq \Phi_k \leq 2\pi$		
$L\,(\Phi_k)$	A_S/Ω_k^2,	$A_S = 3{\cdot}10^{-5}$ m	

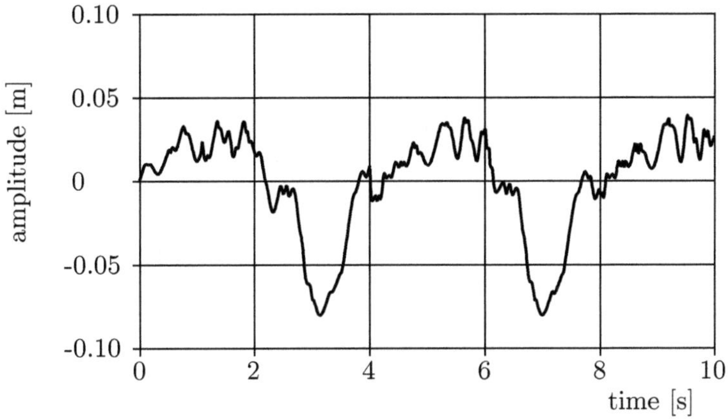

Fig. 7.4. Vertical response of the front wheels of a nonlinear 5-DOF vehicle model during random excitation

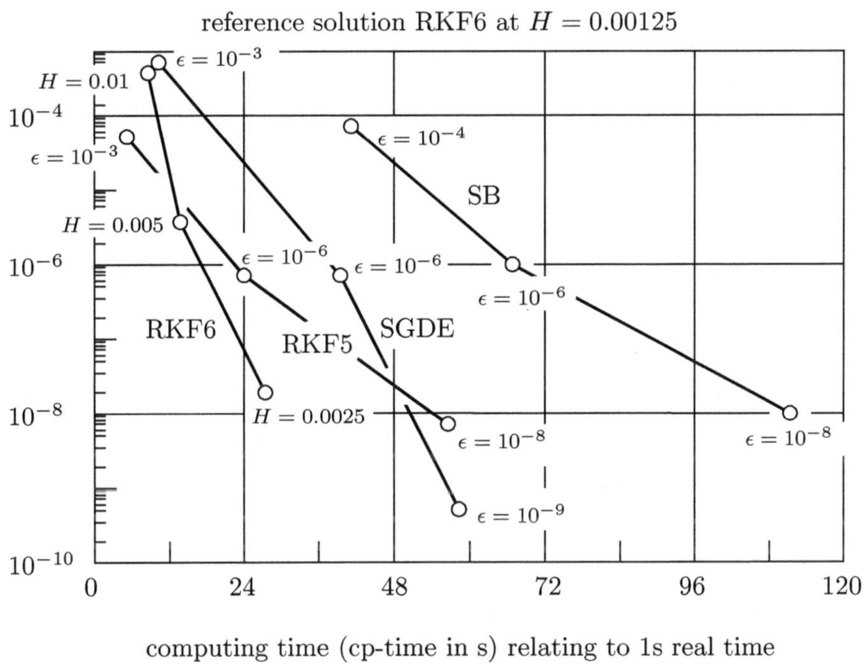

Fig. 7.5. Comparison of time integration codes for a 5-DOF vehicle model during random excitation

The high computation times shown in Fig. 7.2 and Fig. 7.5 are due to the computer technology in the early 1980s. But for the comparison this does not matter.

In a more recent study by Rill and Schiehlen (2009) the efficiency and accuracy of Matlab time integration codes are compared. As benchmark the vertical motion of a 8-DOF vehicle is used with bump and random excitation. Matlab offers seven solvers for technical computing.

ode23 is a one-step solver. Based on an explicit Runge-Kutta (2,3) pair of Bogacki and Shampine. It may be more efficient than *ode45* at crude tolerances and in the presence of mild stiffness.

ode45 is a one-step solver. Based on an explicit Runge-Kutta (4,5) formula, the Dormand-Prince pair. It is a one-step solver - in computing, it needs only the solution at the immediately preceding time point. In general, *ode45* is the best function to apply as a "first try" for most problems.

ode113 Variable order Adams-Bashforth-Moulton PECE solver. It may be more efficient than *ode45* at stringent tolerances and when the ODE function is particularly expensive to evaluate. *ode113* is a multistep solver - it normally needs the solutions at several preceding time points to compute the current solution.

ode15s Variable-order solver based on the numerical differentiation formulas (NDFs). Optionally it uses the backward differentiation formulas, BDFs, (also known as Gear's method). Like ode113, ode15s is a multistep solver. If you suspect that a problem is stiff or if *ode45* failed or was very inefficient, try *ode15s.*

ode23s Based on a modified Rosenbrock formula of order 2. Because it is a one-step solver, it may be more efficient than *ode15s* at crude tolerances. It can solve some kinds of stiff problems for which ode15s is not effective.

ode23t An implementation of the trapezoidal rule using a "free" interpolant. Use this solver if the problem is only moderately stiff and you need a solution without numerical damping.

ode23tb An implementation of TR-BDF2, an implicit Runge-Kutta formula with a first stage that is a trapezoidal rule step and a second stage that is a backward differentiation formula of order 2. Like *ode23s,* this solver may be more efficient than *ode15s* at crude tolerances. For the reference a partially implicit Euler solver is used considering the equations of motion.

As reference a partially implicit Euler solver is used considering the equations of motion (2.96) where $\dot{\boldsymbol{y}}(t)$ is replaced by $\boldsymbol{v}(t)$. With the explicit step

$$\boldsymbol{v}(t+h) = \boldsymbol{v}(t) + h\boldsymbol{M}\left(\boldsymbol{y}\left(t\right)\right)^{-1}\left(\boldsymbol{q}\left(\boldsymbol{y}\left(t\right), \boldsymbol{v}\left(t\right), t\right) - \boldsymbol{k}\left(\boldsymbol{y}\left(t\right), \boldsymbol{v}\left(t\right)\right)\right) , \quad (7.3)$$

the new velocity $\mathbf{v}\left(t+h\right)$ is found where h is the step-size. Further, an implicit step yields the new position

$$\boldsymbol{y}(t+h) = \boldsymbol{y}(t) + h\boldsymbol{v}\,(t+h)\,. \tag{7.4}$$

The partial implicit Euler code is denoted as *ode1m*. The results for step-sizes of $h = 1$ ms and $h = 0.1$ ms are used for the comparison with the Matlab time integration results.

All Matlab solvers use automatic step-size control, i.e., the simulation results are computed with different time intervals. Therefore, the computational efficiency is checked by integral criteria as it is standard in vehicle dynamics for the assessment of the dynamical properties driving comfort and driving safety using root mean square (rms) values or effective (eff) values, respectively,

$$w^{eff} = \sqrt{\frac{1}{T}\int_0^T w^2\,(t)\mathrm{d}t}. \tag{7.5}$$

The computation of this time integral is executed by the solver used for the simulation, too. The criteria chosen are the computation time related to *ode1m* with step-size $h = 1$ ms, the driver's acceleration related to the gravity acceleration of 9.81 m/s^2 and the dynamic wheel loads related to the static wheel loads. The results are summarized in Fig. ??. It turns out that all Matlab codes may meet the required accuracy due to automatic step-size control. However, there are major differences with respect to the computational efficiency. In particular, the implicit codes *ode15s*, *ode23s*, *ode23t*, and *ode23tb* consume extremely large computation times. Most favorable was the explicit code *ode45* based on an explicit Runge-Kutta (4,5) formula. On the other hand, for low accuracies and/or real time simulations the partially implicit Euler code *ode1m* is an interesting alternative.

7.2 Linear Systems

For the analysis of linear vehicle models the efficient tools of the linear vibration and system theory can be used. Then, a numerical simulation is generally not really required.

7.2.1 Stability

Many established methods are available for the stability analysis of linear systems, see e.g. Mueller and Schiehlen (1985). In addition to the eigenvalue criteria there may be used for simple models the criteria based on the characteristic equation (Hurwitz, Routh) and for purely mechanical systems the criteria based on the parameter matrices of the related differential equations of the second order (Thomson and Tait, P.C. Mueller). In vehicle dynamics, however, the mathematical models or differential equations, respectively, are often more complex, and a numerical solution of the eigenvalue problem has to be performed.

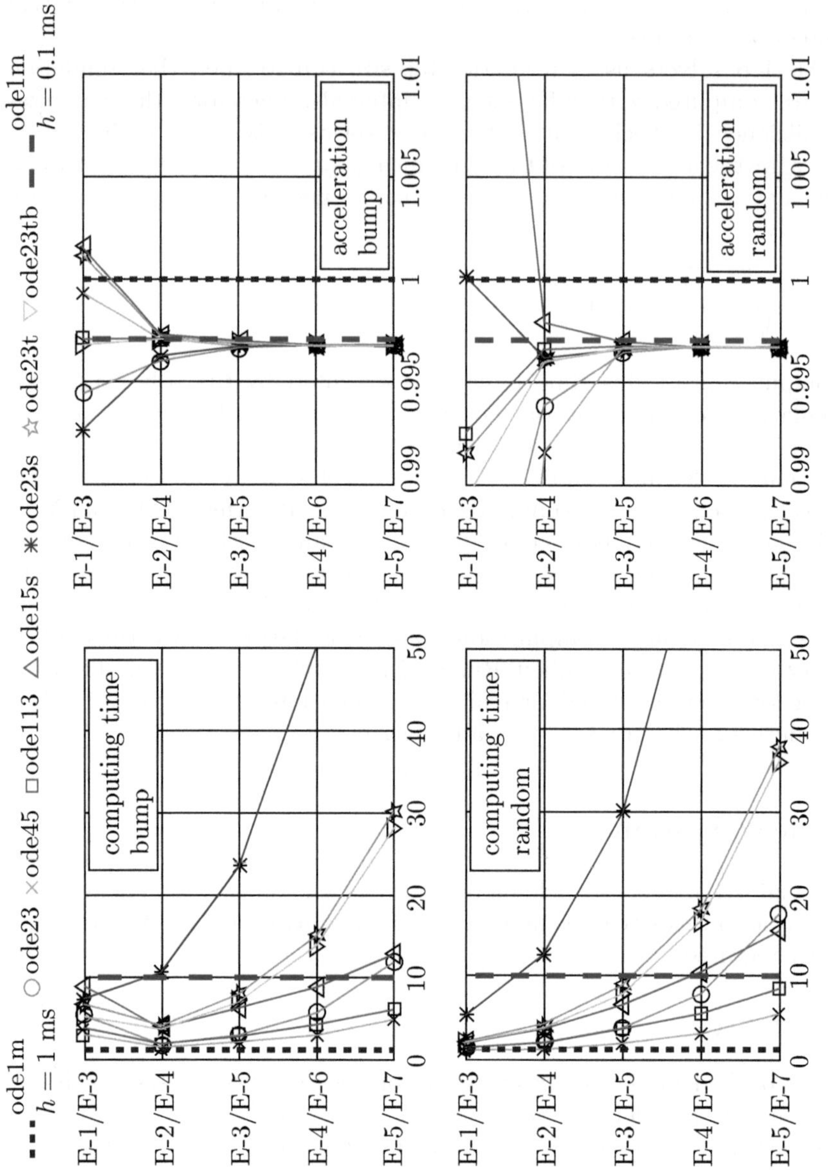

Fig. 7.6. Code comparison for bump and random excitation

The linear system

$$\dot{\boldsymbol{x}} = \boldsymbol{A}\boldsymbol{x}\,, \quad \boldsymbol{x}(0) = \boldsymbol{x}_0 \tag{7.6}$$

with the time-invariant $n \times n$-system matrix $\boldsymbol{A}$ is asymptotically stable if and only if all eigenvalues λ_i have a negative real part Re $\lambda_i < 0$, $i = 1(1)n$. As usual in engineering, in vehicle dynamics asymptotic stability of the system is required. The lateral vibrations of a railway bogie, e.g., shall damp out in time after impulsive excitation.

■ **Problem 7.1 Stability of the hunting motion of a railway wheelset**
According to Klingel's formula (31), (32) in Problem 2.7 a kinematically rolling railway wheelset performs a harmonic undamped oscillation. It has to be shown that the criterium for asymptotic stability is not met.

Solution
The state equations of the kinematics of a railway wheelset read according to (28) of Problem 2.7 as

$$\begin{bmatrix} \gamma' \\ \alpha' \end{bmatrix} = \begin{bmatrix} 0 & \dfrac{1}{r_0} \\ -\dfrac{1}{q} & 0 \end{bmatrix} \begin{bmatrix} \gamma \\ \alpha \end{bmatrix} . \tag{1}$$

The eigenvalues of this system follow form its characteristic equation

$$\det(\lambda \boldsymbol{E} - \boldsymbol{A}) = \det \begin{bmatrix} \lambda & \dfrac{1}{r_0} \\ \dfrac{1}{q} & \lambda \end{bmatrix} = \lambda^2 + \frac{1}{r_0 q} = 0 \tag{2}$$

as

$$\lambda_{1,2} = \pm \mathrm{i} \sqrt{\frac{1}{r_0 q}} . \tag{3}$$

Thus, it yields Re $\lambda_i = 0$, i.e. the railway wheelset is kinematically not asymptotically stable. To achieve asymptotic stability the model has to be extended by dynamical phenomena due to the rolling contact. ■

7.2.2 Frequency Response Analysis

Harmonically excited vehicles are rarely found in practice. However, the sinusoidal excitation is widely used in vehicle testing due to its good repeatability. Then, for comparison a theoretical frequency response analysis is required, too. Furthermore, frequency responses have to be used for the investigation of unbalance phenomena.

An asymptotically stable linear system under harmonic excitation

$$\begin{aligned}\dot{\boldsymbol{x}} &= \boldsymbol{A}\boldsymbol{x} + \boldsymbol{b}^{(1)} \cos \Omega t + \boldsymbol{b}^{(2)} \sin \Omega t \\ &= \boldsymbol{A}\boldsymbol{x} + \boldsymbol{b}\mathrm{e}^{\mathrm{i}\Omega t} + \overline{\boldsymbol{b}}\mathrm{e}^{-\mathrm{i}\Omega t}\end{aligned} \tag{7.7}$$

with the time-invariant $n \times n$-system matrix $\boldsymbol{A}$ and the real $n \times 1$-excitation vectors $\boldsymbol{b}^{(1)}$, $\boldsymbol{b}^{(2)}$ or the complex $n \times 1$-excitation vector $\boldsymbol{b}$ yields the stationary response

$$\begin{aligned}\boldsymbol{x}(t) &= \boldsymbol{g}^{(1)} \cos \Omega t + \boldsymbol{g}^{(2)} \sin \Omega t \\ &= \boldsymbol{g}\mathrm{e}^{\mathrm{i}\Omega t} + \overline{\boldsymbol{g}}\mathrm{e}^{-\mathrm{i}\Omega t}\end{aligned} \tag{7.8}$$

with the corresponding $n \times 1$-vectors $\boldsymbol{g}^{(1)}$, $\boldsymbol{g}^{(2)}$ or $\boldsymbol{g}$ of the frequency response. Here, $\overline{\boldsymbol{b}}$ and $\overline{\boldsymbol{g}}$, respectively, mean the conjugate complex vectors. The corresponding complex $n \times n$-frequency response matrix $\boldsymbol{F}$ reads as

$$\boldsymbol{g} = (\mathrm{i}\Omega \boldsymbol{E} - \boldsymbol{A})^{-1}\boldsymbol{b} = \boldsymbol{F}\boldsymbol{b}\,. \tag{7.9}$$

Then, the amplitude frequency response a_i of the i-th coordinate of the state vector is given by

$$a_i(\Omega) = 2\sqrt{[\mathrm{Re}\, g_i(\Omega)]^2 + [\mathrm{Im}\, g_i(\Omega)]^2}\,. \tag{7.10}$$

For the evaluation of frequency responses often numerical methods are used to avoid the explicit handling of complex quantities what is not so easy.

■ Problem 7.2 Unbalance excitation of wheel vibrations

The vertically moving wheels may perform vibrations excited by unbalances, Fig. 7.7. The frequency response of these vibrations is required.

Solution

The equation of motion of one wheel reads with respect to its equilibrium condition as

$$m\ddot{z} + d\dot{z} + (k_1 + k_2)z = m\varepsilon\Omega^2 \cos \Omega t \tag{1}$$

where $\varepsilon = m_u r/m$ characterizes the magnitude of the unbalance. The state equations read as

$$\begin{bmatrix}\dot{z}\\ \ddot{z}\end{bmatrix} = \underbrace{\begin{bmatrix}0 & 1\\ -\dfrac{k_1+k_2}{m} & -\dfrac{d}{m}\end{bmatrix}}_{\boldsymbol{A}}\begin{bmatrix}z\\ \dot{z}\end{bmatrix} + \underbrace{\begin{bmatrix}0\\ \dfrac{1}{2}\varepsilon\Omega^2\end{bmatrix}}_{\boldsymbol{b}}\mathrm{e}^{\mathrm{i}\Omega t} + \underbrace{\begin{bmatrix}0\\ \dfrac{1}{2}\varepsilon\Omega^2\end{bmatrix}}_{\overline{\boldsymbol{b}}}\mathrm{e}^{-\mathrm{i}\Omega t}\,, \tag{2}$$

and for the frequency response matrix it follows

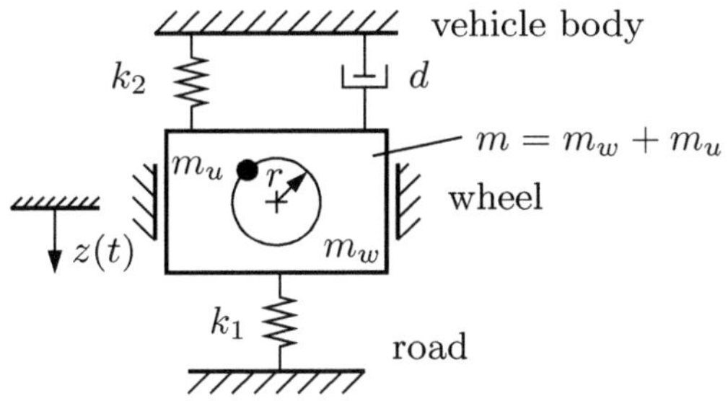

Fig. 7.7. Vehicle wheel with unbalance

$$\boldsymbol{F} = \frac{1}{-\Omega^2 + \mathrm{i}\Omega\frac{d}{m} + \frac{k_1 + k_2}{m}} \begin{bmatrix} \mathrm{i}\Omega + \frac{d}{m} & 1 \\ -\frac{k_1 + k_2}{m} & \mathrm{i}\Omega \end{bmatrix} . \tag{3}$$

The complex frequency response g_1 of the state variable z results in

$$g_1 = \frac{\left(-\Omega^2 + \frac{k_1 + k_2}{m}\right) - \mathrm{i}\Omega\frac{d}{m}}{\left(-\Omega^2 + \frac{k_1 + k_2}{m}\right)^2 + \Omega^2\left(\frac{d}{m}\right)^2} \frac{1}{2}\varepsilon\Omega^2 . \tag{4}$$

Then, it yields for the amplitude frequency response

$$a_1(\Omega) = \frac{\varepsilon\Omega^2}{\sqrt{\left(\frac{k_1 + k_2}{m} - \Omega^2\right)^2 + \left(\frac{d}{m}\Omega\right)^2}} . \tag{5}$$

Figure 7.8 shows frequency responses for different values of the shock absorber characteristic. ■

7.2.3 Random Vibration

The investigation of random vibrations may be performed in the frequency domain using the spectral analysis or in the time domain applying the covariance analysis. Assuming that $\boldsymbol{x}(t) = \boldsymbol{0}$ is an asymptotically stable equilibrium position, then from $\mathrm{E}\{\boldsymbol{w}(t)\} = \boldsymbol{0}$ it follows $\mathrm{E}\{\boldsymbol{x}(t)\} = \boldsymbol{0}$, i. e. the mean value of the state vector vanishes in the steady state. Thus, as essential goal it remains to calculate the characteristic variances, e.g. the variance $\sigma_{\overline{a}}^2 = \mathrm{E}\left\{\overline{a}^2(t)\right\}$ of the frequency-weighted acceleration $\overline{a}(t)$.

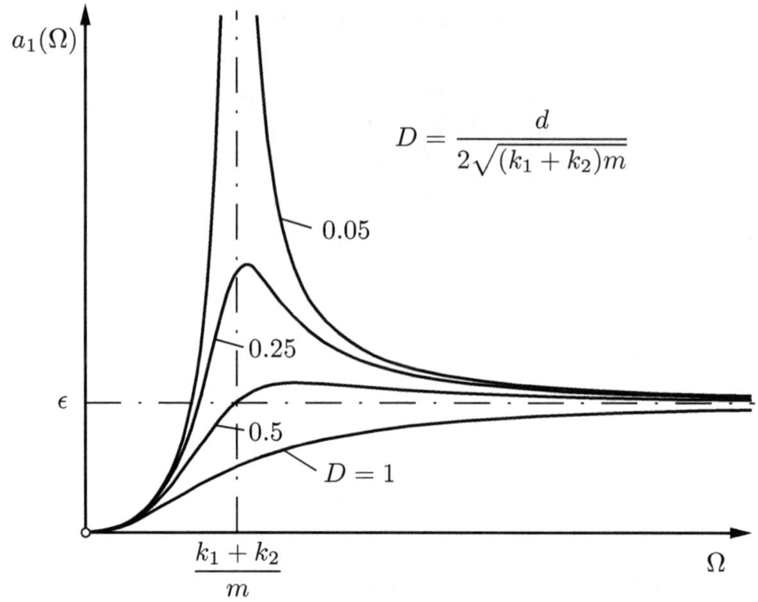

Fig. 7.8. Amplitude frequency response of a vertical wheel vibration excited by unbalance

7.2.3.1 Spectral Density Analysis

The spectral density (PSD) is a characteristic quantity of a random process in the frequency domain. In particular, for linear systems the spectral density of the input and output processes are related to each other by its frequency response. However, one should keep in mind that the ride comfort and safety criteria depend primarily on variances and not on spectral densities.

The $n \times n$-spectral density matrix $\boldsymbol{S}_x(\omega)$ of the entire vehicle system described by the state equations (5.28), where $\boldsymbol{A} = \mathbf{const}$ and $\boldsymbol{B} = \mathbf{const}$ are assumed, follows for an excitation by the white noise process $\boldsymbol{w}(t) \sim (\mathbf{0}, \boldsymbol{Q}_W)$ as

$$\boldsymbol{S}_x(\omega) = (\mathrm{i}\omega\boldsymbol{E} - \boldsymbol{A})^{-1}\boldsymbol{B}\boldsymbol{Q}_W\boldsymbol{B}^{\mathrm{T}}(-\mathrm{i}\omega\boldsymbol{E} - \boldsymbol{A})^{-\mathrm{T}} . \tag{7.11}$$

Thus, the spectral density matrix $\boldsymbol{S}_x(\omega)$ is gained from the input intensity matrix $\boldsymbol{Q}_W$ by matrix multiplications with the frequency response matrix $\boldsymbol{F}_x = (i\omega\boldsymbol{E} - \boldsymbol{A})^{-1}$. Since the state vector $\boldsymbol{x}(t)$ and the system matrix $\boldsymbol{A}$ often can be partitioned it is possible to get partial results by the spectral density matrices of the road excitation, the suspension system and the chassis, respectively. This may reduce the numerical effort considerably. The scalar spectral density of the scalar acceleration $a(t)$, given by (7.1), reads

$$S_a(\omega) = (\boldsymbol{c}^{\mathrm{T}}\boldsymbol{F}_x(\omega)\boldsymbol{B} + \boldsymbol{d}^{\mathrm{T}})\boldsymbol{Q}_W\,(\boldsymbol{B}^{\mathrm{T}}\boldsymbol{F}_x^{\mathrm{T}}(-\omega)\boldsymbol{c} + \boldsymbol{d}) \tag{7.12}$$

where the $n \times n$ -frequency response matrix $\boldsymbol{F}_x$ has been used again. Equation (7.12) shows that the spectral density is a quadratic expression, i. e., the sum of quantities leads to additional terms in the result. Using the spectral density (7.12), the frequency weighting required due to (6.15) can be performed very easily. It remains

$$S_{\overline{a}}(\omega) = \alpha^2 \left|F\ (\omega)\right|^2 S_a \ . \tag{7.13}$$

The variance follows from the spectral density by integration over an infinite interval,

$$\sigma_{\overline{a}}^2 = \int_{-\infty}^{\infty} S_{\overline{a}}(\omega)\, \mathrm{d}\omega \ . \tag{7.14}$$

The numerical computation of the spectral density (7.14) is generally not difficult. However, one has to deal with complex matrices, many approximation points and a very large integration interval. This can lead to numerical errors and considerable computation times. The integration (7.14) is avoided applying the covariance analysis as shown in the next section.

Special care requires the spectral analysis of multi-axle vehicles. The excitation at different axles is given by an excitation function with time delay,

$$\begin{aligned} \boldsymbol{B}\boldsymbol{w}(t) &= \sum_{i=1}^{m} \boldsymbol{B}_i \zeta_i(t) \ , \\ \zeta_i(t) &= \zeta(t - t_i) \ , \quad 0 = t_1 < t_2 < ... < t_m \ , \\ t_i &= \frac{l_i}{v} \ , \quad i = 1, ..., m \ , \end{aligned} \tag{7.15}$$

where l_i denotes the distance between the front axle, and the i-th axle and $v = \text{const}$ is the vehicle speed. An example of an excitation with time delay is shown in Fig. 7.1.The spectral density matrix of a two-axle vehicle under random excitation reads

$$\begin{aligned} \boldsymbol{S}_x(\omega) = (\mathrm{i}\omega \boldsymbol{E} - \boldsymbol{A})^{-1} [&\boldsymbol{B}_1 \boldsymbol{S}_\zeta \boldsymbol{B}_1^{\mathrm{T}} + \boldsymbol{B}_2 \boldsymbol{S}_\zeta \boldsymbol{B}_2^{\mathrm{T}} + \mathrm{e}^{\mathrm{i}\omega(t_1 - t_2)} \boldsymbol{B}_2 \boldsymbol{S}_\zeta \boldsymbol{B}_1^{\mathrm{T}} \\ &+ \mathrm{e}^{\mathrm{i}\omega(t_2 - t_1)} \boldsymbol{B}_1 \boldsymbol{S}_\zeta \boldsymbol{B}_2^{\mathrm{T}}] (-\mathrm{i}\omega \boldsymbol{E} - \boldsymbol{A})^{-\mathrm{T}} \ . \end{aligned} \tag{7.16}$$

In frequency domain, the time delay $(t_2 - t_1)$ results in additional terms, weighted by the exponential function.

7.2.3.2 Covariance Analysis

In contrast to spectral density analysis, the covariance analysis yields directly the variances that are required for the assessment of the vehicle performance. The covariance matrix of the entire vehicle system follows from an algebraic equation, the so-called Lyapunov matrix equation, and integrations are not required. An essential prerequisite of the covariance analysis is a white noise input process. This can always be achieved by modeling the random vehicle excitation by means of a shape filter.

The Lyapunov matrix equation corresponding to the state equations (5.28) of the entire vehicle system reads

$$\boldsymbol{A}\boldsymbol{P}_x + \boldsymbol{P}_x\boldsymbol{A}^{\mathrm{T}} + \boldsymbol{B}\boldsymbol{Q}_W\boldsymbol{B}^{\mathrm{T}} = \boldsymbol{0} \tag{7.17}$$

where $\boldsymbol{P}_x = \mathrm{E}\{\boldsymbol{x}\,\boldsymbol{x}^{\mathrm{T}}\}$ denotes the symmetric $n \times n$-covariance matrix and $\boldsymbol{Q}_W$ is the $r \times r$-intensity matrix of the white noise input. An extensive derivation of (7.17) can be found in Mueller and Schiehlen (1985). Stable numerical procedures to solve the Lyapunov matrix equation are given in Smith (1968) and Kreisselmeier (1972), and are available in Matlab, too. As a result one gets the variances of all state variables of the system. However, the computation of the ride comfort requires the variance of the frequency weighted acceleration. Thus, the covariance analysis has to be applied to the extended system (7.2). The covariance matrix of the extended state vector $\tilde{\boldsymbol{x}}$ reads

$$\boldsymbol{P}_{\tilde{x}} = \begin{bmatrix} \boldsymbol{P}_x & \boldsymbol{P}_{x\overline{v}} \\ \boldsymbol{P}_{\overline{v}x} & \boldsymbol{P}_{\overline{v}} \end{bmatrix} \tag{7.18}$$

where $\boldsymbol{P}_{x\overline{v}} = \boldsymbol{P}_{\overline{v}x}^{\mathrm{T}}$ holds. The $s \times 1$-vector process $\overline{\boldsymbol{v}}(t)$ yields according to (6.20) immediately the frequency weighted acceleration and its variance

$$\sigma_{\overline{a}}^2 = \alpha^2 \overline{\boldsymbol{h}}^{\mathrm{T}} \boldsymbol{P}_{\overline{v}} \overline{\boldsymbol{h}} \,. \tag{7.19}$$

This is the relation corresponding to (7.14). The investigation of multi-axle vehicles is also possible by the covariance analysis. The theory needed is available, as shown in Mueller et al. (1980). In contrast to spectral analysis, the covariance analysis can also be applied to nonstationary and nonlinear problems without difficulties.

■ Problem 7.3 Random vibrations of a single wheel

A wheel guided in vertical direction is excited by the road roughness and performs random vibrations, see Fig. 7.9. The variance of the dynamic wheel load variations has to be investigated applying the covariance analysis.

Solution

The road roughness is characterized by white velocity noise $\dot{\zeta}(t) \sim (0, q)$. The equation of motion of the wheel reads

$$m\ddot{z} + d\dot{z} + (k_1 + k_2)z = k_1\zeta(t) \,, \tag{1}$$

where $\zeta(t)$ is a scalar random process. After differentiation with respect to time the corresponding state equation has the form

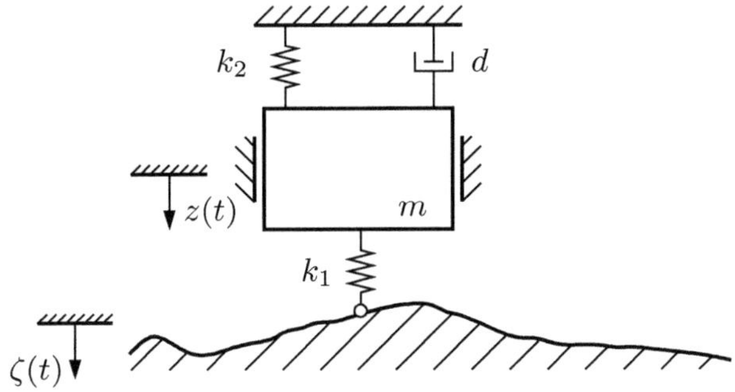

Fig. 7.9. Vehicle wheel subject to random excitation

$$\underbrace{\begin{bmatrix} \ddot{z} \\ \dddot{z} \end{bmatrix}}_{\ddot{\boldsymbol{x}}} = \underbrace{\begin{bmatrix} 0 & 1 \\ -\kappa_{12} & -\delta \end{bmatrix}}_{\boldsymbol{A}} \underbrace{\begin{bmatrix} \dot{z} \\ \ddot{z} \end{bmatrix}}_{\dot{\boldsymbol{x}}} + \underbrace{\begin{bmatrix} 0 \\ \kappa_1 \end{bmatrix}}_{\boldsymbol{B}} \dot{\zeta}(t) , \qquad \ddot{\boldsymbol{x}} = \boldsymbol{A} \dot{\boldsymbol{x}} + \boldsymbol{B} w(t) , \tag{2}$$

where the abbreviations $\delta = d/m$, $\kappa_{12} = (k_1 + k_2)/m$, $\kappa_1 = k_1/m$ have been used and the random excitation is given by a white velocity noise process. Then, the Lyapunov equation reads

$$\boldsymbol{A P}_{\dot{x}} + \boldsymbol{P}_{\dot{x}} \boldsymbol{A}^{\mathrm{T}} + \boldsymbol{B q B}^{\mathrm{T}} = \boldsymbol{0} , \tag{3}$$

where the 2×2-covariance matrix of the first derivative $\dot{\boldsymbol{x}}$ of the state vector is introduced,

$$\boldsymbol{P}_{\dot{x}} = \begin{bmatrix} P_{11} & P_{12} \\ P_{12} & P_{22} \end{bmatrix} = \begin{bmatrix} P_{\dot{z}\dot{z}} & P_{\dot{z}\ddot{z}} \\ P_{\dot{z}\ddot{z}} & P_{\ddot{z}\ddot{z}} \end{bmatrix} . \tag{4}$$

The solution of (3) yields the following matrices

$$\boldsymbol{A P}_{\dot{x}} = \begin{bmatrix} P_{12} & P_{22} \\ -\kappa_{12} P_{12} - \delta P_{12} & -\kappa_{12} P_{12} - \delta P_{22} \end{bmatrix} , \tag{5}$$

$$\boldsymbol{B q B}^{\mathrm{T}} = \begin{bmatrix} 0 & 0 \\ 0 & \kappa_1^2 q \end{bmatrix} . \tag{6}$$

Due to the symmetry of the covariance matrix from (3) it follow altogether three linear equations for the three unknowns,

$$\begin{bmatrix} 0 & 2 & 0 \\ -\kappa_{12} & -\delta & 1 \\ 0 & 2\kappa_{12} & 2\delta \end{bmatrix} \begin{bmatrix} P_{11} \\ P_{12} \\ P_{22} \end{bmatrix} = \begin{bmatrix} 0 \\ 0 \\ \kappa_1^2 q \end{bmatrix} , \tag{7}$$

having the solutions

$$P_{11} = \frac{k_1^2}{2d(k_1+k_2)}q\,, \quad P_{12} = 0\,, \quad P_{22} = \frac{k_1^2}{2dm}q\,. \tag{8}$$

The weighting vector $\boldsymbol{b}$ of the dynamic wheel loads reads according to (1)

$$f_{Dyn} = k_1(z-\zeta) = \underbrace{\begin{bmatrix} -d & -m \end{bmatrix}}_{\boldsymbol{b}^{\mathrm{T}}} \underbrace{\begin{bmatrix} \dot{z} \\ \ddot{z} \end{bmatrix}}_{\dot{\boldsymbol{x}}}\,, \tag{9}$$

where $k_2 \ll k_1$ has been regarded. The variance σ_f^2 of the dynamic wheel load variation reads

$$\sigma_f^2 = \boldsymbol{b}^{\mathrm{T}}\boldsymbol{P}_{\dot{x}}\boldsymbol{b} = \left[\frac{k_1 d}{2} + \frac{k_1^2 m}{2d}\right] q\,. \tag{10}$$

It turns out, that σ_f^2 becomes a minimum if the normalized damping coefficient $D = d/(2\sqrt{k_1 m}) = 0.5$ is chosen. In this case of optimal damping it follows $\sigma_f^2 = k_1\sqrt{k_1 m}\, q$. This result shows that small wheel masses m and small tire stiffnesses k_1 are reducing the dynamic wheel load variations. However, there are design limits for these parameters due to the durability of wheel and tire.

The assumption $k_2 \ll k_1$ is necessary in this example, since an idealized white velocity noise process has been considered. In the general case of colored noise this assumption is not required. ■

7.3 Nonlinear Systems

Nonlinearities occur by all events in vehicle dynamics if the amplitudes are strongly increasing due to self-excitation or forced excitation, respectively. Typical examples include nonlinear characteristics of suspension devices to avoid hard impacts in vertical direction. The lateral motion of a railway wheelset is also strongly nonlinear if large displacements occur resulting in lateral impacts of the wheel flange. Therefore, some approximation methods for nonlinear vibrations are discussed. In contrary to numerical simulations, approximation methods provide only qualitative results.

7.3.1 Harmonic Linearization

Harmonic linearization in an approximation procedure for free self-excited oscillations and harmonically forced nonlinear vibrations. The state equations (5.29) are now decomposed in a linear and a nonlinear part,

$$\dot{\boldsymbol{x}}(t) = \boldsymbol{A}\boldsymbol{x}(t) + \boldsymbol{f}(\boldsymbol{x}) + \boldsymbol{b}_c \cos\Omega t + \boldsymbol{b}_s \sin\Omega t\,. \tag{7.20}$$

Further, it is assumed that a periodic solution with the period T exists

$$\boldsymbol{x}(t) = \boldsymbol{x}(t+T)\ . \tag{7.21}$$

where two cases are distinguished,

- vanishing excitation, T unknown,
- harmonic excitation, $T = 2\pi/\Omega$ known.

The periodic solution is now approximated by a harmonic solution

$$\boldsymbol{x}(t) \approx \boldsymbol{x}_h(t) = \boldsymbol{x}_c \cos \Omega t + \boldsymbol{x}_s \sin \Omega t\ . \tag{7.22}$$

The nonlinearities are assumed to show odd characteristics

$$\boldsymbol{f}(\boldsymbol{x}) = -\boldsymbol{f}(-\boldsymbol{x})\ , \tag{7.23}$$

and shall be described by a linear ansatz at the best

$$\boldsymbol{f}(\boldsymbol{x}) \approx \boldsymbol{F}_h \boldsymbol{x}\ . \tag{7.24}$$

Thus, the remaining error

$$\boldsymbol{e}(t) = \boldsymbol{f}(\boldsymbol{x}_h(t)) - \boldsymbol{F}_h \boldsymbol{x}_h(t) \tag{7.25}$$

has to be minimized

$$\frac{\partial}{\partial F_{hij}} \int_0^T \mathbf{e}^{\mathrm{T}} \mathbf{e} \mathrm{d}t = 0\ , \quad i,j = 1(1)n\ . \tag{7.26}$$

With this condition, the non-unique equivalent $n \times n$-coefficient matrix

$$\boldsymbol{F}_h = \boldsymbol{F}_h(\boldsymbol{x}_c, \boldsymbol{x}_s; \Omega)\ . \tag{7.27}$$

can be found. Then, the linearized substituted system has the form

$$\boldsymbol{x}(t) = (\boldsymbol{A} + \boldsymbol{F}_h)\boldsymbol{x}(t) + \boldsymbol{b}_c \cos \Omega t + \boldsymbol{b}_s \sin \Omega t\ . \tag{7.28}$$

By inserting of the approximated solution (7.22) in (7.28) one gets conditions for the vectors $\boldsymbol{x}_s$, $\boldsymbol{x}_c$ and the frequency Ω if not known.

The coefficients of the equivalent matrix $\boldsymbol{F}_h$ are available in tables for many nonlinearities, see e.g. Magnus et al. (2008). The evaluation of the unknown $\boldsymbol{x}_s$, $\boldsymbol{x}_c$ and Ω is not unique and has to be handled in each case individually.

■ Problem 7.4 Harmonic linearization of self-excited vibrations
The development of nonlinear hunting motions of a railway wheelset can be explained qualitatively by an instability for small amplitudes and a strong

damping for large amplitudes due to the wheel flange. The resulting hunting represents a limit cycle as it is found with the Van der Pol differential equation. Evaluate the limit cycle by harmonic linearization.

Solution
The Van der Pol differential equation reads as

$$\ddot{y} + (-2\delta + \varepsilon y^2)\dot{y} + y = 0 \,. \tag{1}$$

where δ is the positive real part of the eigenvalues of the corresponding linear system ($\varepsilon = 0$). On the other hand large amplitudes of $y(t)$ result in a strong damping to $\varepsilon > 0$.

The state equations read as

$$\underbrace{\begin{bmatrix} \dot{y} \\ \ddot{y} \end{bmatrix}}_{\dot{\boldsymbol{x}}} = \underbrace{\begin{bmatrix} 0 & 1 \\ -1 & 2\delta \end{bmatrix}}_{\boldsymbol{A}} \underbrace{\begin{bmatrix} y \\ \dot{y} \end{bmatrix}}_{\boldsymbol{x}} + \underbrace{\begin{bmatrix} 0 \\ -\varepsilon y^2 \dot{y} \end{bmatrix}}_{\boldsymbol{f}(x)} , \tag{2}$$

where the nonlinear 2×1-vector function $\boldsymbol{f}(\boldsymbol{x})$ is replaced by the linear approximation

$$\boldsymbol{f}(\boldsymbol{x}) = \underbrace{\begin{bmatrix} F_{11} & F_{12} \\ F_{21} & F_{22} \end{bmatrix}}_{\boldsymbol{F}_h} \underbrace{\begin{bmatrix} y \\ \dot{y} \end{bmatrix}}_{\boldsymbol{x}} \,. \tag{3}$$

With the harmonic solution (7.22) based on the vectors

$$\boldsymbol{x}_c = \begin{bmatrix} a \\ 0 \end{bmatrix} , \quad \boldsymbol{x}_s = \begin{bmatrix} 0 \\ -a\Omega \end{bmatrix} \tag{4}$$

one gets from (7.26)

$$F_{11} = F_{12} = F_{21} = 0 \,, \quad F_{22} = -\frac{\varepsilon a^2}{4} \,. \tag{5}$$

Inserting (2) to (5) in (7.28), it follows

$$a = \sqrt{\frac{8\delta}{\varepsilon}} \,, \quad \Omega = 1 \,. \tag{6}$$

The limit cycle has the amplitude $A = \sqrt{8\delta/\varepsilon}$ and the frequency is $\Omega = 1$. In Fig. 7.10 some trajectories are shown. ■

■ Problem 7.5 Harmonic linearization of a forced oscillator
The suspension springs of a vehicle show a nonlinear progressive characteristic to avoid impacts by large displacements. This behaviour is modeled by an

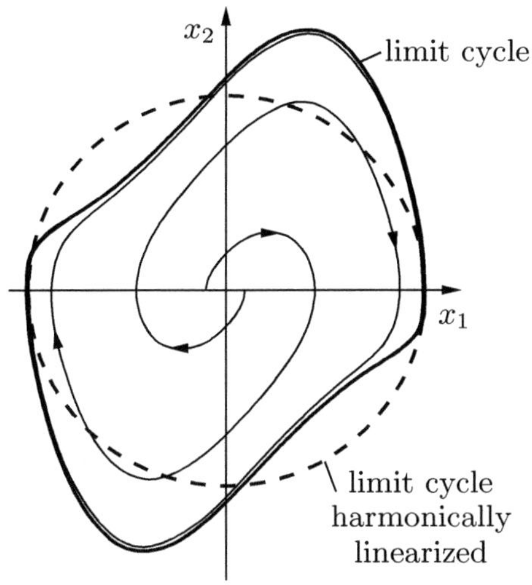

Fig. 7.10. Limit cycle of the Van der Pol differential equation

oscillator with a cubic spring what results in a Duffing differential equation. Evaluate the frequency response by the method of harmonic linearization.

Solution

The Duffing differential equation reads as

$$\ddot{y} + y + \alpha y^3 = b \cos \Omega t \,. \tag{1}$$

According to the ansatz function (7.22) and (7.24) one gets

$$y_h = A \cos \Omega t \tag{2}$$

and

$$y^3 \approx k_h y \tag{3}$$

From (7.26) it follows

$$\frac{\partial}{\partial k_h} \int_0^T (y_h^3 - k_h y_h)^2 \mathrm{d}t = 0 \tag{4}$$

or

$$k_h = \frac{3}{4} A^2 \,. \tag{5}$$

Thus, the differential equation (1) is substituted by a linear system where the eigenfrequency depends on the amplitude A of the oscillator,

$$\ddot{y} + \left(1 + \frac{3}{4}\alpha A^2\right) y = b \cos \Omega t \, . \tag{6}$$

For the evaluation of the still unknown amplitude A, (2) is inserted in (6). Then, it remains the relation

$$\frac{3}{4}\alpha A^3 + (1 - \Omega^2)A - b = 0 \, . \tag{7}$$

This polynomial of the third order has a maximum of three real roots which result in a non-unique frequency response $A(\Omega)$, Fig. 7.11. The resonance of the undamped system (1) at $\Omega = 1$ does not longer exist, it remains a finite amplitude $A(\Omega = 1) = 4b/3a$. It turns out that the frequency response of nonlinear systems is characterized by a bent backbone. ■

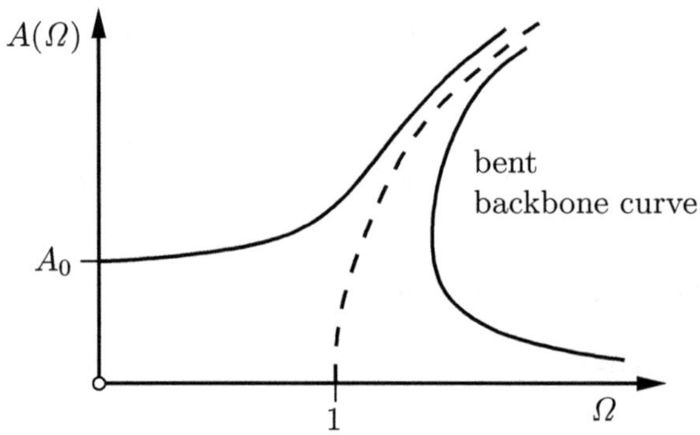

Fig. 7.11. Frequency response of Duffing differential equation

7.3.2 Statistical Linearization

The statistical linearization deals with the qualitative analysis of nonlinear stochastically forced systems. The stochastic differential equations read according to (7.20) as

$$\dot{\boldsymbol{x}}(t) = \boldsymbol{A}\boldsymbol{x}(t) + \boldsymbol{f}(\boldsymbol{x}(t)) + \boldsymbol{B}\boldsymbol{w}(t) \, , \tag{7.29}$$

where $\boldsymbol{w}(t)$ is a stationary, Gaussian and zero mean process, $E\{\boldsymbol{w}(t)\} = 0$.

The assumption is that there exists an also Gaussian and stationary solution process

$$\boldsymbol{x}(t) \approx \boldsymbol{x}_{st}(t) \, , \quad \mathrm{E}\{\boldsymbol{x}_{st}\boldsymbol{x}_{st}^{\mathrm{T}}\} = \boldsymbol{P}_x \, , \tag{7.30}$$

and that the nonlinearities have an odd characteristic (7.23). Then, the nonlinearity is approximated by a linear ansatz,

$$\boldsymbol{f}(\boldsymbol{x}) \approx \boldsymbol{F}_{st}\boldsymbol{x} \tag{7.31}$$

and the expected value

$$\mathrm{E}\{\boldsymbol{e}_{st}^{\mathrm{T}}\boldsymbol{e}_{st}\} = \mathrm{Min} \tag{7.32}$$

of the resulting error

$$\boldsymbol{e}_{st} = \boldsymbol{f}(\boldsymbol{x}_{st}(t)) - \boldsymbol{F}_{st}\boldsymbol{x}_{st}(t) \tag{7.33}$$

shall be minimized. After some lengthly evaluations one gets

$$\boldsymbol{F}_{st} = \boldsymbol{F}_{st}(\boldsymbol{P}_x)\ . \tag{7.34}$$

The linearized substituted system has the form

$$\dot{\boldsymbol{x}}(t) = (\boldsymbol{A} + \boldsymbol{F}_{st})\boldsymbol{x}(t) + \boldsymbol{B}\boldsymbol{w}(t)\ . \tag{7.35}$$

It turns out that for the determination of the approximated solution process always the complete covariance matrix $\boldsymbol{P}_x$ is required. This means that the covariance analysis is most adequate for nonlinear systems. Further details are reported by Mueller et al. (1980).

7.3.3 Investigation of Linearized Systems

The methods presented in Sect. 7.2 for linear system can be applied to the linearized system (7.28) and (7.35), too. However, the stability is theoretically not guaranteed. Nevertheless one gets often useful results.

The linearized state equations (7.28) of self-excited systems $\boldsymbol{b}_c = \boldsymbol{b}_s = 0$ hold always a pair of purely imaginary eigenvalues $\lambda = \pm\mathrm{i}\omega$ and therefore, they are not asymptotically stable. But it is possible to check at least the orbital stability of the limit cycle by considering the system behavior in the limit cycle's neighborhood. For this purpose the amplitudes of the solution ansatz (7.22) are varied by a parameter $\mu = 1 + \varepsilon$, $\varepsilon \ll 1$. Orbital stability is at hand if the following conditions are fulfilled:

$$\begin{array}{lll}
\mu = 1\ : & \lambda_{1,2} = \pm\mathrm{i}\omega\ ; \quad \text{limit cycle} & \\
 & \mathrm{Re}\ \lambda_i < 0\ , \quad i = 3(1)n\ ; & \\
(1-\varepsilon) < \mu < 1\ : & \mathrm{Re}\ \lambda_{1,2} > 0\ , \quad \mathrm{Re}\lambda_i < 0\ , & i = 3(1)n\ ; \\
1 < \mu < (1+\varepsilon)\ : & \mathrm{Re}\ \lambda_i < 0\ , \quad i = 1(1)n\ . &
\end{array}$$

Obviously these conditions are based on plausibility considerations only.

The frequency response analysis of harmonically linearized systems results in an amplitude-dependent frequency response matrix

$$\boldsymbol{F}(\Omega) = [\mathrm{i}\Omega\boldsymbol{E} - \boldsymbol{A} - \boldsymbol{F}_h(\boldsymbol{x}_c, \boldsymbol{x}_s; \Omega)]^{-1} \tag{7.36}$$

which is known in control theory as describing function, too. A simple case of frequency response is given in Problem 7.5.

The investigation of random vibrations of statistically linearized systems requires the covariance matrix of the solutions process. Starting with the spectral density matrix $\boldsymbol{S}_w$ of the excitation one gets the spectral density matrix of the solution

$$\boldsymbol{S}_x(\omega) = \boldsymbol{F}(\omega; \boldsymbol{P}_x)\boldsymbol{B}\boldsymbol{S}_w\boldsymbol{B}^{\mathrm{T}}\boldsymbol{F}^{\mathrm{T}}(-\omega; \boldsymbol{P}_x)\ . \tag{7.37}$$

On the other hand, it yields for the covariance matrix

$$\boldsymbol{P}_x = \int_{-\infty}^{+\infty} \boldsymbol{S}_x(\omega; \boldsymbol{P}_x)\mathrm{d}\omega \tag{7.38}$$

a relation which can be solved only iteratively by numerical methods

$$\boldsymbol{P}_x^{(i+1)} = \int_{-\infty}^{\infty} \boldsymbol{S}_x(\omega; \boldsymbol{P}_x^{(i)})\mathrm{d}\omega\ , \quad \lim_{i\to\infty} \boldsymbol{P}_x^{(i)} = \boldsymbol{P}_x\ . \tag{7.39}$$

This procedure is computationally costly since the iteration (7.39) involves an improper integral.

The covariance analysis of a statistically linearized system results in a nonlinear Lyapunov matrix equation as

$$[\boldsymbol{A} + \boldsymbol{F}_{st}(\boldsymbol{P}_x)]\,\boldsymbol{P}_x + \boldsymbol{P}_x\,[\boldsymbol{A} + \boldsymbol{F}_{st}(\boldsymbol{P}_x)]^{\mathrm{T}} + \boldsymbol{B}\boldsymbol{Q}_W\boldsymbol{B}^{\mathrm{T}} = \boldsymbol{0} \tag{7.40}$$

where $\boldsymbol{w}(t) \sim (\boldsymbol{0}, \boldsymbol{Q}_W)$ means a white noise process. This equation has also to be solved numerically but it is a purely algebraic equation. Moreover, the convergence is proven so that covariance analysis provides theoretically well-grounded results.

The approximation methods for nonlinear systems assume that the system show a quasi-linear dynamical behavior. In particular, periodical solutions are searched. But nonlinear system may show strongly irregular, chaotic behavior, too. For their analysis more recently developed methods of nonlinear dynamics are required, see, e.g., Moon (1987) and Kreuzer (1987). Due to the nonlinearities and the high number of degrees of freedom in vehicles, see True (2007), undoubtedly chaotic vibrations may occur. However, the distinction between stochastically excited random vibrations and nonlinear chaotic vibration is theoretically intricate and experimentally hardly possible.

7.4 Optimization Problems

Optimal behavior of a vehicle system is achieved by the proper choice of design variables. Since the driver does not have any influence on the condition of the guideway the design variables have to be chosen from the parameters p_j, $j = 1(1)s$, of the vehicle and the suspension and guidance devices. These parameters are summarized in an $s \times 1$-vector

$$\boldsymbol{p} = [p_1, p_2, \ldots, p_s]^{\mathrm{T}} \tag{7.41}$$

and the state equations (5.28) are rewritten in the form

$$\dot{\boldsymbol{x}}(t) = \boldsymbol{A}(\boldsymbol{p}, t)\boldsymbol{x}(t) + \boldsymbol{B}(\boldsymbol{p}, t)\boldsymbol{w}(t) \ . \tag{7.42}$$

As performance criteria for an optimization the assessment criteria introduced in Chap. 6 are at hand. These criteria can be classified in instationary processes like driving stability and stationary processes like driving comfort and safety on uneven guideways. Thus, a general cost function reads as

$$\boldsymbol{J} = \frac{1}{T}\int_0^T \boldsymbol{x}_{instat}^{\mathrm{T}} \boldsymbol{Q}_{instat} \boldsymbol{x}_{instat} \mathrm{d}t + \frac{1}{T}\int_0^T \boldsymbol{x}_{stat}^{\mathrm{T}} \boldsymbol{Q}_{stat} \boldsymbol{x}_{stat} \mathrm{d}t \tag{7.43}$$

where $\boldsymbol{Q}$ are the corresponding $n \times n$-assessment matrices and T means the period considered. Inserting (7.42) into (7.43) it remains

$$\boldsymbol{J} = \boldsymbol{J}(\boldsymbol{p}) = \mathrm{Min} \ . \tag{7.44}$$

This optimization condition can be usually only computationally evaluated. In addition to (7.44) also constraints of the parameters have to be considered.

The optimization of mechanical systems was primarily developed in structural dynamics, see, e.g., Haug and Arora (1979). The multicriteria optimization problems were considered in detail in Stadler (1988). Contributions to the parameter optimization of multibody systems are due Bestle (1994). An typical application of these methods was presented by Wimmer and Rauh (1996).

Furthermore, genetic algorithms may be used to the design optimization of vehicle systems, see e.g., Baumal et al. (1998).

8

Longitudinal Motions

The longitudinal or forward motion, respectively, is most important for all vehicles representing the desired motion for any transportation task. In particular, the performance of a vehicle depends on the driving performance achieved by its propulsion system resulting in the longitudinal motion what will be discussed in the following. In this chapter only road vehicles are considered particularly challenging by the tire elasticity. From the many references the books by Gillespie (1992), Mitschke and Wallentowitz (2004) and Wong (2001) are mentioned.

8.1 Elastic Wheel

An essential component of drive and brake systems is the wheel with an elastic tire which has been treated in detail in Sect. 3.4.4. The important relations for the longitudinal motion are summarized once again in this Section. The contact force f_{t1} of the elastic wheel depends according to (3.138) on the longitudinal slip s where the definitions on the last column of Table 3.9 are used. Then, it yields

$$f_{t1} = \overline{\varphi}_1(s) f_n \ , \tag{8.1}$$

where $\overline{\varphi}_1$ is the rolling contact coefficient and f_n means the normal force. The longitudinal slip is distinguished by the kind of motion:

$$\begin{aligned} &\text{Driving slip} \qquad s_A = \frac{\nu_1}{\nu_1 - 1} \ , \\ &\text{Braking slip} \qquad s_B = \nu_1 \end{aligned} \tag{8.2}$$

where $\nu_1 = 1 - \omega r_e / v_c$, see also (3.108). Then, it yields

$$0 \leq s_{A,B} \leq 1 \ . \tag{8.3}$$

The special cases included are $s_{A,B} = 0$ for pure rolling, $s_A = 1$ for wheelspinning (pure slipping) and $s_B = 1$ for wheel locking (pure sliding). The differences in the rolling contact coefficient for driving and braking are neglected here, $\overline{\varphi}_A(s_A) = \overline{\varphi}_B(s_B) = \overline{\varphi}_1(s_{A,B})$. A typical rolling contact coefficient in dependency of the slip s is shown in Fig. 3.9. Moreover, the rolling contact coefficient depends on many factors:

- tire type,
- tire profile,
- road surface,
- surface condition (wet, dry),
- speed,
- wheel load,
- lateral force.

Therefore, the rolling contact coefficient has to be measured experimentally. Nevertheless, the models presented in Sect. 3.4.4 are most helpful as well as Pacejka's magic formula, see Pacejka (2002). As a result tire characteristics as shown in Fig. 8.1 are found.

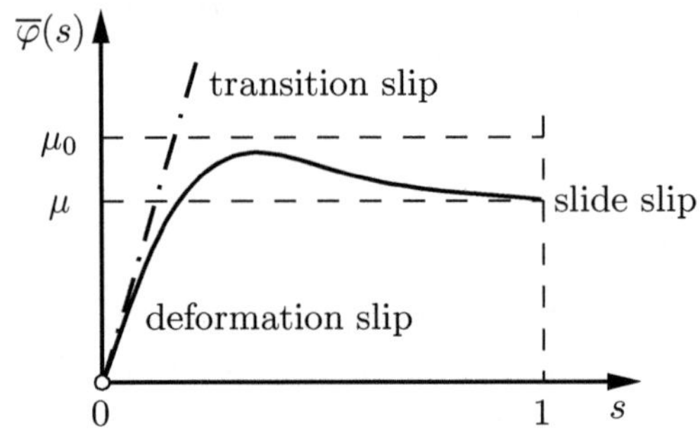

Fig. 8.1. Rolling contact coefficient $\overline{\varphi}$ in dependency of slip s

It follows once again for the elastic tire:

- forces can not be transmitted without slip.

Standard driving conditions are characterized by a slip of $3 - 10\%$, i.e., the slip is due to the deformation of the tire but there is no sliding on the road surface. In Table 8.1 some characteristic values of the sticking and sliding friction coefficient μ_0 and μ are summarized which limit the maximal value of the rolling contact coefficient. The rolling contact coefficients are subject to major variations resulting in the well-known problems for safe braking. Therefore, it is desirable to use the maximum rolling contact coefficient if required. For this purpose anti-lock braking systems (ABS) and acceleration slip regulators (ASR) have been developed which ensure the utilization of

Table 8.1. Friction coefficients between tire and road

Road	Static friction coefficient μ_0	Dynamic friction coefficient μ
Asphalt and concrete (dry)	0.8 - 1.0	0.75
Asphalt (wet)	0.5 - 0.7	0.45 - 0.6
Concrete (wet)	0.8	0.7
Gravel	0.6	0.55
Snow	0.2	0.15
Ice	0.1	0.07

the maximum rolling contact coefficient by control engineering means. More recently both systems have been included in an electronic stability program (ESP). For engineering details see Bosch (2007), Leiber and Czinczel (1979), Burckhardt (1986).

■ Problem 8.1 Control process of a vehicle wheel

A braked wheel with an elastic tire is subject to a torque $M_C = -M(t)$. For the control design of the relative velocity v_P in the contact point, a linear model shall be used.

Solution

For the solution of the given task the equations of motion (3.51) of the slipping wheel is applied. As position coordinates the absolute motion x_C of the center of mass and the relative motion x_P of the contact point P are introduced. Then, it yields according to (8.2) for the kinematics considering the negative velocity v_p in the contact point

$$v_C = \dot{x}_C \ , \quad \omega = \frac{1}{r} v_C + \frac{1}{r} v_P \ . \tag{1}$$

Inserted in (3.43) and (3.44) it follows

$$\ddot{x}_C = \dot{v}_C = \frac{1}{m} f_t(v_C, v_P) \ , \tag{2}$$

$$\ddot{x}_P = \dot{v}_P = -\frac{mr^2 + I_C}{mI_C} f_t(v_C, v_P) - \frac{r}{I_C} M(t) \ . \tag{3}$$

Obviously, the equations of motion (2), (3) are strongly nonlinear due to the contact force $f_t(v_C, v_P)$. The typical control process is highly nonlinear, too.

However, during the very beginning of the braking event only deformation slip occurs and it remains for the contact force

$$f_t = \frac{v_P}{v_C} k f_n \tag{4}$$

where k is a constant coefficient. Initially, the velocity changes Δv_C are small

$$v_C = v_{C0} + \Delta v_C \;, \quad |\Delta v_C| \ll v_{C0} \;, \\ |v_P| \ll v_{C0} \tag{5}$$

where v_{C0} is the initial velocity of the wheel. Then, (4) can be linearized, and one gets from (3) a scalar PT$_1$-control process for the velocity in the contact point

$$\dot{v}_P + \frac{mr^2 + I_C}{mI_C}\frac{kf_n}{v_{C0}} v_P = -\frac{r}{I_C} M(t) \;. \tag{6}$$

For this PT$_1$-control process a standard control design could be used. But such a control design is not feasible for ABS braking systems. The strong nonlinearities and the very high reliability requirements are handled digitally with logical circuits. ■

For the longitudinal motion of the elastic wheel another consequence of the elasticity has to be considered and that is the rolling resistance or rolling friction, respectively. The steady rolling process of the viscoelastic rubber tire results in a certain energy loss. The consumption of energy results in a rolling resistances torque M_R proportional to the normal force or wheel load, respectively, see Fig. 8.2. It yields for a constant speed, $v = const.$, $\omega = const.$,

$$M_R = ef_n = r\overline{\varphi}_r f_n \;, \quad f_x = \frac{M_R}{r} = -W_R \;. \tag{8.4}$$

The rolling resistance can be interpreted differently: either as the displacement e of the action line of the normal force f_n or the torque M_R, both resulting in a resistance force W_R acting on the car body. Typical values for passenger cars are $\overline{\varphi}_r = e/r = 0.01 - 0.02$.

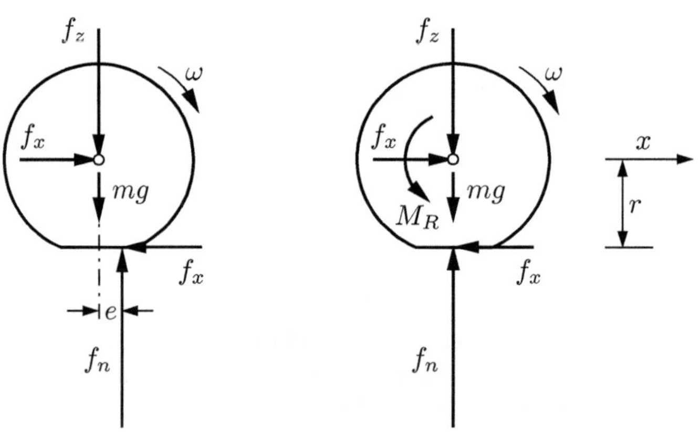

Fig. 8.2. Free wheel diagram with applied torque M_R and constraint force f_x in the bearing acting as resistance force W_R on the car body

8.2 Entire Vehicle

The equations of motion in longitudinal direction are generated now for a two-axle automobile as shown in Fig. 8.3. According to the free body principle the vehicle is represented by three bodies in the vertical plane. The vehicle model has three degrees of freedom if the suspension of the front and rear axle are neglected. The generalized coordinates chosen are the forward motion x and the rotational motions φ_V, φ_H of both wheels of the front and rear axle, respectively. In the front and rear bearings two reaction forces are acting at each axle named as F_{xV}, F_{zV}, F_{xH}, F_{zH}. The road is loaded by the normal forces N_V and N_H due to the radially rigid wheels. Applied forces are acting as aerodynamic forces and torques W_L, F_L, M_L on the vehicle body, as driving or braking torques M_V, M_H, respectively, between the vehicle body and the axles as well as tangential forces T_V, T_H on the wheels due to the tangentially elastic tires. The rolling resistance is considered by a displacement of the normal forces N_V, N_H, representing applied forces, too. The climbing resistance is characterized by the inclination angle α of the road.

Using the geometrical dimensions defined in Fig. 8.3 Newton's and Euler's equations of the plane motion read as follows.

Vehicle body

$$m\ddot{x} = -mg\sin\alpha - W_L - F_{xV} - F_{xH} \ , \tag{8.5}$$

$$0 = mg\cos\alpha - F_L - F_{zV} - F_{zH} \ , \tag{8.6}$$

$$0 = -M_L + M_V + M_H - (h-r)(F_{xV} + F_{xH}) + l_V F_{zV} - l_H F_{zH} \ . \tag{8.7}$$

Front axle

$$m_V\ddot{x} = -m_V g\sin\alpha + F_{xV} + T_V \ , \tag{8.8}$$

$$0 = F_{zV} - N_V \ , \tag{8.9}$$

$$I_V\ddot{\varphi}_V = M_V - rT_V - e_V N_V \ . \tag{8.10}$$

Rear axle

$$m_H\ddot{x} = -m_H g\sin\alpha + F_{xH} + T_H \ , \tag{8.11}$$

$$0 = F_{zH} - N_H \ , \tag{8.12}$$

$$I_H\ddot{\varphi}_H = M_H - rT_H - e_H N_H \ . \tag{8.13}$$

The mass of the vehicle body is denoted m, the mass of both wheels at each axle is m_V, m_H, and I_V, I_H are the corresponding moments of inertia. There are nine equations (8.5) to (8.13) available for the nine unknown of this model, namely x, φ_V, φ_H, F_{xV}, F_{zV}, F_{xH}, F_{zH}, N_V, N_H. Thus, the dynamics of the vehicle are completely determined.

The assumption of a constant slip $s_{AV} = s_{AH} = s_A = \text{const}$, $s_{BV} = s_{BH} = s_B$ results in additional constraints for driving or braking, respectively,

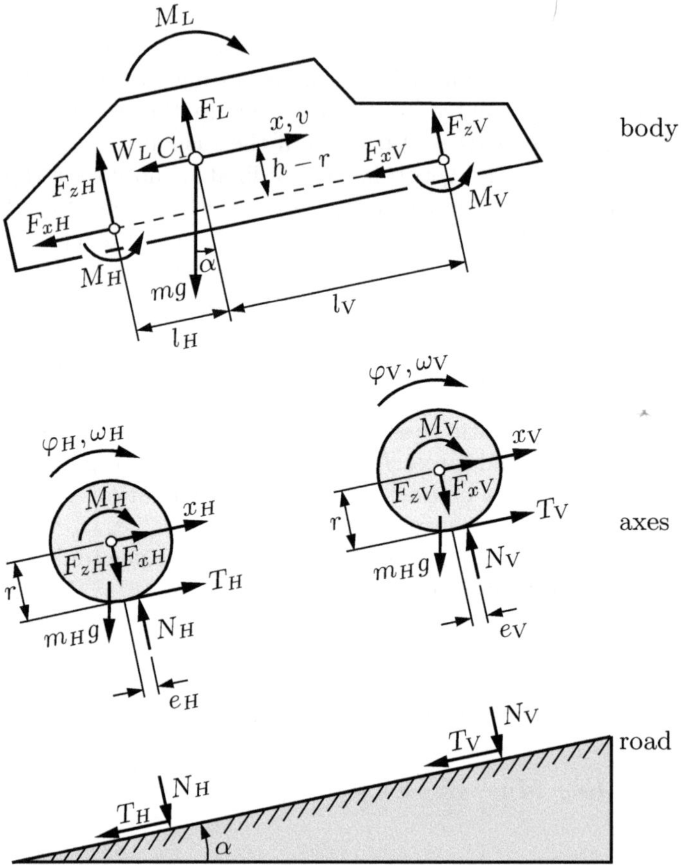

Fig. 8.3. Planar model of an automobile

$$\dot{\varphi}_V = \dot{\varphi}_H = \frac{\dot{x}}{r(1-s_A)} \qquad \text{driving}\,, \tag{8.14}$$

$$\dot{\varphi}_V = \dot{\varphi}_H = \frac{\dot{x}}{r}(1-s_B) \qquad \text{braking}\,, \tag{8.15}$$

with the consequence that the tangential forces T_V and T_H are reaction forces which can be eliminated. With (8.8) and (8.10) or (8.11) and (8.13), respectively, one gets for the axles for

driving

$$\left(m_{V,H} + \frac{I_{V,H}}{r^2}\frac{1}{1-s_A}\right)\ddot{x} = -m_{V,H}\,g\sin\alpha + F_{xV,H} + \frac{1}{r}M_{V,H} - W_{RV,H}\,, \tag{8.16}$$

braking

$$\left(m_{V,H} + \frac{I_{V,H}}{r^2}(1 - s_B)\right)\ddot{x} = -m_{V,H}g\sin\alpha + F_{xV,H} - \frac{1}{r}M_{V,H} - W_{RV,H}\ , \tag{8.17}$$

where the rolling resistance force W_R characterizes the rolling resistance according to (8.4). Furthermore, the braking torques have a negative sign. Due to the slip the generalized inertia of the driven wheel is increasing while it is decreasing for the braked wheel. The slip supports the stopping power of the vehicle in a natural manner.

Further, for vanishing slip $s = 0$ it follows from (8.5), (8.16) or (8.17) the fundamental equation of the longitudinal motion as

$$\left(m + m_V + m_H + \frac{I_V + I_H}{r^2}\right)\ddot{x} = \frac{M_V + M_H}{r} - W_L - W_R - G\sin\alpha\ , \tag{8.18}$$

where $W_R = W_{RV} + W_{RH}$ means the total rolling resistance of all four wheels and $G = (m + m_V + m_H)\,g$ represents the total weight of the vehicle.

8.3 Aerodynamic Forces and Torques

The aerodynamic forces and torques are most essential for the driving performance of the automobiles at medium and high speeds. The aerodynamics are also important for energy conservation and economic efficiency of cars.

The aerodynamic resistance is originating from three sources:

- shape resistance due to the turbulence of the air flow at the rear of the vehicle with a strong influence of the design of the vehicle (85%)
- friction resistance due to the shear flow at the car body depending on its surface (10%)
- internal resistance due to the flow through the car body (5%). The inner flow is required for the cooling of the engine and the air ventilation of the passenger compartment.

The aerodynamic resistance and therefore the aerodynamic forces are mainly resulting form the turbulent flow and they are proportional to the dynamic pressure

$$p_L = \frac{1}{2}\rho v_L^2 \tag{8.19}$$

where ρ is the air density and v_L is the velocity of the air relative to the car. The aerodynamic forces and torques depend on dimensionless coefficients c, the characteristic area A of the vehicle and the wheel base l.

The air resistance reads as

$$W_L = c_W A p_L \ , \tag{8.20}$$

the lifting force as

$$F_L = c_A A p_L \ , \tag{8.21}$$

and the aerodynamic torque is given by

$$M_L = c_M A l p_L \ . \tag{8.22}$$

Furthermore, for the lateral motion the side force

$$S_L = c_N A p_L \tag{8.23}$$

and the yaw torque

$$M_{Lz} = c_{Mz} A l p_L \tag{8.24}$$

have to be considered. The coefficients c depend as already mentioned strongly on the body design and the direction of the air flow relative to the vehicle. Some air resistance coefficient c_W for head-on flow are presented in Table 8.2. These coefficients are found experimentally in wind tunnels or evaluated by computational fluid dynamics (CFD) software. All new vehicles are subject to a thorough optimization to reduce the air resistance. As shown by (8.20) the air resistance force depends not only on the coefficients c_W often used in commercials but also on the front surface A of the vehicle.

Table 8.2. Air resistance coefficients of different vehicles types

Type of value	Air resistance coefficient c_W
Passenger car	0.3 - 0.4
Bus	0.6 - 0.7
Truck	0.6 - 1.0
Motor cycle	0.5 - 1.0

8.4 Driving and Braking Torques

The acceleration or driving torque, respectively, and the braking torque are acting on the front and/or rear axle of the vehicle depending on its design,

$$M_{V,H} = M_{AccelerationV,H} - M_{BrakingV,H} \ . \tag{8.25}$$

The braking torque M_B originates form the brake system and it is controlled by the driver via the brake pedal. Therefore, the braking torque $M_B(t)$ is an arbitrary function of time t.

The acceleration torque M_A is originating from the engine and it is transferred by the transmission to the axles. In the drivetrain the transmission ratios i_D and i_G of differential and gearbox as well as the moments of inertia I_G and I_M of gearbox and engine have to be regarded. For the engine an averaged constant moment of inertia is chosen. Further, the degree of efficiency of the transmission is an important characteristic. The degree of efficiency is around $\eta = 0.98$ for manual transmissions while $\eta = 0.85$ and more is achieved by automatic transmissions. For a front drive vehicle, e.g., the driving torque reads as

$$M_{AV} = \eta i_D i_G M_M(\dot{x}) - [(i_D i_G)^2 I_M + (i_D)^2 I_G]\frac{\ddot{x}}{r} \, . \tag{8.26}$$

The engine torque M_M depends on the engine speed as shown in Fig. 8.4. The engine speed ω_M is related to the vehicle speed $\dot{x}$ as

$$\omega_M = i_D i_G \frac{\dot{x}}{r} \, . \tag{8.27}$$

Thus, the fundamental equation of the longitudinal motion (8.18) can be rewritten as

$$\hat{M}\ddot{x} = \hat{A} - \hat{B} - G\sin\alpha - W_L - W_R \tag{8.28}$$

using the following abbreviations:
generalized inertia

$$\hat{M} = m + m_V + m_H + \frac{I_V + I_H}{r^2} + i_D^2\frac{I_G}{r^2} + i_D^2 i_G^2\frac{I_M}{r^2} \, ,$$

driving force

$$\hat{A} = \frac{1}{r}\eta i_D i_G M_M(\dot{x}) \, ,$$

total braking force

$$\hat{B} = \frac{1}{r}(M_{BV} + M_{BH}) \, ,$$

vehicle weight

$$G = (m + m_V + m_H)g \, ,$$

air resistance force

$$W_L = W_L(\dot{x}^2)$$

rolling resistance force

$$W_R = \overline{\varphi}_r G\cos\alpha \, .$$

It is to be mentioned that (8.18) is subject to numerous assumptions where the vanishing tire slip and the constant moment of inertia of the engine are just two of them.

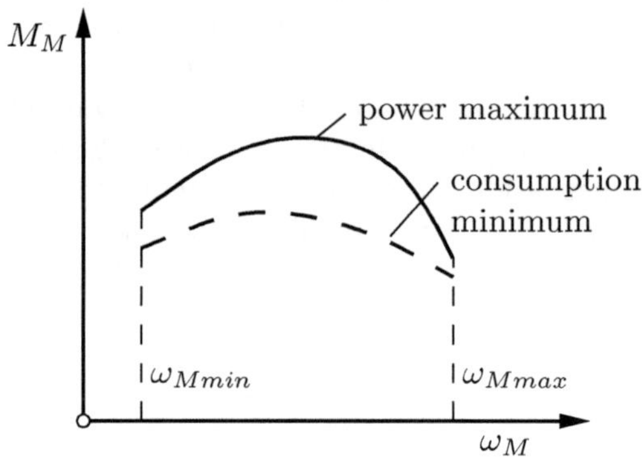

Fig. 8.4. Engine torque M_M as a function of the engine speed ω_M for a combustion engine

■ Problem 8.2 Acceleration of an automobile

For the evaluation of the acceleration of an automobile on a horizontal road the following data are given.

Total mass	$m + m_V + m_H = 1200\,\text{kg}$,
Moment of inertia at one axle	$I_V = I_H = 1.8\,\text{kg}\,\text{m}^2$,
Moment of inertia of transmission	$I_G = 0$,
Moment of inertia of engine	$I_M = 0.25\,\text{kg}\,\text{m}^2$,
Transmission ratio (high gear)	$i_D i_G = 4$,
Wheel radius	$r = 0.3\,\text{m}$,
Engine torque	$M_M = 300\,\text{Nm}$,
Engine speed	$\omega_M = 360\,\text{s}^{-1}$,
Degree of efficiency	$\eta = 0.9$,
Braking forces	$B = 0$,
Inclination of road	$\alpha = 0$,
Front area	$A = 2\,\text{m}^2$,
Air density	$\rho = 1.2\,\text{kg/m}^3$,
Air resistance coefficient	$c_W = 0.3$,
Rolling resistance coefficient	$\overline{\varphi}_r = 0.02$.

Solution

The solution follows from the fundamental equation (8.28). The quantities required are computed as follows.

The vehicle speed is according to (8.27)

$$v = \dot{x} = \frac{\omega_M r}{i_D i_G} = 97\,\text{km/h}\,. \tag{1}$$

The air resistance force W_L is given by (8.18), (8.20) and (1)

$$W_L = \frac{1}{2} c_W A \rho v_L^2 = 262\,\mathrm{N}\,. \tag{2}$$

For the rolling resistance force it yields according to (8.4)

$$W_R = \overline{\varphi}_r (m + m_V + m_H) g = 235\,\mathrm{N}\,. \tag{3}$$

Further, one gets the driving force

$$\hat{A} = \frac{1}{r} \eta i_D i_G M_M = 3600\,\mathrm{N} \tag{4}$$

and the generalized inertia as

$$\hat{M} = 1284\,\mathrm{kg}\,. \tag{5}$$

Thus, for the acceleration it yields

$$a = \ddot{x} = \frac{1}{\hat{M}} (\hat{A} - W_L - W_R) = 2.4\,\mathrm{m/s^2}\,. \tag{6}$$

It turns out that the rotating part of engine and wheels have only a small influence on the acceleration in a high gear. ■

8.5 Driving Performance

In this section some quantities characterizing the driving performance are presented.

The *maximum speed* is defined for a horizontal road. With $\alpha = 0$, $\hat{B} = 0$ and $\ddot{x} = 0$ it follows from (8.28) an equation for the maximum speed

$$\hat{A}(v_{max}) = W_R + W_L(v_{max}^2)\,. \tag{8.29}$$

A graphical representation is shown in the driving performance diagram, Fig. 8.5, which is obtained from the motor characteristics, Fig. 8.4, and five different gear or transmission ratios, respectively.

The *gradability* is found for small grade angles $p = \tan(\alpha) \approx \sin(\alpha)$, $\cos(\alpha) \approx 1$ with $\ddot{x} = 0$, $\hat{B} = 0$ also from (8.28) as

$$p_{max} = \frac{\hat{A}_{max}}{G} - \frac{W_L(v(\hat{A}_{max}))}{G} - \overline{\varphi}_r\,. \tag{8.30}$$

The maximal available driving force $\hat{A}_{max}$ and the corresponding speed $v(\hat{A}_{max})$ can be found from the driving performance diagram, too.

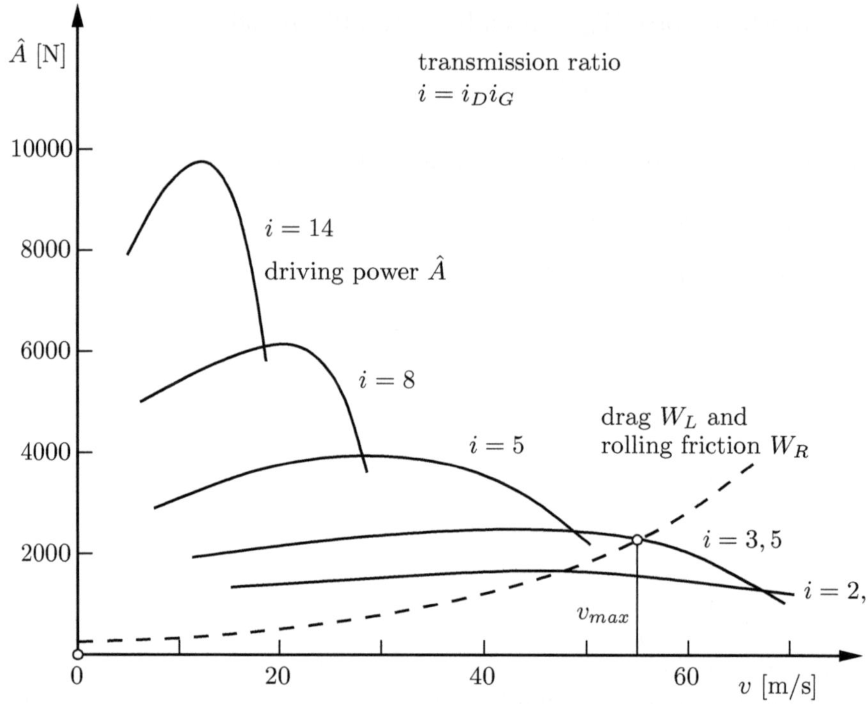

Fig. 8.5. Driving performance diagram, driving and resistance forces depending on the vehicle speed v for different transmission ratios

For the evaluation of the *traction limits* for maximum acceleration and maximum braking the tangential forces $T_{V,H}$ have to be determined in addition. For this purpose the equations (8.10) and (8.13) with (8.14), (8.15) and (8.28) are at hand. Neglecting the moments of inertia of the wheels it follows immediately form (8.10) or (8.13), respectively, in the case of braking for the tangential forces

$$T_{V,H} = -\hat{B}_{V,H} - W_{RV,H} \ , \tag{8.31}$$

where only the braking forces $\hat{B}_{V,H} = M_{BV,H}/r$ appear. The maximum tangential forces are determined by the maximal rolling contact coefficient $\overline{\varphi}_{max}$ and the wheel load at the front and rear axle F_{zV}, F_{zH} known from (8.6) and (8.7),

$$T_{V,H,max} \leq \overline{\varphi}_{max} F_{zV,H} \ . \tag{8.32}$$

Corresponding results can be derived for the acceleration of the vehicle. Then, it turns out that four wheel drive improves the traction limits what is most valuable for driving in winter time.

The *mileage* of a car is also an important feature of a vehicle. As shown in Fig. 8.4 the consumption minimum of a combustion engine is different from its maximum power. A low mileage is achieved for low and medium engine speeds. This properties can be considered in the transmission design. Often a vehicle is more economic in the fifth gear as in the fourth gear which may deliver the maximum speed, see Fig. 8.5.

9

Lateral Motions

The lateral motion of vehicles is discussed separately for road and rail vehicles. Due to different constraints both kinds of vehicles require dynamical approaches different from each other. In contrary to automobiles equipped with actively operated steering, railway vehicles have a passive lateral guidance. For automobiles driving stability is achieved or improved, respectively, by interplay of the driver and the vehicle.

9.1 Handling of Road Vehicles

The analysis of automobile handling is an extremely complex problem. But the fundamental handling behavior of road vehicles can already be understood by the strongly simplified model published by Riekert and Schunck (1940) today also denoted as bicycle model. This model is based on the following assumptions:

- constant vehicle speed, vanishing longitudinal acceleration
- vanishing width of the vehicle, no roll motion
- constant wheel loads, no heave and pitch motion
- small displacements, linear tire forces
- wheels without inertia.

Such a simplified vehicle model is depicted in Fig. 9.1.

9.1.1 Elastic Wheel

For the road vehicle handling the elastic wheel is the most important component, treated in detail in Sect. 3.4.4. Here, the essential relations are reviewed once again. The lateral force

$$S = \overline{\varphi}_2(\alpha) f_n \tag{9.1}$$

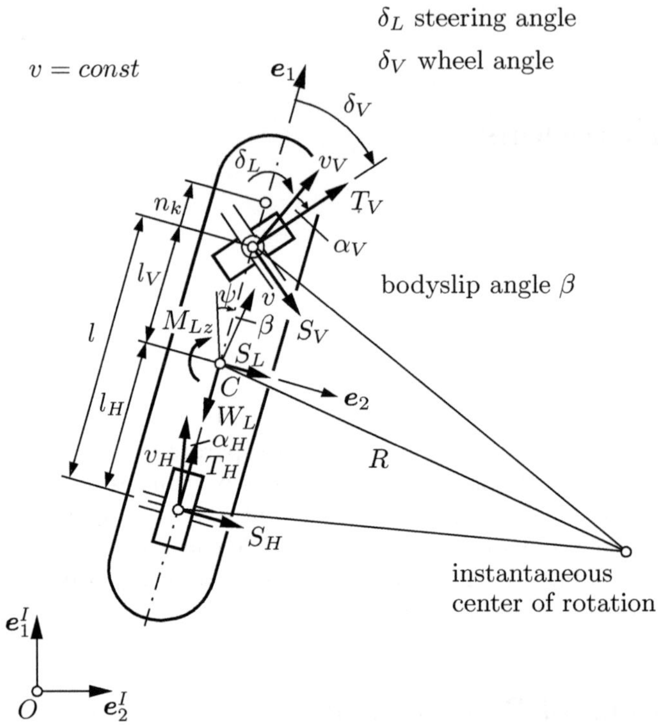

Fig. 9.1. Simplified vehicle model or bicycle model, respectively

is depending on the slip angle α or the lateral slip ν_2, respectively, see (3.112) as well as on thc normal wheel load f_n. The characteristic of the rolling contact coefficient $\overline{\varphi}_2$ is shown in Fig. 9.2. The action point of the lateral force f_{t2} is displaced to the rear by the trail n_S according to (3.130). For small slip angles $\alpha \ll 1$ and constant normal load $f_n = \text{const}$ it follows from Fig. 9.2 a lateral force linearly depending on the slip angle and a constant trail

$$S = k_S \alpha \,, \quad n_S = \text{const} \tag{9.2}$$

where k_S is the lateral force coefficient or cornering stiffness, respectively. According to (3.115) and (3.130) it yields $k_S = 2k_2a^2$ and $n_S = a/3$, respectively. The lateral force coefficient and the trail are usually found from experiments. Characteristic values are given by Mitschke and Wallentowitz (2004). For a rough estimate $k_S = 40\,\text{kN/rad}$ and $n_S = 3\,\text{cm}$ represent characteristic numbers for medium-size automobiles.

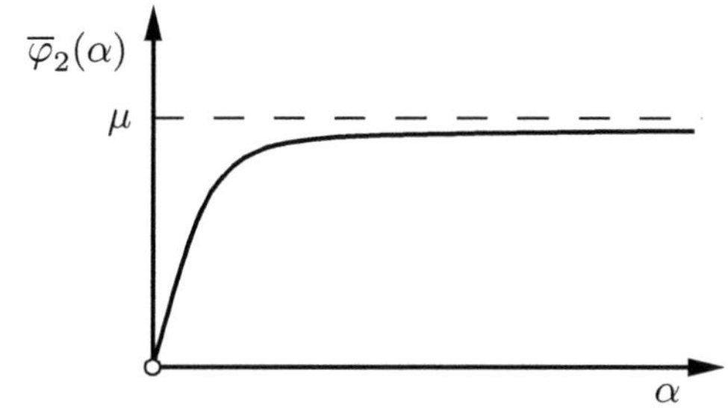

Fig. 9.2. Rolling contact coefficient $\overline{\varphi}_2$ of a tire depending on the slip angle α

9.1.2 Vehicle Model

The whole vehicle is represented by the model shown in Fig. 9.1 running on elastic wheels without inertia. Thus, a system consisting of one rigid body is given performing a motion in a horizontal plane. The vehicle speed is assumed to be constant, $v = \text{const}$, what is considered as a nonholonomic constraint. For the evaluation of the equations of motion a vehicle-fixed moving reference frame $\{C, \boldsymbol{e}_v\}$ is used resulting in a simple representation of the tire forces.

As generalized coordinates for the description of the $f = 3$ degrees of freedom in the horizontal plane the coordinates x,y of the center of mass C in the inertial frame $\{0, \boldsymbol{e}_v^I\}$ and the yaw angle ψ are introduced. Due to the nonholonomic constraint

$$v = \sqrt{\dot{x}^2 + \dot{y}^2} = \sqrt{v_1^2 + v_2^2} = \text{const} \tag{9.3}$$

there remain $g = 2$ generalized velocities, the body slip angle β and the yaw angular velocity $\dot{\psi}$. The body slip angle β characterizes the $\boldsymbol{e}_2$-component of the vehicle speed vector and represents therefore a normalized velocity. Then, the kinematic relations of the plane motion in the inertial frame read as

$$\left.\begin{aligned} \dot{x} &= v\cos(\psi + \beta), \\ \dot{y} &= v\sin(\psi + \beta), \\ \dot{\psi} &= \dot{\psi}, \end{aligned}\right\} \tag{9.4}$$

while one gets in the moving vehicle-fixed frame

$$\left.\begin{aligned} v_1 &= v\cos\beta, \\ v_2 &= v\sin\beta, \\ \omega_3 &= \dot{\psi} \end{aligned}\right\} \tag{9.5}$$

Obviously, both representations (9.4) and (9.5) comply with the constraint (9.3).

The accelerations are now computed in the moving vehicle-fixed frame according to Sect. 2.2.3 resulting in

$$\begin{bmatrix} a_1 \\ a_2 \\ a_3 \end{bmatrix} = \begin{bmatrix} \dot{v}\cos\beta - v\dot{\beta}\sin\beta \\ \dot{v}\sin\beta + v\dot{\beta}\cos\beta \\ 0 \end{bmatrix} + \begin{bmatrix} 0 & -\dot{\psi} & 0 \\ +\dot{\psi} & 0 & 0 \\ 0 & 0 & 0 \end{bmatrix} \begin{bmatrix} v\cos\beta \\ v\sin\beta \\ 0 \end{bmatrix} . \tag{9.6}$$

Considering $v = \text{const}$ and a small body slip angle $\beta \ll 1$, it remains

$$\left.\begin{aligned} a_1 &= -v\beta\dot{\psi}, \\ a_2 &= v(\dot{\psi} + \dot{\beta}), \\ \alpha_3 &= \ddot{\psi}, \end{aligned}\right\} \tag{9.7}$$

where the first two equations represent the lateral acceleration supplemented by the yaw angle acceleration. In matrix notation,

$$\begin{bmatrix} a_1 \\ a_2 \\ \alpha_3 \end{bmatrix} = \underbrace{\begin{bmatrix} 0 & 0 \\ v & 0 \\ 0 & 1 \end{bmatrix}}_{\overline{\boldsymbol{L}}} \underbrace{\begin{bmatrix} \dot{\beta} \\ \ddot{\psi} \end{bmatrix}}_{\dot{\boldsymbol{z}}} + \begin{bmatrix} -v\beta\dot{\psi} \\ v\dot{\psi} \\ 0 \end{bmatrix} , \tag{9.8}$$

the 3×2-Jacobian matrix $\overline{\boldsymbol{L}}$ appears and the 2×1-vector $\boldsymbol{z}(t)$ of the generalized velocities is introduced.

In the wheel contact points the lateral velocities v_{2V} and v_{2H} are found which follow with small slip angles $\alpha_{V,H}$ and a front wheel with small angle δ_V according to Fig. 9.1 as

$$v_{2V} = (\delta_V - \alpha_V)v = v\beta + l_V\dot{\psi} , \tag{9.9}$$

$$v_{2H} = -\alpha_H v = v\beta - l_H\dot{\psi} . \tag{9.10}$$

Then, one gets immediately the slip angles at the front axle V and the rear axle H as

$$\alpha_V = \delta_V - \beta - \frac{l_V\dot{\psi}}{v} , \tag{9.11}$$

$$\alpha_H = -\beta + \frac{l_H\dot{\psi}}{v} . \tag{9.12}$$

The front wheel angle δ_V depends on the steering angle δ_L where only the elasticity of the steering assembly is considered. The equilibrium of the torque at the front wheel results in

$$c_L\left(\frac{\delta_L}{i_L} - \delta_V\right) = (n_k + n_S)S_V , \tag{9.13}$$

where c_L means the generalized torsion spring coefficient. Further, i_L is the transmission ratio of the steering gear, n_k the suspension trail, n_S the trail of the tire and S_V the lateral tire force. Thus, on the right hand side of (9.13) the tire reset torque of the vehicle is found. Neglecting the steering elasticity, $c_L \to \infty$, it follows from (9.13)

$$\delta_V = \frac{1}{i_L}\delta_L \,, \tag{9.14}$$

i.e., the front wheel angle δ_V can be chosen directly as input variable for the handling of the vehicle.

In the contact points of the wheels there are acting longitudinal and lateral tire forces $T_{V,H}$ and $S_{V,H}$. Furthermore, the vehicle is subject to the air resistance force W_L, lateral aerodynamic force S_L and the aerodynamic torque M_{Lz}. For small slip angles it yields in addition

$$T_{V,H} \ll N \,, \quad S_{V,H} \ll N \tag{9.15}$$

where N is the vertical tire load.

The Newton-Euler equations read with (9.8) and the mentioned forces and torques neglecting quadratically small terms as

$$\begin{bmatrix} 0 & 0 \\ mv & 0 \\ 0 & I \end{bmatrix} \begin{bmatrix} \dot\beta \\ \ddot\psi \end{bmatrix} + \begin{bmatrix} -mv\dot\psi\beta \\ mv\dot\psi \\ 0 \end{bmatrix} = \begin{bmatrix} T_V + T_H - W_L \\ S_V + S_H + S_L \\ S_V l_V - S_H l_H + M_{Lz} \end{bmatrix} . \tag{9.16}$$

Here, m is the mass and I the moment of the inertia of the vehicle with respect to the yaw axis. The equations of motion follow from Jourdain's principle (2.81) what means in this case a left multiplication of (9.16) with the transposed Jacobian matrix $\overline{\boldsymbol{L}}^{\mathrm{T}}$ from (9.8):

$$\begin{bmatrix} mv^2 & 0 \\ 0 & I \end{bmatrix} \begin{bmatrix} \dot\beta \\ \ddot\psi \end{bmatrix} + \begin{bmatrix} mv^2\dot\psi \\ 0 \end{bmatrix} = \begin{bmatrix} v(S_V + S_H + S_L) \\ S_V l_V - S_H l_H + M_{Lz} \end{bmatrix} . \tag{9.17}$$

The reaction force related to the nonholonomic constraint (9.3) reads from (9.16) as

$$T_V + T_H = W_L - mv\dot\psi\beta \,, \tag{9.18}$$

where an allocation of the reaction force $T_V + T_H$ to the front and rear axle is not possible due to the simplicity of the model chosen.

Considering the linear elastic characteristic of the tires with respect to the small lateral forces (9.15) assumed, it yields

$$S_{V,H} = k_{S\,V,H}\,\alpha_{V,H} \,. \tag{9.19}$$

Inserting the slip angles (9.11) and (9.12) one gets from (9.17) and (9.19) the differential equations of the Riekert-Schunck model as

$$m v\dot{\beta} + (k_{SV} + k_{SH})\beta + \left(mv + \frac{k_{SV}l_V - k_{SH}l_H}{v}\right)\dot{\psi} = S_L + k_{SV}\delta_V \,, \tag{9.20}$$

$$I\ddot{\psi} + \frac{1}{v}(k_{SV}l_V^2 + k_{SH}l_H^2)\dot{\psi} + (k_{SV}l_V - k_{SH}l_H)\beta = M_{LZ} + k_{SV}l_V\delta_V \,. \tag{9.21}$$

These equations of motion are very useful for a fundamental analysis for steady-state cornering and driving stability of an automobile.

9.1.3 Steady-state Cornering

For the steady-state cornering on a circular path with vanishing aerodynamic forces and torques a relation between the body slip angle and the radius of the path can be evaluated. The corresponding assumptions are

$$S_L = M_{LZ} = 0 \,, \tag{9.22}$$

$$\ddot{\psi} = \dot{\beta} = 0 \,, \quad \dot{\psi}_0 = \text{const} \,, \quad \beta_0 = \text{const} \,. \tag{9.23}$$

Comparing to (9.8) it is obvious that (9.23) means stationary behavior, indeed. The vehicle is travelling with constant speed according to (9.3) and with constant angular velocity $\dot{\psi}_0$ due to (9.23). Thus, the radius R of the circular path of the center of mass C follows from

$$R = \frac{v}{\dot{\psi}_0} \tag{9.24}$$

according to Fig. 9.1.

From (9.20) and (9.21) one gets together with (9.22) and (9.23) the linear, inhomogeneous system of equations

$$(k_{SV} + k_{SH})\beta_0 + \left(mv + \frac{k_{SV}l_V - k_{SH}l_H}{v}\right)\dot{\psi}_0 = k_{SV}\delta_V \,, \tag{9.25}$$

$$(k_{SV}l_V - k_{SH}l_H)\beta_0 + \frac{k_{SV}l_V^2 + k_{SH}l_H^2}{v}\dot{\psi}_0 = k_{SV}l_V\delta_V \,. \tag{9.26}$$

After multiplication of (9.25) with l_V, subtraction of (9.26) and consideration of (9.24) it remains for the body slip angle

$$\beta_0 = \frac{l_H}{R}\left(1 - \frac{l_V}{k_{SH}l_H l}mv^2\right) \tag{9.27}$$

where $l = l_V + l_H$ is the wheel base.

The first term characterizes the behavior of a vehicle with rigid wheels while the second term shows the influence of the elastic wheel with the cornering stiffness k_{SH}. Inserting (9.27) in (9.25) one gets after some manipulations

$$\delta_V = \frac{l}{R}\left(1 + \frac{k_{SH}l_H - k_{SV}l_V}{k_{SV}k_{SH}l^2}mv^2\right) . \tag{9.28}$$

The steering behavior undergoes a noticeable change increasing with the square of the speed v if the term $k_{SH}l_H - k_{SV}l_V$ is not vanishing. In vehicle dynamics this phenomenon is called understeer or oversteer, respectively. For

$$k_{SH}l_H > k_{SV}l_V \tag{9.29}$$

understeering occurs, i.e., the steering angle δ_V required from the driver is larger than for neutral steering. On the other hand,

$$k_{SH}l_H < k_{SV}l_V \tag{9.30}$$

means the contrary, i.e., a smaller steering angle δ_V is sufficient. Oversteer vehicles are more easy to drive and less effort of the driver is required, they appear to be more favourable at a first glance.

9.1.4 Driving Stability

The driving stability is analyzed for a vehicle running straight ahead, $\delta_V = 0$, and vanishing aerodynamics $S_L = M_{LZ} = 0$. Then, it remains from (9.20) and (9.21) a homogeneous system of differential equations for the generalized velocities denoted in matrix form as

$$\begin{bmatrix} \dot{\beta} \\ \ddot{\psi} \end{bmatrix} = \underbrace{\begin{bmatrix} -\dfrac{k_{SV}+k_{SH}}{mv} & -1-\dfrac{k_{SV}l_V - k_{SH}l_H}{mv^2} \\ -\dfrac{k_{SV}l_V - k_{SH}l_H}{I} & -\dfrac{k_{SV}l_V^2 + k_{SH}l_H^2}{Iv} \end{bmatrix}}_{\boldsymbol{A}} \begin{bmatrix} \beta \\ \dot{\psi} \end{bmatrix} . \tag{9.31}$$

The driving stability depends on the eigenvalues of the system matrix $\boldsymbol{A}$, see Sect. 7.2.1. The eigenvalues are computed via the characteristic equation

$$\det(\lambda E - A) = \lambda^2 + a_1\lambda + a_2 = 0 \tag{9.32}$$

with the characteristic coefficients

$$a_1 = \frac{k_{SV}+k_{SH}}{mv} + \frac{k_{SV}l_V^2 + k_{SH}l_H^2}{Iv} \tag{9.33}$$

and

$$a_2 = \frac{k_{SV}k_{SH}l^2}{Imv^2}\left(1 + \frac{k_{SH}l_H - k_{SV}l_V}{k_{SV}k_{SH}l^2}mv^2\right) . \tag{9.34}$$

The eigenvalues of (9.32) are

$$\lambda_{1,2} = -\frac{a_1}{2} \pm \sqrt{\frac{a_1^2}{4} - a_2} \,. \tag{9.35}$$

Asymptotic stability is guaranteed according to Sect. 7.2.1 if all eigenvalues have negative real parts. Then, the characteristic coefficients have to meet the Hurwitz conditions

$$a_1 > 0 \,, \quad a_2 > 0 \,. \tag{9.36}$$

For the vehicle under consideration $a_1 > 0$ is always meet, see (9.33). The second condition is for understeer vehicles (9.29) also meet while for oversteer vehicles (9.30) a critical vehicle speed v_{crit} exists which shall not to be exceeded,

$$v_{crit}^2 = \frac{1}{m} \frac{k_{SV} k_{SH} l^2}{k_{SV} l_V - k_{SH} l_H} \,. \tag{9.37}$$

This means that oversteer vehicles may be unstable. Therefore, today's vehicles are designed to be slightly understeering and, thus, naturally stable. Higher steering effort by the driving is the price to be paid for a reliable driving stability.

In addition to the driving stability naturally designed, by electronic devices the stability behavior of vehicle can be achieved artificially. The mechanical design is then supplemented by a control system device, see e.g., Kiencke and Nielsen (2005) and Rajamani (2006).

■ Problem 9.1 Driving stability of a road vehicle

With the assumptions of the Riekert-Schunck or bicycle model, respectively, a rough estimation of the driving stability of an automobile can be obtained. The tires may have the same cornering stiffness $k_{SV} = k_{SH} = 80\,\text{kN/rad}$ at the front and the rear axle. The critical speed has to be evaluated depending on the position of the center of mass. The wheelbase is $l = 3\,\text{m}$, the mass of the vehicle is given as $m = 1200\,\text{kg}$.

Solution

The position of the center of mass measured from the front axle is introduced as independent variable. Then, it yields

$$l_H = l - l_V \,. \tag{1}$$

Inserting (1) in (9.37) one gets with $k_{SV} = k_{SH} = k_S$

$$v_{crit} = \sqrt{\frac{k_S l}{m} \frac{1}{2(l_V/l) - 1}} \,. \tag{2}$$

For the given data the factor $\sqrt{k_S l/m} = \sqrt{200}\,\text{m/s} = 14.14\,\text{m/s}$ is found. The result is shown in Fig. 9.3. It is obvious that the critical speed may be low for rear loaded vehicles. That is the reason why today almost all automobiles are designed with front engine. ■

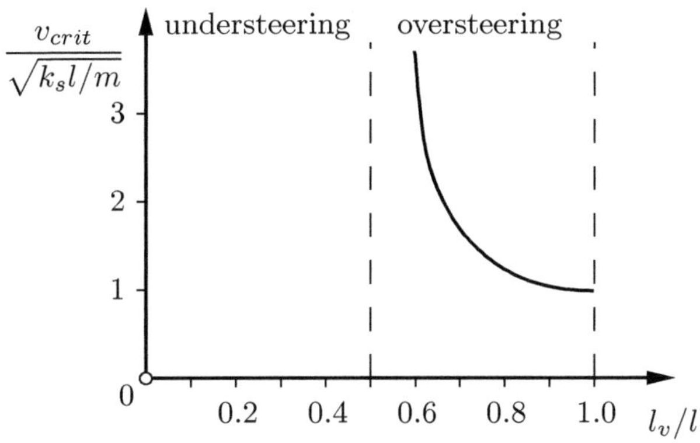

Fig. 9.3. Critical speed v_{crit} of an automobile depending on the position l_V/l of the center of mass

9.1.5 Experimental Studies

The complex dynamics of lateral motions of road vehicles requires in addition to design and simulation also testing and validation. For this purpose all major automobile manufactures are using special test tracks, see e.g. ATP Automotive Testing Papenburg (2009). Moreover, there are special measurement devices for vehicle dynamics testing available like the dSPACE tools, see Nishimura and Ohoha (2007). Such sophisticated tools enable the test engineer to perform measurement, calibration, and bypassing tasks with a minimal workload.

The question of checking of simulation results found by theoretical models is discussed in simulation engineering under the heading of validation, often supplemented by model verification. With strongly increasing cost in automotive testing on one hand, and the steadily increasing computational power of simulation tools on the other hand, it is a very responsible engineering task to find a good compromise between simulation and testing.

9.2 Driving Stability of Railways

A lateral motion of a railway vehicle occurs already in the case of pure rolling due to the hunting motion of the wheelsets. The corresponding differential equations have been analyzed in Problem 7.2. Due to the more general contact phenomena between wheel and rail wheelsets show slip and may perform lateral motions independent from the longitudinal motion. The fundamental stability behavior of railways can be already discussed for a single wheelset. For this purpose the equations of motion found in Problem 3.5 will be used and analyzed in detail.

9.2.1 Equation of Motion of a Railway Wheelset

The equations of motion (27) of Problem 3.5 are simplified for a stability analysis as follows. The contact force coefficients are referred to the weight $G = mg$ of the wheelset, they are equal with respect to the 1- and 2-direction and neglected in the 3-direction. Then, it yields

$$f_{11} = \frac{1}{2}hmg\,, \quad f_{22} = \frac{1}{2}hmg\,, \quad f_{23} = f_{33} = 0 \tag{9.38}$$

with a rolling contact coefficient h. The moment of inertia related to the 1-axis is approximated by $I_1 = ma^2$, i.e., the wheel inertia is concentrated in its center of mass what is feasible according to Problem 2.8. The moment of inertia related to the 2-axis is introduced as $I_2 = \frac{1}{2}mr_0^2$. Then, it remains

$$\begin{bmatrix} 1 & 0 \\ 0 & a^2 \end{bmatrix} \begin{bmatrix} \ddot{y} \\ \ddot{\gamma} \end{bmatrix} + \begin{bmatrix} \frac{hg}{v_0} & -\frac{r_0 v_0}{2a}\delta_0 \\ \frac{r_0 v_0}{2a}\delta_0 & \frac{hga^2}{v_0} \end{bmatrix} \begin{bmatrix} \dot{y} \\ \dot{\gamma} \end{bmatrix} + \begin{bmatrix} g\frac{\delta_0}{a} & -hg \\ \frac{hga\delta_0}{r_0} & 0 \end{bmatrix} \begin{bmatrix} y \\ \gamma \end{bmatrix} = 0\,. \tag{9.39}$$

Due to the assumption introduced the mass of the wheelset disappears in the equations of motion (9.39) what makes the stability analysis more simple.

9.2.2 Stability of a Free Wheelset

The motion behavior of a free wheeset without bogie and vehicle body is studied for different speeds. It turns out that the wheelset is an inherent unstable component which can fulfil its task only in concurrence within the whole railway vehicle.

For very small speeds only the rolling contact has to be considered. Then, it remains from the equations of motion (9.39) only the terms directly depending on the contact forces

$$\begin{bmatrix} \frac{hg}{v_0} & 0 \\ 0 & \frac{hga^2}{v_0} \end{bmatrix} \begin{bmatrix} \dot{y} \\ \dot{\gamma} \end{bmatrix} + \begin{bmatrix} 0 & -hg \\ \frac{hga\delta_0}{r_0} & 0 \end{bmatrix} \begin{bmatrix} y \\ \gamma \end{bmatrix} = 0\,. \tag{9.40}$$

The characteristic equation of this system reads as

$$\det \begin{bmatrix} \frac{\lambda}{v_0 a} & -\frac{1}{a} \\ \frac{\delta_0}{r_0} & \frac{\lambda a}{v_0} \end{bmatrix} (hga)^2 = 0 \tag{9.41}$$

or

$$\lambda^2 + \frac{\delta_0}{r_0 a} v_0^2 = 0 \,. \tag{9.42}$$

Considering the abbreviations (23) of Problem 2.7, one gets

$$\lambda_{1,2} = \pm i \sqrt{\frac{1}{r_0 q}} v_0 \,, \tag{9.43}$$

what is exactly the frequency of the hunting motion, see (3) in Problem 7.1. In contrary to Problem 2.7 and Problem 7.1 the time was not normalized so that in (9.43) still the vehicle speed v_0 appears.

For small speeds the weight and the contact forces define the motion. The corresponding terms of the equation of motion (9.39) are

$$\begin{bmatrix} \frac{hg}{v_0} & 0 \\ 0 & \frac{hga^2}{v_0} \end{bmatrix} \begin{bmatrix} \dot{y} \\ \dot{\gamma} \end{bmatrix} + \begin{bmatrix} \frac{g\delta_0}{a} & -hg \\ \frac{hga\delta_0}{r_0} & 0 \end{bmatrix} \begin{bmatrix} y \\ \gamma \end{bmatrix} = 0 \,. \tag{9.44}$$

According to the Hurwitz-Kriterium (9.36) asymptotic stability is given. The gravitational force in concurrence with the contact forces result in energy dissipation and a damped hunting motion occurs. The corresponding normalized damping coefficient reads as

$$D = \frac{\frac{\delta_0 v_0}{ha}}{2\sqrt{\frac{\delta_0 v_0^2}{r_0 a}}} = \frac{1}{2}\sqrt{\frac{\delta_0 r_0}{h^2 a}} \tag{9.45}$$

and for the eigenfrequency it yields

$$\omega_D = \omega_0 \sqrt{1 - D^2} \,, \tag{9.46}$$

where ω_0 is the eigenfrequency in the undamped case.

For medium speeds the motion is determined by gravitation, rolling contact and inertia. However, the gyroscopic forces are still neglected. Then, characteristic equation reads as

$$\det \begin{bmatrix} \lambda^2 + \frac{hg}{v_0}\lambda + \frac{g\delta_0}{a} & -hg \\ \frac{hg\delta_0}{r_0 a} & \lambda^2 + \frac{hg}{v_0}\lambda \end{bmatrix} a^2 = 0 \,, \tag{9.47}$$

or

$$\lambda^4 + \underbrace{2\frac{hg}{v_0}}_{a_1}\lambda^3 + \underbrace{\left(\frac{h^2g^2}{v_0^2} + \frac{g\delta_0}{a}\right)}_{a_2}\lambda^2 + \underbrace{\frac{hg^2\delta_0}{av_0}}_{a_3}\lambda + \underbrace{\frac{h^2g^2\delta_0}{r_0 a}}_{a_4} = 0\,. \tag{9.48}$$

According to the Hurwitz criterium, see e.g. Mueller and Schiehlen (1985) the conditions on the characteristic coefficients are for asymptotic stability

$$a_1 > 0\,, \quad a_2 > 0\,, \quad a_3 > 0\,, \quad a_4 > 0\,, \tag{9.49}$$

and

$$H_3 = a_1a_2a_3 - a_3^2 - a_1^2a_4 > 0\,. \tag{9.50}$$

The conditions (9.49) are completely met for the characteristic coefficients in (9.48). Therefore, only the Hurwitz determinant (9.50) has to be investigated. From

$$H_3 = \frac{h^2g^4\delta_0}{av_0^2}\left[\frac{2gh^2}{v_0^2} + \frac{\delta_0}{a} - \frac{4h^2}{r_0}\right] > 0 \tag{9.51}$$

it follows the critical speed

$$v_0^2 < v_{crit}^2 = \frac{gr_0}{2(1-D^2)}\,, \tag{9.52}$$

where the normalized damping coefficient (9.45) was used as an abbreviation. For the critical speed (9.52) the wheelset is operating at its vibration boundary, the critical frequency is

$$\omega_{crit}^2 = \frac{a_3}{a_1} = \frac{g\delta_0}{2a}\,. \tag{9.53}$$

For high speeds only the gravitational, gyroscopic and inertia forces are considered. The limited contact forces are neglected here. The equations of motion read now as

$$\begin{bmatrix} 1 & 0 \\ 0 & a^2 \end{bmatrix}\begin{bmatrix} \ddot{y} \\ \ddot{\gamma} \end{bmatrix} + \begin{bmatrix} 0 & -\frac{r_0v_0}{2a}\delta_0 \\ \frac{r_0v_0}{2a}\delta_0 & 0 \end{bmatrix}\begin{bmatrix} \dot{y} \\ \dot{\gamma} \end{bmatrix} + \begin{bmatrix} \frac{g\delta_0}{a} & 0 \\ 0 & 0 \end{bmatrix}\begin{bmatrix} y \\ \gamma \end{bmatrix} = 0\,. \tag{9.54}$$

The corresponding characteristic equation is

$$\lambda^4 + \left(\frac{r_0^2v_0^2}{4a^4}\delta_0^2 + \frac{g\delta_0}{a}\right)\lambda^2 = 0\,. \tag{9.55}$$

The eigenvalues are obtained as

$$\lambda_{1,2}^2 = 0\,, \quad \lambda_{3,4}^2 = -\left(\frac{r_0^2v_0^2}{4a^4}\delta_0^2 + \frac{g\delta_0}{a}\right)\,. \tag{9.56}$$

The wheelset is not asymptotically stable due to the double zero eigenvalue. Moreover the wheelset is unstable due to the double nullity of the characteristic matrix, see e.g. Mueller and Schiehlen (1985). At very high speeds the wheelset which is a rigid body behaves like a gyro. The nutation frequency is found as $\omega_N = (r_0\delta_0/2a^2)v_0$ for $v_0 > ga$.

Thus, it turns out that a railway wheelset is a dynamical critical component. Only in concurrence of several wheelsets in bogies under the vehicle body itself the indispensable asymptotic behavior is achieved. Obviously the design of railway vehicles for very high speeds is a challenging task for vehicle engineers. The latest world record of a high-speed railway vehicle was established by the French railways SNCF on 3 April 2007 with 574.8 km/h or 160 m/s respectively on the new track Strassbourg-Paris.

■ Problem 9.2 Stabilization of railway wheelsets

For the following numbers the critical parameters of a railway wheelset have to be evaluated. Engineering measures to be taken for the stabilization of wheelsets shall be discussed. The number given: $a = 0.75\,\mathrm{m}$, $\delta_0 = 26\,\mathrm{grd} = 0.45\,\mathrm{rad}$, $h = 100$.

Solution

A railway wheelset is characterized by the following parameters for increasing speed. The frequency of the hunting motion is according to (9.43)

$$\omega_S = \sqrt{\frac{\delta_0}{r_0 a}} v_0 \quad \text{with} \quad \sqrt{\frac{\delta_0}{r_0 a}} = 1.09 \left[\frac{1}{\mathrm{m}}\right] . \tag{1}$$

It has to be pointed out that a pure hunting motion does occur only for very low speeds v_0 without any slipping in the contact point.

Considering gravity one gets the normalized damping coefficient (9.45) as

$$D = \frac{1}{2h}\sqrt{\frac{\delta_0 r_0}{a}} = 0.0027 , \tag{2}$$

what means a very low damping. Thus, the eigenfrequency in the damped hunting motion remains unchanged

$$\omega_D \approx \omega_S . \tag{3}$$

The critical speed (9.52) is found as

$$v_{crit} = \sqrt{\frac{1}{2} g r_0} = 1.56\,\mathrm{m/s} = 5.63\,\mathrm{km/h} . \tag{4}$$

A free wheelset is getting unstable for a very low speed and it is of no use for engineering applications. The nutation frequency of the wheelset follows from (9.56) for $v_0 > ga$ as

$$\omega_N = \frac{r_0 \delta_0}{2a^2} v_0 \quad \text{with} \quad \frac{r_0 \delta_0}{2a^2} = 0.2 \left[\frac{1}{\text{m}}\right] . \tag{5}$$

This unacceptable behavior of a free wheelset is completely changed by its assembly in bogies under the vehicle body. Essential features are:

1. The weight of the vehicle body boosts the normal force in the contact patch and the rolling contact forces are getting much higher.
2. The viscoelastic coupling of the wheelsets in the bogie, and the bogies with the vehicle body as well, results in a strongly stabilizing effect. The dynamical behavior of such designs has been investigated in detail, e.g., by De Pater (1987), Wickens (1987) and Iwnicki (2006).

■

10

Vertical Motions

The vertical motions of road and rail vehicles can be treated jointly again. The investigations of the vertical motion provide a basis for the optimization of thc driving comfort, and the driving safety as far as road vehicles are concerned. The contact forces in wheel contact patches most important for the longitudinal and lateral motions are directly affected by the vertical motion according to the rolling contact laws. The fundamental principles of a vehicle suspension can be completely studied by a model with two vertical degrees of freedom. However, for a more detailed analysis more complex models can be used, too. For magnetically levitated vehicles the active suspension control has to be considered, too. For wheeled vehicles actively controlled suspensions are also of increasing interest.

10.1 Principles of Vehicles Suspension

The main tasks of a vehicle suspension are twofold. On the one hand the vehicle body has to be carried on constant height without vibrational accelerations while on the other hand the vehicle wheels should follow the uneven profile of the road without any delay resulting in a constant wheel load. Both requirements are inconsistent with each other, and, therefore, the problem may be solved by frequency decoupling. Then, the eigenfrequency of the body should be as low as possible while the eigenfrequency of the wheels has to be as high as possible. However, there are design constraints which restrain the frequency decoupling, like the limited relative motion space between of the wheels and the body. This means that an optimal design has to consider the engineering constraints, and the design process has to include the randomly uneven road in addition to the vehicle structure.

The fundamentals of vehicle suspensions are presented for a car model constrained to vertical motions only, Fig. 10.1, which offers frequency decoupling by two degrees of freedom and includes random excitation by road unevenness. Such a model is often denoted as quarter car model, too. With respect

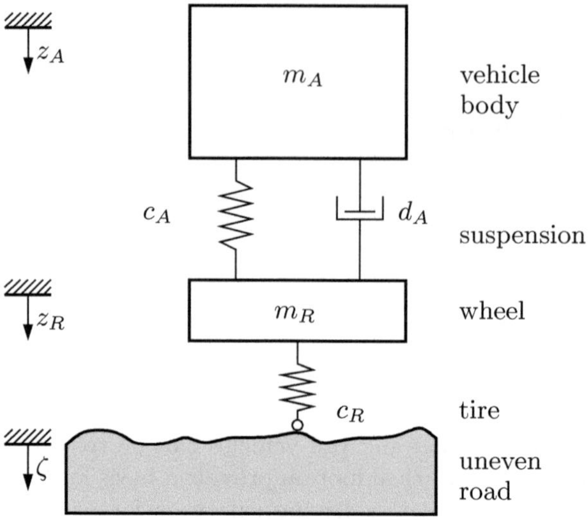

Fig. 10.1. Vehicle model for the analysis of vertical vibrations

to the equilibrium condition the linear equations of motion read as

$$\begin{aligned} &\begin{bmatrix} m_A & 0 \\ 0 & m_R \end{bmatrix} \begin{bmatrix} \ddot{z}_A \\ \ddot{z}_R \end{bmatrix} + \begin{bmatrix} d_A & -d_A \\ -d_A & d_A \end{bmatrix} \begin{bmatrix} \dot{z}_A \\ \dot{z}_R \end{bmatrix} \\ &+ \begin{bmatrix} c_A & -c_A \\ -c_A & c_A + c_R \end{bmatrix} \begin{bmatrix} z_A \\ z_R \end{bmatrix} = \begin{bmatrix} 0 \\ c_R \end{bmatrix} \zeta(t) \end{aligned} \tag{10.1}$$

where m_A is the body mass, m_R is the generalized mass of all wheels as well as d_A, c_A and c_R represent the generalized damping coefficient, body spring coefficient and tire spring coefficient of all wheels. For the derivation of these equations of motion see Problem 2.12. As excitation a white velocity noise with intensity q is assumed,

$$\dot{\zeta}(t) \sim (0, q) \; . \tag{10.2}$$

Using the abbreviations

$$a = \frac{c_A}{m_A}, \quad b = \frac{c_A}{m_R}, \quad c = \frac{c_R}{m_R}, \quad d = \frac{d_A}{m_A}, \quad e = \frac{d_A}{m_R} \tag{10.3}$$

one gets from (10.1) the state equations which have been differentiated once to obtain an velocity excitation of the vehicle,

$$\underbrace{\begin{bmatrix} \ddot{z}_A \\ \ddot{z}_R \\ \dddot{z}_A \\ \dddot{z}_R \end{bmatrix}}_{\ddot{\boldsymbol{x}}} = \underbrace{\begin{bmatrix} 0 & 0 & 1 & 0 \\ 0 & 0 & 0 & 1 \\ -a & a & -d & d \\ b & -b-c & e & -e \end{bmatrix}}_{\boldsymbol{A}} \underbrace{\begin{bmatrix} \dot{z}_A \\ \dot{z}_R \\ \ddot{z}_A \\ \ddot{z}_R \end{bmatrix}}_{\dot{\boldsymbol{x}}} + \underbrace{\begin{bmatrix} 0 \\ 0 \\ 0 \\ c \end{bmatrix}}_{\boldsymbol{B}} \underbrace{\dot{\zeta}(t)}_{\boldsymbol{w}(t)} . \tag{10.4}$$

For the assessment of the vehicles performance driving comfort, driving safety and suspension travel are used. It yields for the vertical acceleration characterizing the driving comfort

$$a_A = \ddot{z}_A = \begin{bmatrix} 0\,0\,1\,0 \end{bmatrix} \dot{\boldsymbol{x}} = \boldsymbol{c}^{\mathrm{T}} \dot{\boldsymbol{x}} , \tag{10.5}$$

and for the dynamical wheel load affecting the driving safety

$$f = c_R(\zeta - z_R) = m_A \ddot{z}_A + m_R \ddot{z}_R = \begin{bmatrix} 0\,0\,m_A\,m_R \end{bmatrix} \dot{\boldsymbol{x}} = \boldsymbol{e}^{\mathrm{T}} \dot{\boldsymbol{x}} \tag{10.6}$$

and for the suspension travel

$$\begin{aligned} s = z_A - z_R &= \frac{1}{c_A}(-m_A \ddot{z}_A - d_A(\dot{z_A} - \dot{z_R})) \\ &= \begin{bmatrix} -\frac{d_A}{c_A} & \frac{d_A}{c_A} & -\frac{m_A}{c_A} & 0 \end{bmatrix} \dot{\boldsymbol{x}} = \boldsymbol{g}^{\mathrm{T}} \dot{\boldsymbol{x}} \end{aligned} \tag{10.7}$$

where the 4×1-weighting vectors $\boldsymbol{c}$, $\boldsymbol{e}$, $\boldsymbol{g}$ are introduced. Then, it remains for the standard deviations of acceleration and wheel load

$$\sigma_a^2 = \boldsymbol{c}^{\mathrm{T}} \boldsymbol{P} \boldsymbol{c} = P_{33} \tag{10.8}$$

and

$$\sigma_f^2 = \boldsymbol{e}^{\mathrm{T}} \boldsymbol{P} \boldsymbol{e} = m_A^2 P_{33} + 2 m_A m_R P_{34} + m_R^2 P_{44} \tag{10.9}$$

as well as the standard deviation the suspension travel

$$\sigma_s^2 = \boldsymbol{g}^{\mathrm{T}} \boldsymbol{P} \boldsymbol{g} = \frac{1}{c_A^2} \left[d_A^2 \left(P_{11} - 2P_{12} - P_{22} \right) + 2 d_A m_A \left(P_{13} - P_{23} \right) + m_A^2 P_{33} \right] \tag{10.10}$$

with the stationary 4×4-covariance matrix

$$\boldsymbol{P} = \mathrm{E}\{\dot{\boldsymbol{x}}(t) \dot{\boldsymbol{x}}^{\mathrm{T}}(t)\} = \text{const} . \tag{10.11}$$

The problem is reduced now to the evaluation of the covariance matrix (10.11) of the system (10.4). For this purpose the covariance analysis is most adequate offering analytical solutions for low order systems in contrary to the power spectral density approach. The corresponding Lyapunov matrix equation (7.17) reads

$$\boldsymbol{AP} + \boldsymbol{PA}^{\mathrm{T}} + \boldsymbol{B}q\boldsymbol{B}^{\mathrm{T}} = \boldsymbol{0}\,. \tag{10.12}$$

For the solution a matrix polynomial is applied as outlined by Mueller and Schiehlen (1985). This polynomial reads for $n = 4$ as follows

$$\boldsymbol{P} = \frac{1}{2\det \boldsymbol{H}} \sum_{k=0}^{3} H_{k+1,1} \sum_{m=0}^{2k} (-1)^m \boldsymbol{A}_m \boldsymbol{Q} \boldsymbol{A}_{2k-m}^{\mathrm{T}} \tag{10.13}$$

where

$$\boldsymbol{H} = \begin{bmatrix} a_1 & 1 & 0 & 0 \\ a_3 & a_2 & a_1 & 1 \\ 0 & a_4 & a_3 & a_2 \\ 0 & 0 & 0 & a_4 \end{bmatrix} \tag{10.14}$$

is the 4×4-Hurwitz matrix also known from stability theory and $H_{k+1,1}$ are the corresponding scalar cofactors. Further,

$$a_i\,, \quad i = 1(1)4 \tag{10.15}$$

mean the characteristic coefficients of the system matrix $\boldsymbol{A}$. In addition, the 4×4 - auxiliary matrices are defined by

$$\boldsymbol{A}_m = \boldsymbol{A}\boldsymbol{A}_{m-1} + a_m \boldsymbol{E}\,, \quad m = 0(1)6 \tag{10.16}$$

where it yields $\boldsymbol{A}_p = \boldsymbol{0}$, $p = 4(1)6$. The 4×4-intensity matrix reads as

$$\boldsymbol{Q} = q\boldsymbol{B}\boldsymbol{B}^{\mathrm{T}}\,. \tag{10.17}$$

Some intermediate results are listed for the following quantities. Characteristic equation:

$$\det(\lambda \boldsymbol{E} - \boldsymbol{A}) = \lambda^4 + (d+e)\lambda^3 + (a+b+c)\lambda^2 + cd\lambda + ac = 0\,. \tag{10.18}$$

Determinant of Hurwitz matrix:

$$\det \boldsymbol{H} = a_4 H_3 = ac(c^2 de)\,. \tag{10.19}$$

Cofactors of Hurwitz matrix:

$$\left.\begin{aligned} H_{11} &= a_4(a_2 a_3 - a_1 a_4) = ac^3 d\,, \\ H_{21} &= -a_3 a_4 = -ac^2 d\,, \\ H_{31} &= a_1 a_4 = ac(d+e)\,, \\ H_{41} &= -(a_1 a_2 - a_3) = -(a+b)(d+e) - ce\,. \end{aligned}\right\} \tag{10.20}$$

Auxiliary matrices multiplied by 4×1-matrix $\boldsymbol{B}$:

$$\left.\begin{aligned} \boldsymbol{A}_0\boldsymbol{B} &= \boldsymbol{B} = c[0\;0\;0\;1]^{\mathrm{T}}\,, \\ \boldsymbol{A}_1\boldsymbol{B} &= \boldsymbol{A}\boldsymbol{B} + a_1\boldsymbol{B} = c[0\;1\;d\;d]^{\mathrm{T}}\,, \\ \boldsymbol{A}_2\boldsymbol{B} &= \boldsymbol{A}\boldsymbol{A}_1\boldsymbol{B} + a_2\boldsymbol{B} = c[d\;d\;a\;a]^{\mathrm{T}}\,, \\ \boldsymbol{A}_3\boldsymbol{B} &= \boldsymbol{A}\boldsymbol{A}_2\boldsymbol{B} + a_3\boldsymbol{B} = c[a\;a\;0\;0]^{\mathrm{T}}\,, \\ \boldsymbol{A}_4\boldsymbol{B} &= \boldsymbol{0}\,. \end{aligned}\right\} \tag{10.21}$$

Then, the elements of the covariance matrix required for the assessments (10.8) to (10.10) are easily found as

$$P_{11} = \frac{q}{2e}(a+b)\,, P_{12} = \frac{q}{2e}(a+b-c)\,, P_{13} = 0\,, \tag{10.22}$$

$$P_{22} = \frac{q}{2e}(a+b-2c+\frac{c(c+d^2)}{a})\,, P_{23} = \frac{q}{2e}cd\,, \tag{10.23}$$

$$P_{33} = \frac{q}{2de}[cd^3 + a^2(d+e)]\,, \tag{10.24}$$

$$P_{34} = \frac{q}{2de}[cd(d^2 - a) + a^2(d+e)]\,, \tag{10.25}$$

$$P_{44} = \frac{q}{2de}[(a-c)^2 d + cd^3 + a^2 e]\,. \tag{10.26}$$

Inserting (10.22) to (10.26) in (10.8) to (10.10), and eliminating the abbreviations (10.3), it remain the final results

$$\sigma_a^2 = \frac{q}{2}\left[\frac{c_R d_A}{m_A^2} + \frac{c_A^2(m_A+m_R)}{d_A m_A^2}\right]\,, \tag{10.27}$$

$$\begin{aligned} \sigma_f^2 = \frac{q}{2}\Bigg[&\left(1+\frac{m_R}{m_A}\right)^3 \frac{c_A^2 m_A}{d_A} + \left(1+\frac{m_R}{m_A}\right)^2 c_R d_A \\ &- 2\left(1+\frac{m_R}{m_A}\right)\frac{c_A c_R m_R}{d_A} + \frac{c_R^2 m_R}{d_A}\Bigg]\,, \end{aligned} \tag{10.28}$$

$$\sigma_s^2 = \frac{q}{2}\left(\frac{m_A+m_R}{d_A}\right)\,. \tag{10.29}$$

Thus, analytical explicit solutions are found which can be used for optimization of the original design parameters.

The optimization will be performed for a compact passenger car with a body mass $m_A = 1200$ kg. Then, there remain four design variables with the following reference values:

Wheel mass	m_R	$= 80$ kg ,
Body spring	c_A	$= 30000$ N/m ,
Tire spring	c_R	$= 320000$ N/m ,
Shock absorber	d_A	$= 4800$ Ns/m .

These design variables are generalized quantities. e.g., the wheel mass m_R includes not only the masses of the four wheels but also the corresponding moments of inertia of the control arms of the wheel suspensions.

For the discussion of the sensitivity with respect to the design variables the following specific quantities are useful.

$$\frac{\sigma_a^2(c_A, d_A)}{\sigma^2_{a\ reference}} \quad \text{Driving comfort by suspension strut}$$

$$\frac{\sigma_f^2(c_A, d_A)}{\sigma^2_{f\ reference}} \quad \text{Driving safety by suspension strut}$$

$$\frac{\sigma_s^2(c_A, d_A)}{\sigma^2_{s\ reference}} \quad \text{Suspension travel by suspension strut}$$

$$\frac{\sigma_a^2(c_R, m_R)}{\sigma^2_{a\ reference}} \quad \text{Driving comfort by axle and tire}$$

$$\frac{\sigma_f^2(c_R, m_R)}{\sigma^2_{f\ reference}} \quad \text{Driving safety by axle and tire}$$

$$\frac{\sigma_s^2(c_R, m_R)}{\sigma^2_{s\ reference}} \quad \text{Suspension travel by axle and tire}$$

Thus, the original four-dimensional optimization problem is reduced to two two-dimensional optimization problems which are evaluated graphically for driving comfort, driving safety and suspension travel, Fig. 10.2 to Fig. 10.4, where the height of the box represents driving comfort, driving safety, and suspension travel, respectively, of the reference values. For the strut parameters the following tendencies are found.

1. Large and small damper coefficients are harmful, for medium damping comfort and safety are optimal.
2. Small spring and damper coefficients improve the comfort considerably.

However, soft springs result in large variations of the equilibrium position due to body mass variations which may be only compensated by a level control system.

3. The safety is insensitive to the spring coefficient.
4. The reference value of the damping coefficient yields optimal safety.

The damping coefficient with optimal comfort deteriorates the safety essentially. Since the efficiency of a damper decreases with time, drivers feel in used cars more comfortable forgetting about safety.

5. The suspension travel is independent of the spring coefficient and requires a damping coefficient as less as possible.

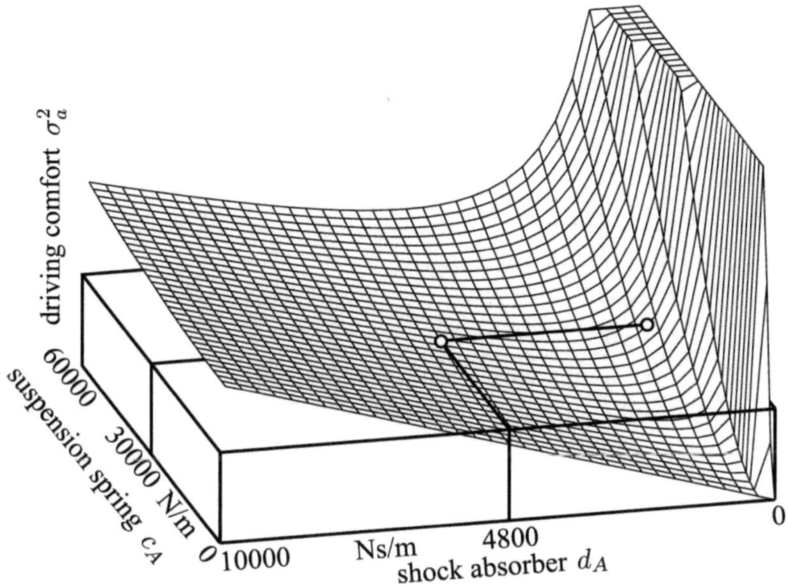

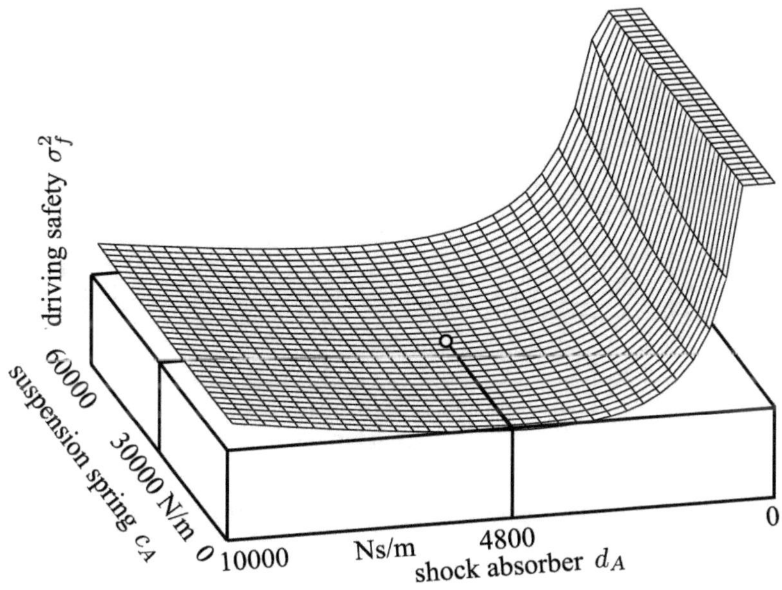

Fig. 10.2. Driving comfort and driving safety by suspension strut variables

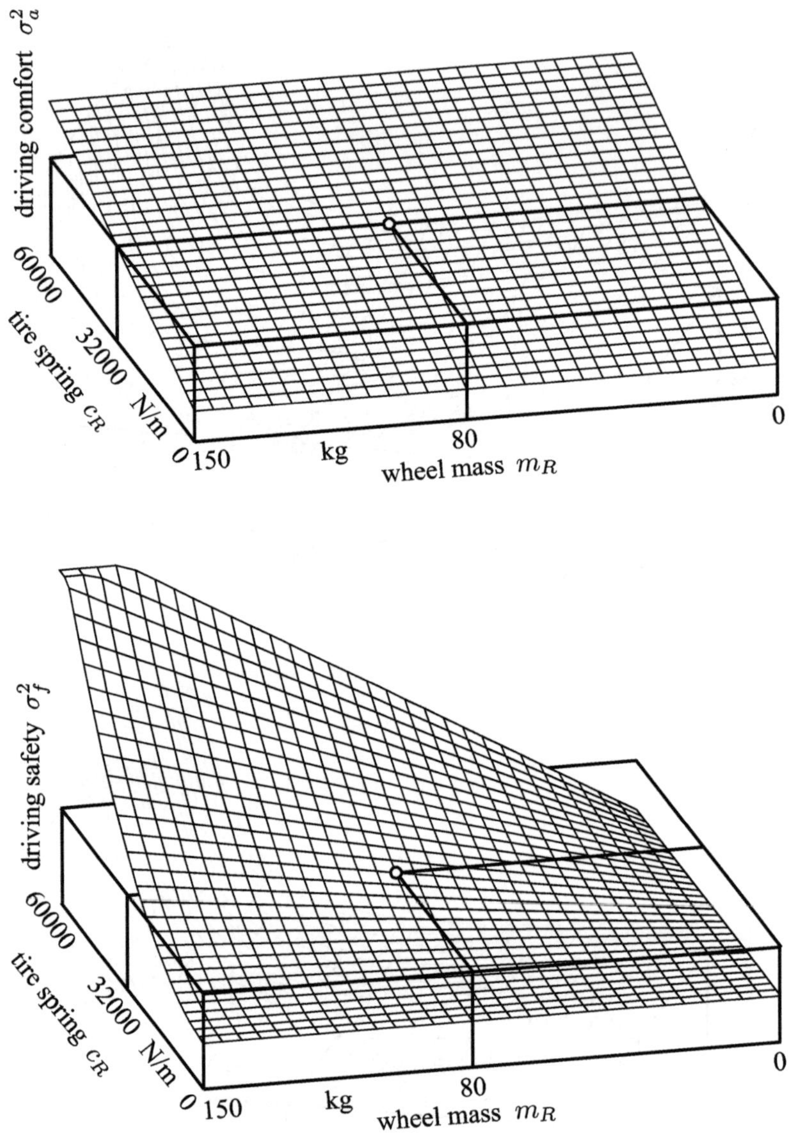

Fig. 10.3. Driving comfort and driving safety by axle and tire design variables

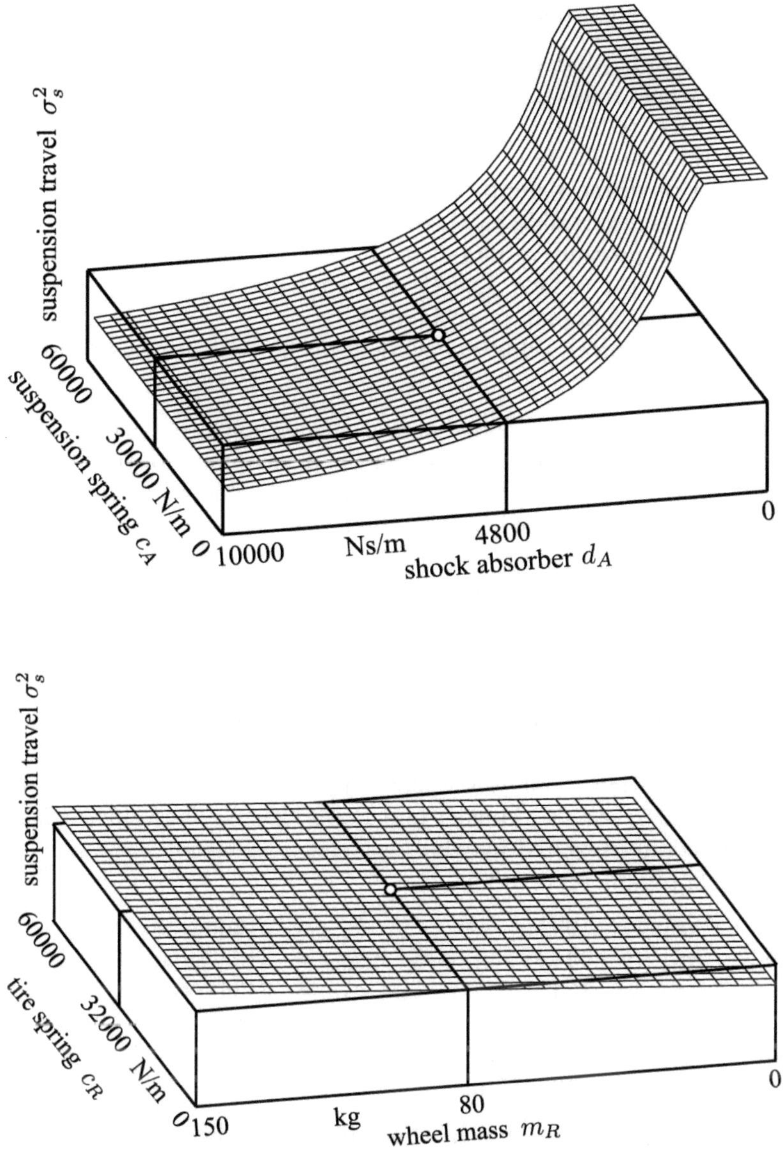

Fig. 10.4. Suspension travel by suspension strut variables, and axle and tire design variables

The suspension travel is characterized by an optimum on the boundary of the design parameter space.

For axle and tire parameters also some general tendencies are observed.

1. The comfort is independent of the wheel mass.
2. Small wheel mass improves safety essentially. The generalized mass can be also reduced by a suitable suspension design. Nevertheless, the reductions are limited.
3. Small tire stiffness improves comfort and safety. The air pressure in the tires is responsible for the tire stiffness and can not reduced essentially. Low pressure results in strong wear of the tire and weak lateral guidance of the vehicle. Therefore, the reference value of the tire spring coefficient is a boundary optimum.
4. The suspension travel is independent of the tire spring and insensitive to the tire mass.

In summary, the reference values of the suspension are close to an optimum which was found empirically during the development of the automobile in more than a century. Further progress will be achieved in the future by mechatronic system design where the stochastic methods presented will be most helpful.

Some results for an active suspension system are presented by Schiehlen et al. (2007).

10.2 Random Vibrations of a Two Axle Vehicle

An extended covariance analysis is now presented for the two axle vehicle shown in Fig. 10.5, see also Mueller and Popp (1979). The vehicle with four degrees of freedom is excited by white velocity noise again.

The vehicle parameters read as

$$\begin{aligned}
m_1 &= 100\,\text{kg} & k_4 &= 51208\,\text{N/m}\\
m_2 &= 54\,\text{kg} & d_3 &= 3198\,\text{Ns/m}\\
m_3 &= 1247\,\text{kg} & d_4 &= 3434\,\text{Ns/m}\\
J &= 1945\,\text{kgm}^2 & a &= 1.2\,\text{m}\\
k_1 &= k_2 = 343350\,\text{N/m} & b &= 1.3\,\text{m}\\
k_3 &= 47382\,\text{N/m} & l &= a+b = 2.5\,\text{m}
\end{aligned}$$

Small motions are assumed and the time lag $t_2 = l/v$ between front and rear axle excitation has to be taken into account.

The generalized coordinates are summarized in the position vector $\boldsymbol{y}(t) = [z_1(t)\ \ z_2(t)\ \ z_3(t)\ \ \beta(t)]^{\mathrm{T}}$. Then, the equations of motion read as

$$\begin{aligned}
&\boldsymbol{M}\ddot{\boldsymbol{y}}(t) + \boldsymbol{D}\dot{\boldsymbol{y}}(t) + \boldsymbol{K}\boldsymbol{y}(t) = \boldsymbol{s}_1\zeta_1(t) + \boldsymbol{s}_2\zeta_2(t)\ ,\\
&\zeta_1(t) = \zeta(t)\ , \quad \zeta_2(t) = \zeta(t-t_2)\ ,
\end{aligned} \tag{10.30}$$

where $\boldsymbol{M}$, $\boldsymbol{D}$, $\boldsymbol{K}$ characterize the inertia, damping and stiffness matrix, respectively, and $\boldsymbol{s}_1, \boldsymbol{s}_2$ are the excitation input vectors,

$$
\begin{aligned}
\boldsymbol{M} &= \mathbf{diag}[m_1 \;\; m_2 \;\; m_3 \;\; J] \, , \\
\boldsymbol{s}_1 &= [k_1 \;\; 0 \;\; 0 \;\; 0]^{\mathrm{T}} \, , \\
\boldsymbol{s}_2 &= [0 \;\; k_2 \;\; 0 \;\; 0]^{\mathrm{T}} \, ,
\end{aligned}
\tag{10.31}
$$

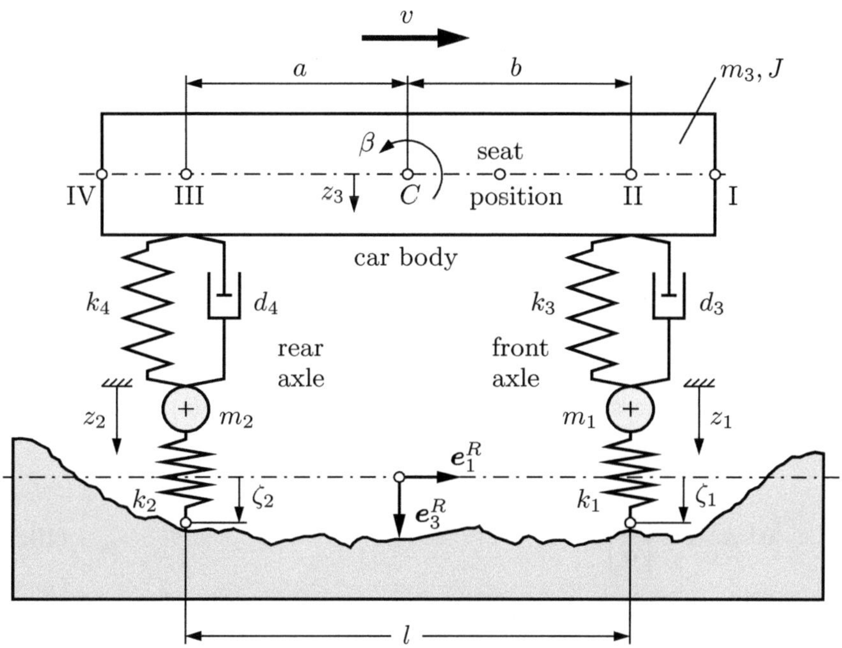

Fig. 10.5. Two axle vehicle model traveling on rough road

$$
\boldsymbol{D} = \begin{bmatrix} d_3 & 0 & -d_3 & bd_3 \\ 0 & d_4 & -d_4 & -ad_4 \\ -d_3 & -d_4 & d_3 + d_4 & -bd_3 + ad_4 \\ bd_3 & -ad_4 & -bd_3 + ad_4 & b^2 d_3 + a^2 d_4 \end{bmatrix} ,
\tag{10.32}
$$

$$
\boldsymbol{K} = \begin{bmatrix} k_1 + k_3 & 0 & -k_3 & bk_3 \\ 0 & k_2 + k_4 & -k_4 & -ak_4 \\ -k_3 & -k_4 & k_3 + k_4 & -bk_3 + ak_4 \\ bk_3 & -ak_4 & -bk_3 + ak_4 & b^2 k_3 + a^2 k_4 \end{bmatrix} .
\tag{10.33}
$$

From (10.30) the vehicle state space representation can be obtained,

$$\dot{\boldsymbol{x}}(t) = \boldsymbol{A}\boldsymbol{x}(t) + \boldsymbol{b}_1\zeta_1(t) + \boldsymbol{b}_2\zeta_2(t) = \boldsymbol{A}\boldsymbol{x}(t) + \boldsymbol{B}\boldsymbol{\xi}\,, \tag{10.34}$$

$$\boldsymbol{x}(t) = \begin{bmatrix} \boldsymbol{y}(t) \\ \dot{\boldsymbol{y}}(t) \end{bmatrix}, \quad \boldsymbol{\xi}(t) = \begin{bmatrix} \zeta_1(t) \\ \zeta_2(t) \end{bmatrix}, \quad \boldsymbol{A} = \begin{bmatrix} \boldsymbol{0} & \boldsymbol{E} \\ -\boldsymbol{M}^{-1}\boldsymbol{K} & -\boldsymbol{M}^{-1}\boldsymbol{D} \end{bmatrix},$$

$$\boldsymbol{b}_1 = \begin{bmatrix} \boldsymbol{0} \\ \boldsymbol{M}^{-1}\boldsymbol{s}_1 \end{bmatrix}, \quad \boldsymbol{b}_2 = \begin{bmatrix} \boldsymbol{0} \\ \boldsymbol{M}^{-1}\boldsymbol{s}_2 \end{bmatrix}, \quad \boldsymbol{B} = \begin{bmatrix} \boldsymbol{b}_1\ \boldsymbol{b}_2 \end{bmatrix}\,. \tag{10.35}$$

In order to use white noise vehicle excitation, $\dot{\zeta}(t) \equiv w(t)$, (10.35) has to be differentiated,

$$\ddot{\boldsymbol{x}}(t) = \boldsymbol{A}\dot{\boldsymbol{x}}(t) + \boldsymbol{B}\boldsymbol{w}(t)\,. \tag{10.36}$$

Here, the autocorrelation matrix of the white noise vector process $\boldsymbol{w}$ reads

$$\boldsymbol{R}_w(\tau) = \mathrm{E}\{\boldsymbol{w}(t)\boldsymbol{w}^{\mathrm{T}}(t-\tau)\} = q_{\dot{\zeta}} \begin{bmatrix} \delta(\tau) & \delta(\tau+t_2) \\ \delta(\tau-t_2) & \delta(\tau) \end{bmatrix}. \tag{10.37}$$

Thus, the covariance matrix $\boldsymbol{P}_{\dot{x}}$ follows in the steady state from the extended algebraic Lyapunov equation

$$\boldsymbol{A}\boldsymbol{P}_{\dot{x}} + \boldsymbol{P}_{\dot{x}}\boldsymbol{A}^{\mathrm{T}} + \boldsymbol{Q} = \boldsymbol{0}, \quad \boldsymbol{Q} = q_{\dot{\zeta}} \left[\boldsymbol{b}_1\boldsymbol{b}_1^{\mathrm{T}} + \boldsymbol{b}_2\boldsymbol{b}_2^{\mathrm{T}} + e^{\boldsymbol{A}t_2}\boldsymbol{b}_1\boldsymbol{b}_2^{\mathrm{T}} + \boldsymbol{b}_2\boldsymbol{b}_1^{\mathrm{T}}e^{\boldsymbol{A}^{\mathrm{T}}t_2}\right]. \tag{10.38}$$

From $\boldsymbol{P}_{\dot{x}}$ the variances σ_a^2 of the car body acceleration $\ddot{z}_P(t)$ at any position P with coordinate $x_P = \boldsymbol{a}^{\mathrm{T}}\boldsymbol{y}$ can immediately be calculated by

$$\sigma_a^2 = [\boldsymbol{0}^{\mathrm{T}}\ \boldsymbol{a}^{\mathrm{T}}]\boldsymbol{P}_{\dot{x}} \begin{bmatrix} \boldsymbol{0} \\ \boldsymbol{a} \end{bmatrix}. \tag{10.39}$$

A similar expression holds for the variances σ_f^2 of the wheel loads $f_i = k_i(\zeta_i - y_i)$, $i = 1, 2$, but here (10.30) has to be used to get the weighting vector of the displacements as shown in Sect. 10.1. Fig. 10.6 shows some numerical results where a smooth road excitation ($q_{\dot{\zeta}} = 3.14 \cdot 10^{-6} v\ \mathrm{m}^2/\mathrm{s}$, solid line) and a rough road excitation ($q_{\dot{\zeta}} = 24.7 \cdot 10^{-6} v\ \mathrm{m}^2/\mathrm{s}$, dashed line) is assumed.

In Fig. 10.6 a) the standard deviation σ_f/f_{Stat} of the load variation of the front and rear wheel is plotted against speed v where f_{Stat} denotes the static load. It increases with $\sqrt{v}$ since the excitation intensity is proportional to v. Even in the worst case (rough road, front wheel, maximum speed) the safety margin R^* according to (6.28) takes the value $R^* = 1 - \sigma_F/f_{Stat} \approx 0.75$ indicating a safe ride.

In Fig. 10.6 b) the standard deviation σ_a/g of the car body acceleration versus the vehicle length is given, which clearly shows a minimum for the center of mass position at different speeds. Neglecting the time lag between front and rear excitation (dotted line) can lead to incorrect results especially near the seat position.

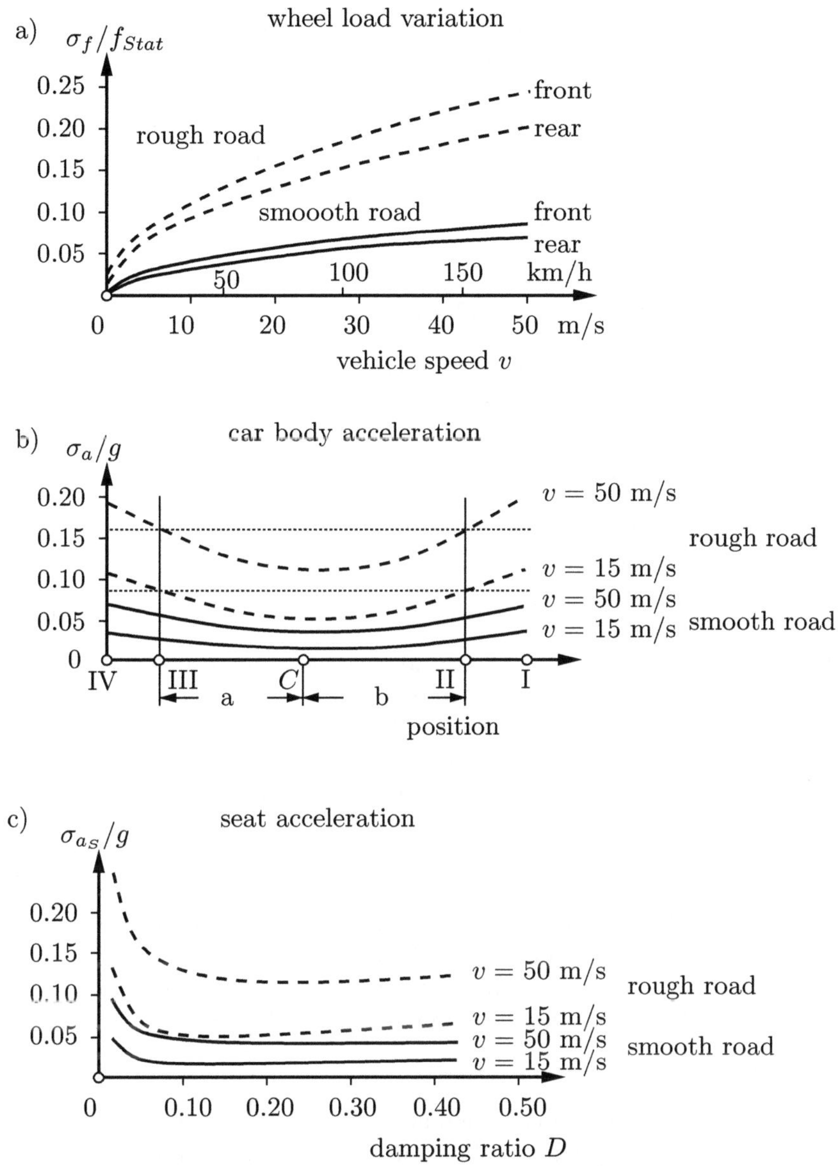

Fig. 10.6. Wheel load variation and vertical accelerations of the two axle vehicle. The dotted line in b) does not consider the time lag between front and rear axle.

The covariance analysis has also an important application in optimization problems. Fig. 10.6 c) shows a simple example, where the influence of the nondimensional damping ratio $D = D_i \equiv d_i/(2\sqrt{k_i \overline{m}_i})$, $i = 3, 4$, $(\overline{m}_3 = m_3 a/l, \overline{m}_4 = m_3 b/l)$ on the standard deviation of the seat acceleration, σ_{as}/g, is studied. Best results are obtained in the range $0.1 \leq D \leq 0.25$. However, tradeoffs considering the wheel load variation call for higher damping. For a more detailed ride comfort evaluation the perception measure K has to be calculated. A rough approximate estimation using the thumb rule $a_{max} \leq 0.5$ m/s^2 $\approx 0.05g$ shows that the worst case (rough road, maximum speed, maximum damping) certainly results in a bad ride comfort.

10.3 A Complex Vehicle Model

For the refinement of advanced automobiles a three body model is not sufficient. Therefore, a complex vehicle model is analyzed adopted from Problem 2.14.

The vehicle model proposed by Kreuzer and Rill (1982) consists of the componentes listed in Table 10.1. The model is shown in Fig. 10.7, and the corresponding data are found in Table 10.2, see also Hirschberg (1981). The state equations read as

$$\dot{\boldsymbol{x}}_F = \boldsymbol{A}\boldsymbol{x}_F + \boldsymbol{B}_1\boldsymbol{x}_W(t) + \boldsymbol{B}_2\boldsymbol{x}_W(t - t_2) , \tag{10.40}$$

where the state vector consists of $n = 40$ variables composed by $2 * 19 = 38$ body state variables and by 2 state variables describing the two serial spring-damper devices according to Table 3.1. These devices are used to reduce the engine vibrations without increasing the noise in the car body. The excitation of the vehicle is due to the left and right road profile, both of them with a time delay of

$$t_2 = l/v \tag{10.41}$$

between front and rear axle as discussed by (7.15). The road unevenness is modeled by colored noise processes as shown in (5.14) and (5.15).

The assessment is based on the vertical acceleration at the point (U, V) related to the vehicle body and the four dynamic wheel loads,

$$\sigma_a = \sigma_a(U, V) , \quad \sigma_{fi} , \quad i = 1(1)4 . \tag{10.42}$$

The result of the computation by Kreuzer and Rill (1982) using the covariance analysis, see Sect. 7.2.3.2, is shown in Fig. 10.8. In the middle of the car body, i.e. for the driver and the left passenger, the driving comfort is high. The dynamical wheel load variation is larger at the rear axle than at the front axle. This is due to the front wheel drive of the vehicle chosen.

A parameter optimization can be performed only numerically for such a complex vehicle model as indicated in Section 7.4. Specific results for the design of the suspension dynamics can be found by such a complex model.

Table 10.1. Parts of the vehicle model

Components	Mass points	Rigid bodies	Constraints	Degrees of freedom	Generalized coordinates
Front wheels	2	-	4	2	$Z1$, $Z2$
Chassis	-	1	3	3	ZK, AK, BK
Engine	-	1	1	5	XM, YM, ZM, AM, BM
Drive shaft	-	2	10	2	ZG, BG
Rear axle	-	1	3	3	ZT, AT, BT
	-	2	10	2	$BS1$, $BS2$
Exhaust pipe	1	-	2	1	ZA
Driver	1	-	2	1	ZF
Total	4	7	35	19	

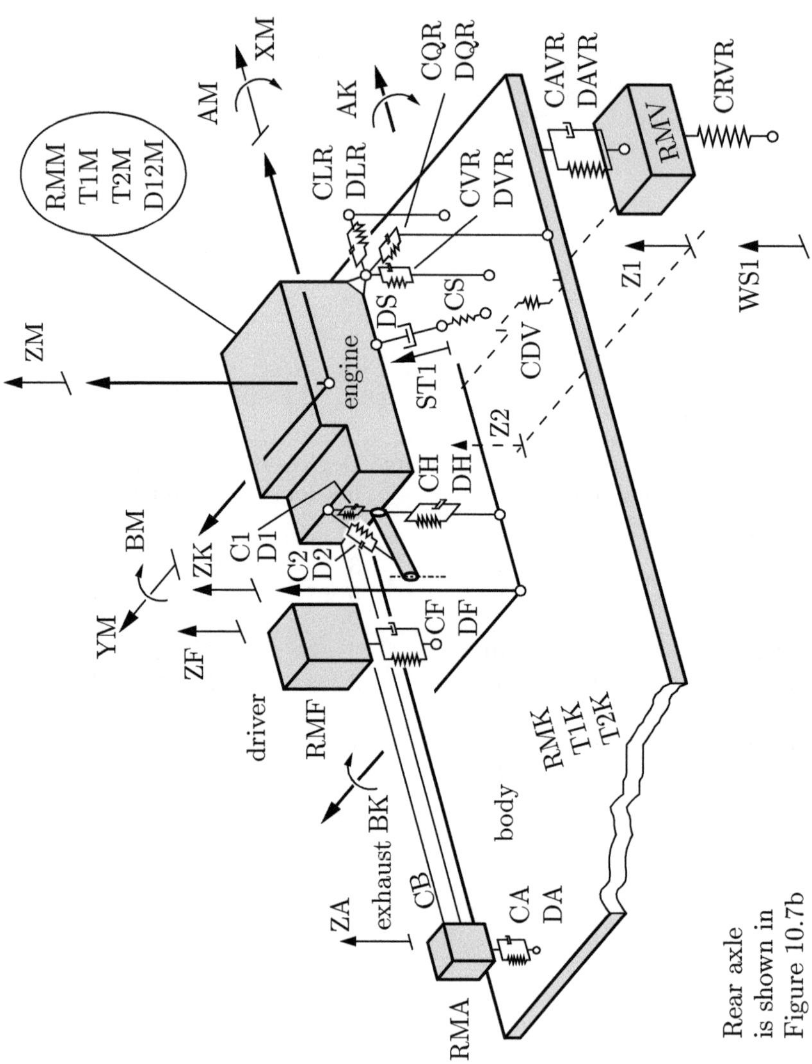

Fig. 10.7. a) Complex vehicle model proposed by Kreuzer and Rill (1982), chassis

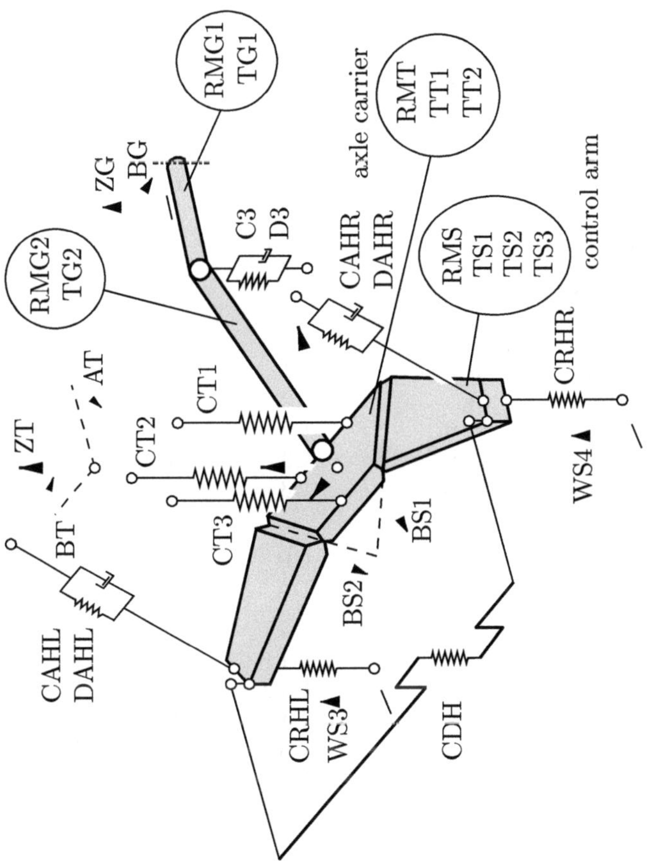

Fig. 10.7. b) Complex vehicle model proposed by Kreuzer and Rill (1982), rear axle

Table 10.2. Parameters of the vehicle model

		Unit			Unit
Engine					
RMM	= 293	kg	T2M	= 31.09	kg m^2
T1M	= 9.175	kg m^2	D12M	= 0.6024	kg m^2
Engine mounting					
CS	= 580000	N/m	DQR	= 20	Ns/m
DS	= 2900	Ns/m	DQL	= 20	Ns/m
CLR	= 95000	N/m	CVR	= 183000	N/m
CLL	= 95000	N/m	CVL	= 183000	N/m
DLR	= 20	Ns/m	DVR	= 20	Ns/m
DLL	= 20	Ns/m	DVL	= 20	Ns/m
CQR	= 113000	N/m	CH	= 88000	N/m
CQL	= 113000	N/m	DH	= 25	Ns/m
Chassis					
RMK	= 1130	kg	CAHL	= 52800	N/m
T1K	= 504	kg m^2	DAVR	= 1880	Ns/m
T2K	= 1840	kg m^2	DAVL	= 1880	Ns/m
CAVR	= 10700	N/m	DAHR	= 5700	Ns/m
CAVL	= 10700	N/m	DAHL	= 5700	Ns/m
CAHR	= 52800	N/m			
Drive shaft					
RMG1	= 4.8	kg	D1	= 40	Ns/m
RMG2	= 7.2	kg	C2	= 51	Nm/rad
TG1	= 0.78	kg m^2	D2	= 0.00006	Nms/rad
TG2	= 2.3	kg m^2	C3	= 40000	N/m
C1	= 6380000	N/m	D3	= 30	Ns/m
Exhaust pipe					
RMA	= 8	kg	CB	= 2400	N/m
CA	= 67600	N/m	DA	− 210	Ns/m
Rear axle					
RMT	= 25	kg	CT1	= 250000	N/m
RMS	= 40.2	kg	CT2	= 250000	N/m
TT1	= 1.5	kg m^2	CT3	= 250000	N/m
TT2	= 2	kg m^2	CRHR	= 175000	N/m
TS1	= 0.8	kg m^2	CRHL	= 175000	N/m
TS2	= 0.5	kg m^2	CDH	= 5500	N/m
TS3	= 0.4	kg m^2			
Front wheels					
RMV	= 42.5	kg	CRVL	= 150000	N/m
CRVR	= 150000	N/m	CDV	= 20000	N/m
Driver					
RMF	= 60	kg	DF	= 360	Ns/m
CF	= 54000	N/m			

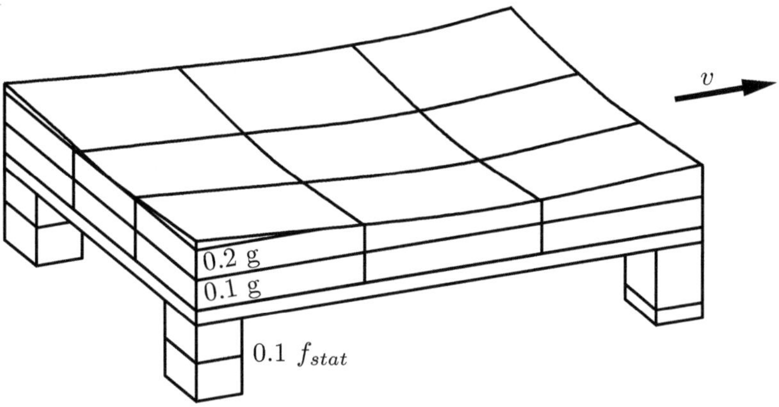

Fig. 10.8. Driving comfort σ_a depending on the horizontal position on the car body and dynamical loads σ_f of the four wheels related to the static wheel load f_{stat}

10.4 Magnetically Levitated Vehicles

Magnetically levitated (maglev) vehicles are an alternative to wheeled railway vehicles in particular for high speeds. While wheels have to provide propulsion, guidance and suspension simultaneously, magnetically levitated vehicles are separating these tasks. Then, each mode of operation can be controlled separately offering much more freedom to the designer and more comfort to the passenger. Maglev vehicles called Transrapid have been developed in Germany during the last three decades, see Transrapid International (2009).

Special consideration requires the active control of the vertical motion of maglev vehicles which is excited by the elastic displacement of the elevated guideway due to the vehicles weight and perturbations of the guideway. The state equations (5.26) of the vertical vibrations of a vehicle on a flexible guideway have to be analyzed in detail where the choice of the controller for the suspension system with magnetic actuators according to Sect. 3.2 is an important engineering task. For the assessment of maglev vehicles the human body acceleration or the human perception of vibration according to Sect. 6.2, respectively, as well as the size of the air gap between the magnets on the vehicle and the rail at the guideway are the essential design criteria. Small air gaps result in damages of the magnets by hitting to rail due to vertical vibrations. Large air gaps, on the other hand, reduce the load capacity of the magnet and result to increased weight of the magnetic suspension system. Therefore, the air gap is an important criterium for the controller design, too. A detailed treatment of the vertical dynamics of maglev vehicles is due to Popp (1978), and it is closely related to the notation used in this book. In particular, the concepts of height control providing optimal comfort and gap control guaranteeing a proper operation of the levitation magnet are discussed. The dynamic and control requirements for electro- and maglev suspensions have been discussed by Goodall (2004) with reference to the people mover at Birmingham Airport, a low speed shuttle in operation from 1984 to 1995.

However, for high-speed vehicles the requirements of driving comfort and levitation can not be met simultaneously by control devices only. Therefore, a secondary suspension has to be provided for maglev vehicles, too. The levitation magnets follow by a stiff gap control the guideway as wheels are doing while the vehicle body is supported by a soft suspension system, see Fig. 10.9. This design principle is called in the literature 'magnetic wheel', see Gottzein (1984). Thus, the principle of frequency decoupling and optimal damping well proven for automobiles and railways has been transferred to maglev vehicles. An actively controlled secondary suspension offers additional possibilities which are only weakly coupled to the kind of the primary suspension control chosen.

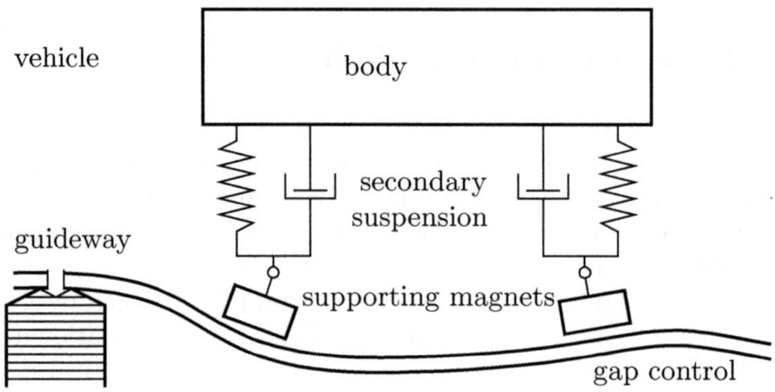

Fig. 10.9. Principle of the magnetic wheel

A

Appendix: Optimal Control of Multivariable Systems

In the following, some basics of controller design of linear multivariable systems are recapitulated without giving derivations. Details can be found in Bryson (2002), Williams II and Lawrence (2007) and Heimann et al. (2007).

A.1 Mathematical Model

The mathematical model of an uncontrolled, time-invariant system Σ reads

$$\dot{\boldsymbol{x}}(t) = \boldsymbol{A}\boldsymbol{x}(t) + \boldsymbol{B}\boldsymbol{u}(t)\;, \quad \boldsymbol{x}(0) = \boldsymbol{x}_0\;, \tag{A.1}$$

$$\boldsymbol{y}(t) = \boldsymbol{C}\boldsymbol{x}(t) \tag{A.2}$$

where the vectors and matrices are defined as

$\boldsymbol{u}$	$r \times 1$ control vector,	$\boldsymbol{A}$	$n \times n$ system matrix,
$\boldsymbol{x}$	$n \times 1$ state vector,	$\boldsymbol{B}$	$n \times r$ input or control matrix,
$\boldsymbol{y}$	$m \times 1$ measurement vector,	$\boldsymbol{C}$	$m \times n$ output or measurement matrix.

Figure A.1 shows the system structure. The equations of the dynamical system (A.1) and the measurement system (A.2) can be read from the summing-point of the integrator and the system output, respectively.

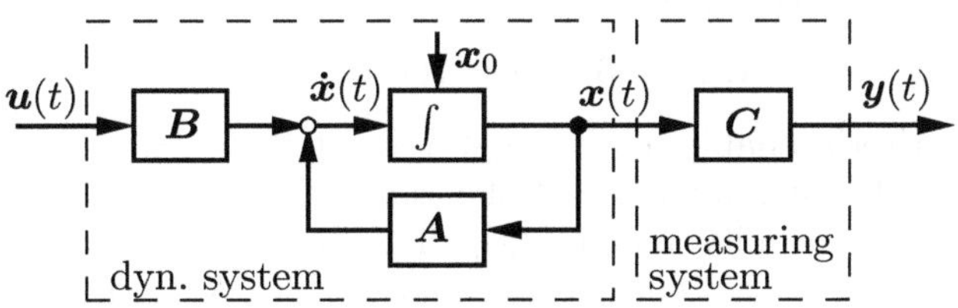

Fig. A.1. System structure

A.2 Task Formulation and Structure Issues

The aim of designing the control is to find a control law $\boldsymbol{u}(t)$ such that the system state $\boldsymbol{x}(t)$ takes a desired state $\boldsymbol{x}_s = \text{const}$ (fix point control) or $\boldsymbol{x}_s = \boldsymbol{x}_s(t)$ (follower control) keeping it regardless of the kind, position and magnitude of disturbances. Two structural questions are to be addressed before designing the control law:

I. Can the dynamical behavior of the system be changed by a control function $\boldsymbol{u}(t)$ in the desired way (is the system controllable)?
II. Can sufficient information on the system be obtained from the measurements (is the system observable)?

I. Controllability
Definition: The system Σ of order n is complete controllable, if there exists for any initial condition $\boldsymbol{x}(0) = \boldsymbol{x}_0$ and for any arbitrary state $\boldsymbol{x}_1$ a finite time $t_1 > 0$ and a input function $\boldsymbol{u}(t)$ defined in the time interval $[0, t_1]$, so that the trajectory starting in $\boldsymbol{x}_0$ reaches $\boldsymbol{x}_1$ at $t = t_1$.

Kalman - criterion:

Σ completely controllable requires

$$\text{rank}\, \boldsymbol{Q}_S = \text{rank}\, [\boldsymbol{B} \vdots \boldsymbol{AB} \vdots \boldsymbol{A}^2\boldsymbol{B} \vdots \cdots \vdots \boldsymbol{A}^{n-1}\boldsymbol{B}] = n\,. \tag{A.3}$$

Hautus - criterion:

Σ completely controllable requires

$$(\lambda_i \boldsymbol{E} - \boldsymbol{A}^{\mathrm{T}})\bar{\boldsymbol{x}}_i = \boldsymbol{0} \Rightarrow \boldsymbol{B}^{\mathrm{T}}\bar{\boldsymbol{x}}_i \neq \boldsymbol{0}\,, \quad i = 1(1)n\,. \tag{A.4}$$

II. Observability
Definition: The system Σ is completely observable, if there exists for any arbitrary initial condition $\boldsymbol{x}(0) = \boldsymbol{x}_0$ a finite time $t_1 > 0$ so that the initial condition $\boldsymbol{x}_0$ can be deduced by the knowledge of the control function $\boldsymbol{u}(t)$ and the measured function $\boldsymbol{y}(t)$ in the time interval $[0, t_1]$.

Kalman - criterion:

Σ completely observable requires

$$\text{rank}\, \boldsymbol{Q}_B = \text{rank}\, [\boldsymbol{C}^{\mathrm{T}} \vdots \boldsymbol{A}^{\mathrm{T}}\boldsymbol{C}^{\mathrm{T}} \vdots \boldsymbol{A}^{T2}\boldsymbol{C}^{\mathrm{T}} \vdots \cdots \vdots \boldsymbol{A}^{Tn-1}\boldsymbol{C}^{\mathrm{T}}] = n\,. \tag{A.5}$$

Hautus - criterion:

Σ completely observable requires

$$(\lambda_i \boldsymbol{E} - \boldsymbol{A})\bar{\boldsymbol{x}}_i = \boldsymbol{0} \Rightarrow \boldsymbol{C}\bar{\boldsymbol{x}}_i \neq \boldsymbol{0}\,, \quad i = 1(1)n\,. \tag{A.6}$$

The Kalman criteria enable a yes/no statement for controllability and observability from the examination of the rank of the controllability matrix $\boldsymbol{Q}_S$ resp. the observability matrix $\boldsymbol{Q}_B$. The Hautus criteria enable additional statements about not controllable and not observable modes, for which $\boldsymbol{B}^{\boldsymbol{T}}\bar{\boldsymbol{x}}_{\boldsymbol{i}} = \boldsymbol{0}$ and $\boldsymbol{C}^{\boldsymbol{T}}\bar{\boldsymbol{x}}_{\boldsymbol{i}} = \boldsymbol{0}$ respectively holds. Here, $\bar{\boldsymbol{x}}_{\boldsymbol{i}}$ are the eigenvectors of the eigenvalues λ_i of system Σ.

A.3 Structure and Properties of Controllers

In the following, linear controllers for linear systems Σ are designed which transfer the state vector $\boldsymbol{x}(t)$ from the initial condition $\boldsymbol{x}(t_0) = \boldsymbol{x}_0 \neq \boldsymbol{0}$ (initial disturbance) to the target state $\boldsymbol{x}_s = \boldsymbol{0}$. This control task corresponds to a complete fixed-point control. Other control tasks can be reduced to or derived by this basic task.

The controller must assure that the controlled system

a) is asymptotically stable,
b) has a certain performance.

This aim can be achieved with the principle of feedback. The deviation of the system output from the target state is countersteered by amplifying and changing its algebraic sign and feeding it back to the system input. This is termed a linear state or output feedback respectively,

$$\boldsymbol{u}(t) = \quad \boldsymbol{K}_x \boldsymbol{x}(t) \, , \tag{A.7}$$

$$\boldsymbol{u}(t) = -\boldsymbol{K}_y \boldsymbol{y}(t) \, , \tag{A.8}$$

depending on the linear feedback of the state vector or the output vector, cp. Fig. A.2 (a) and (b). The constant feedback matrices $\boldsymbol{K}_x$ and $\boldsymbol{K}_y$ are composed by the control gains. The system state $\boldsymbol{x}(t)$ is often not directly available. Therefore, a state estimation $\hat{\boldsymbol{x}}(t)$ must be deduced from measurements $\boldsymbol{y}(t)$ by an observer. Normally a complete state feedback cannot be realized, but a feedback of the state estimation $\hat{\boldsymbol{x}}(t)$, Fig. A.2 c). Firstly the design of the controller is treated followed by the observer design.

A.4 Controller Design

Two methods are discussed to determine the gain matrix $\boldsymbol{K}_x$ for an ideal state feedback. These methods show the power of the feedback principle. The case of an output feedback can be realized with these methods as well, if the measurement matrix $\boldsymbol{C}$ is a regular $n \times n$ matrix. Therefore, the number of measured quantities must be equal to the number of state variables. In this case the feedback matrix reads $\boldsymbol{K}_y = \boldsymbol{K}_x \boldsymbol{C}^{-1}$. If the number of measured quantities is less then the number of state variables, more specific methods of control theory may be used to determine $\boldsymbol{K}_y$.

A.4.1 Controller Design by Pole Assignment

If and only if the system Σ of order n is complete controllable, a state feedback $\boldsymbol{u}(t) = -\boldsymbol{K}_x \boldsymbol{x}(t)$ can be found so that the closed-loop system

$$\dot{\boldsymbol{x}}(t) = \hat{\boldsymbol{A}} \boldsymbol{x}(t) \, , \quad \hat{\boldsymbol{A}} = \boldsymbol{A} - \boldsymbol{B} \boldsymbol{K}_x \tag{A.9}$$

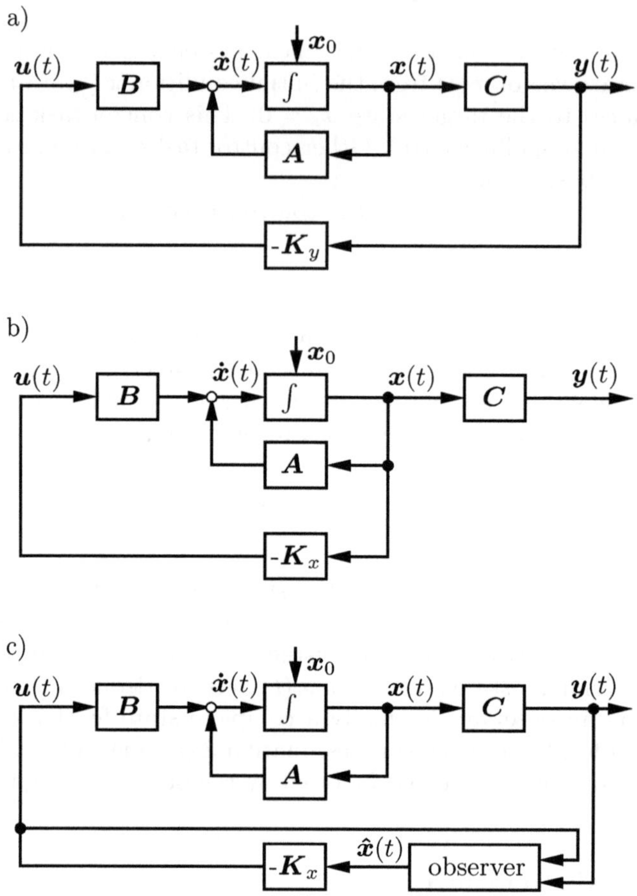

Fig. A.2. Control loops: a) output feedback; b) ideal state feedback; c) state feedback with observer

has arbitrarily prescribed eigenvalues. For complete controllable systems with only one input, $r = 1$,

$$\dot{\boldsymbol{x}}(t) = \boldsymbol{A}\boldsymbol{x}(t) + \boldsymbol{b}u(t) \ , \quad u(t) = -\boldsymbol{k}_x^{\mathrm{T}}\boldsymbol{x}(t) \tag{A.10}$$

and the target eigenvalues $\hat{\lambda}_i, i = 1\,(1)\,n$, or the target characteristic polynomial, respectively,

$$\hat{p}(\lambda) = (\lambda - \hat{\lambda}_1)(\lambda - \hat{\lambda}_2)\ldots(\lambda - \hat{\lambda}_n) = \lambda^n + \hat{a}_1\lambda^{n-1} + \ldots + \hat{a}_{n-1}\lambda + \hat{a}_n \tag{A.11}$$

the feedback vector $\boldsymbol{k}_x^{\mathrm{T}}$ is determined uniquely,

$$\boldsymbol{k}_x^{\mathrm{T}} = \boldsymbol{e}_n^{\mathrm{T}} \boldsymbol{Q}_S^{-1} \hat{p}(\boldsymbol{A}) \,. \tag{A.12}$$

Here $\boldsymbol{e}_n$ is the n-th unit vector, $\boldsymbol{Q}_S$ the controllability matrix and $\hat{p}(\boldsymbol{A})$ the matrix-polynomial built of the system matrix $\boldsymbol{A}$,

$$\boldsymbol{e}_n^{\mathrm{T}} = [0, 0, 0, \ldots, 0, 1] \,, \tag{A.13}$$

$$\boldsymbol{Q}_S = [\boldsymbol{b} \vdots \boldsymbol{A}\boldsymbol{b} \vdots \cdots \vdots \boldsymbol{A}^{n-1}\boldsymbol{b}] \,, \tag{A.14}$$

$$\hat{p}(\boldsymbol{A}) = \boldsymbol{A}^n + \hat{a}_1 \boldsymbol{A}^{n-1} + \ldots + \hat{a}_{n-1}\boldsymbol{A} + \hat{a}_n \boldsymbol{E} \,. \tag{A.15}$$

For complete controllable systems with multiple inputs ambiguous solutions appear in the calculation of the feedback matrix $\boldsymbol{K}_x$. The target eigenvalues have at least to guarantee asymptotic stability, see Sect. 7.2.1.

A.4.2 Optimal Controller Due to a Quadratic Integral Criterion

For the completely controllable system Σ of order n the following cost function is used:

$$J[\boldsymbol{x}(t), \boldsymbol{u}(t)] = \frac{1}{2} \int_0^{\infty} [\boldsymbol{x}^{\mathrm{T}}(t)\boldsymbol{Q}\boldsymbol{x}(t) + \boldsymbol{u}^{\mathrm{T}}(t)\boldsymbol{R}\boldsymbol{u}(t)\mathrm{d}t \rightarrow \mathrm{Min} \tag{A.16}$$

with the weighting matrices $\boldsymbol{Q} = \boldsymbol{Q}^{\mathrm{T}} \geq \boldsymbol{0}, \boldsymbol{R} = \boldsymbol{R}^{\mathrm{T}} > \boldsymbol{0}$. The matrices $\boldsymbol{A}$ and $\boldsymbol{Q}$ must be completely observable $\left(\mathrm{rank}\left[\boldsymbol{Q} \vdots \boldsymbol{A}^{\mathrm{T}}\boldsymbol{Q} \vdots \cdots \vdots \boldsymbol{A}^{Tn-1}\boldsymbol{Q}\right] = n\right)$. Then a unique optimal control

$$\dot{\boldsymbol{u}}(t) = -\boldsymbol{R}^{-1}\boldsymbol{B}^{\mathrm{T}}\boldsymbol{P}\boldsymbol{x}(t) = -\boldsymbol{K}_x\boldsymbol{x}(t) \,, \tag{A.17}$$

exists for the cost function shown above. Herein $\boldsymbol{P} = \boldsymbol{P}^{\mathrm{T}} > \boldsymbol{0}$ is the unique, positive definite solution of the Riccati equation

$$\boldsymbol{A}^{\mathrm{T}}\boldsymbol{P} + \boldsymbol{P}\boldsymbol{A} - \boldsymbol{P}\boldsymbol{B}\boldsymbol{R}^{-1}\boldsymbol{B}^{\mathrm{T}}\boldsymbol{P} + \boldsymbol{Q} = \boldsymbol{0} \,. \tag{A.18}$$

The minimal value of the cost function concludes in

$$J^* = \boldsymbol{x}_0^{\mathrm{T}}\boldsymbol{P}\boldsymbol{x}_0 \,. \tag{A.19}$$

The closed-loop control

$$\dot{\boldsymbol{x}}(t) = \hat{\boldsymbol{A}}\boldsymbol{x}(t) \,, \quad \hat{\boldsymbol{A}} = \boldsymbol{A} - \boldsymbol{B}\boldsymbol{K}_x = \boldsymbol{A} - \boldsymbol{B}\boldsymbol{R}^{-1}\boldsymbol{B}^{\mathrm{T}}\boldsymbol{P} \,, \tag{A.20}$$

is asymptotically stable.

A.4.3 Choice of Poles and Weighting Matrices

The difficulty of both methods for controller design is the appropriate assignment of the poles and the weighting matrices, respectively. It makes sense to choose the weighting matrices as diagonal matrices. The stronger the weighting of a quantity in the cost function, the smaller this quantity will be due to the optimization. Only the weighting ratio of the control variables and the state variables is important. Therefore, $\boldsymbol{R} = \boldsymbol{E}, \boldsymbol{Q} = \mathbf{diag}\,(q_i)$ can be set, where

$$q_i = \frac{u_{max}^2}{(x_{i,max})^2} \tag{A.21}$$

holds as a rule of thumb. Here $u_{max}^2 = \left(\boldsymbol{u}^{\mathrm{T}}\boldsymbol{u}\right)_{max}$ is a measure of the maximum available control energy and $x_{i,max}$ the maximum tolerated value of the i-th state variable.

As a reference for the choice of the poles, the pole configuration for two special cases of cost functions for a scalar control $u\,(r = 1)$ will be given.

a) Mirroring of the uncontrolled system poles. The special case

$$J = \lim_{\rho \to 0} \frac{1}{2} \int_0^\infty \mathrm{e}^{2\gamma t}(\rho \boldsymbol{x}^{\mathrm{T}} \boldsymbol{Q} \boldsymbol{x} + u^2)\mathrm{d}t \to \mathrm{Min} \tag{A.22}$$

results in a pole configuration that yields minimal control energy for a target stability measure γ. Let $\lambda_i, i = 1\,(1)\,n$ be the poles of the uncontrolled system Σ, $k < n$ of them on the left side and the remaining $n - k$ poles on the right side of the line $\mathrm{Re}\lambda = -\gamma$ in the root locus plane. The k poles on the left side of the line $\mathrm{Re}\lambda = -\gamma$ and the $n - k$ control path poles that are mirrored on the line are chosen as poles $\hat{\lambda}_i, i = 1\,(1)\,n$ of the closed-loop control, Litz and Preuss (1977):

$$\begin{aligned} \hat{\lambda}_i &= \lambda_i & & i = 1(1)k\,, \\ & & \text{for} & \\ \hat{\lambda}_i &= -\lambda_i - 2\gamma & & i = k+1(1)n\,. \end{aligned} \tag{A.23}$$

b) Butterworth configuration. The limit case

$$J = \lim_{\rho \to \infty} \frac{1}{2} \int_0^\infty (\rho \boldsymbol{x}^{\mathrm{T}} \boldsymbol{Q} \boldsymbol{x} + u^2)\mathrm{d}t \to \mathrm{Min} \tag{A.24}$$

results in a pole configuration that yields minimal deviation for the controlled system. Using the nomenclature defined in Sect. A.4.1 the following applies for the poles $\hat{\lambda}_i, i = 1(1)n$:

1. All poles have negative real parts.
2. The $k < n$ dominant poles satisfy the equation

$$p(\hat{\lambda}^2) = \boldsymbol{b}^{\mathrm{T}}[\mathrm{adj}(-\hat{\lambda}\boldsymbol{E} - \boldsymbol{A})]^{\mathrm{T}}\boldsymbol{Q}\ \mathrm{adj}(\hat{\lambda}\boldsymbol{E} - \boldsymbol{A})\boldsymbol{b} = 0\,. \tag{A.25}$$

3. The $n - k$ remaining poles are infinitely large and represent a stable Butterworth configuration. They are located in the root locus plane on a circle around the origin with a radius proportional to $\rho^{\frac{n-m}{2}}$ $(\rho \to \infty)$,

$$\hat{\lambda}_i = \lim_{\rho \to \infty} \left[\rho \frac{n-m}{2} \mathrm{e}^{j\psi_i}\right]\,, \quad i = k + 1(1)n\,, \quad j = \sqrt{-1}\,. \tag{A.26}$$

The phase angles Ψ_i follow from

$$\begin{aligned} \Psi_i &= \frac{2m+1}{n-k} \cdot 90^\circ & & n-k \text{ even,} & & \\ & & \text{for} & & m &= 0, 1, 2, \ldots\,, \\ \Psi_i &= \frac{2m+1}{n-k} \cdot 180^\circ & & n-k \text{ uneven,} & & \end{aligned} \tag{A.27}$$

where Ψ_i has to be chosen so that $90^\circ < \Psi_i < 270^\circ$ holds. Explicitly, for $n - m = 1\,(1)\,4$ the results are given in Table A.1. In order to get restricted control quantities and finite actuating energy, the dominant poles are set exactly, the remaining poles approximatively in Butterworth configuration, but with a finite radius.

Table A.1. Results of Eq. (A.27)

$n-k$	1	2	3	4
Ψ_i	$+180^\circ$	$\pm 135^\circ$	$\pm 120^\circ; +180^\circ$	$\pm 112.5^\circ; \pm 157.5^\circ$

A.5 Structure and Properties of Observers

The control laws presented are depending on the state vector $\boldsymbol{x}$, e. g. the optimal control (A.17), but from the measurements only the vector $\boldsymbol{y} = \boldsymbol{C}\boldsymbol{x}$ is available. In many cases the number of measures is smaller than the system order. Then the inversion of the measurement matrix $\boldsymbol{C}$ and a direct representation of the state vector by $\boldsymbol{x} = \boldsymbol{C}^{-1}\boldsymbol{y}$ is not possible. The not measured state variables $\boldsymbol{T}\boldsymbol{x}$ ($\boldsymbol{T}$ row-regular $s \times n$-matrix) can be simulated by the $s \times 1$-vector $\boldsymbol{\xi}\,(t)$ with an observer. The observer is formulated mathematically as an asymptotical estimator for the estimations $\hat{\boldsymbol{x}}\,(t)$ of the state $\boldsymbol{x}\,(t)$, Luenberger (1964).

Estimation:

$$\hat{\boldsymbol{x}}(t) = \boldsymbol{S}_1\boldsymbol{\xi}(t) + \boldsymbol{S}_2\boldsymbol{y}(t) : \lim_{t\to\infty} [\boldsymbol{x}(t) - \hat{\boldsymbol{x}}(t)] = \boldsymbol{0} . \tag{A.28}$$

Observation:

$$\dot{\boldsymbol{\xi}}(t) = \boldsymbol{D}\boldsymbol{\xi}(t) + \boldsymbol{T}\boldsymbol{B}\boldsymbol{u}(t) + \boldsymbol{L}\boldsymbol{y}(t) : \lim_{t\to\infty} [\boldsymbol{\xi}(t) - \boldsymbol{T}\boldsymbol{x}(t)] = \boldsymbol{0} . \tag{A.29}$$

With Eqs. (A.1) and (A.2) the following matrix relations are available:

$$\boldsymbol{S}_1\boldsymbol{T} + \boldsymbol{S}_2\boldsymbol{C} = \boldsymbol{E}_n , \tag{A.30}$$

$$\boldsymbol{D}\boldsymbol{T} - \boldsymbol{T}\boldsymbol{A} = -\boldsymbol{L}\boldsymbol{C} , \tag{A.31}$$

$$\frac{\mathrm{d}}{\mathrm{d}t}(\boldsymbol{\xi} - \boldsymbol{T}\boldsymbol{x}) = \boldsymbol{D}(\boldsymbol{\xi} - \boldsymbol{T}\boldsymbol{x}) , \tag{A.32}$$

with

$$\mathrm{Re}\bar{\lambda}_i(\boldsymbol{D}) < 0 , \quad i = 1(1)s , \quad n - m \leq s \leq n . \tag{A.33}$$

Regarding the dimension s there are two special cases,

a) the minimal observer with $s = n - m$,
b) the complete observer with $s = n$.

For the complete observer, $\boldsymbol{T} = \boldsymbol{E}_n$ applies. From Eqs. (A.30) and (A.31) it follows

$$\boldsymbol{S}_1 + \boldsymbol{S}_2\boldsymbol{C} = \boldsymbol{E}_n , \quad \boldsymbol{D} = \boldsymbol{A} - \boldsymbol{L}\boldsymbol{C} . \tag{A.34}$$

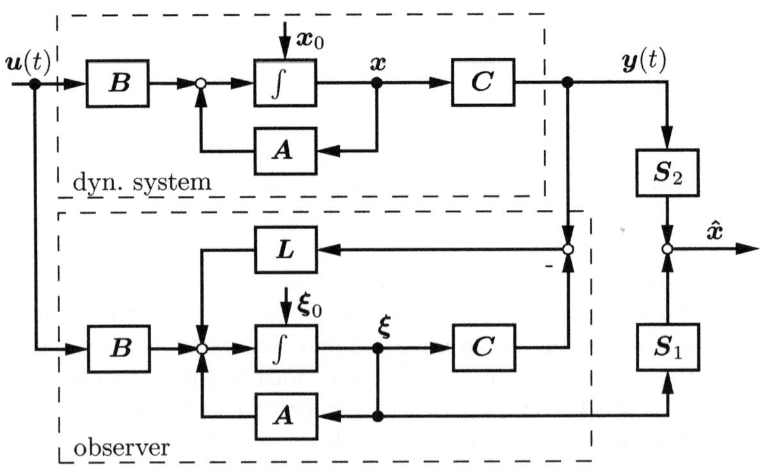

Fig. A.3. Block diagram of a dynamical system with complete observer

Therefore, the observer equations read

$$\hat{\boldsymbol{x}} = \boldsymbol{S}_1\boldsymbol{\xi} + \boldsymbol{S}_2\boldsymbol{y} ,$$

$$\dot{\boldsymbol{\xi}} = \underbrace{(\boldsymbol{A} - \boldsymbol{L}\boldsymbol{C})}_{\boldsymbol{D}}\boldsymbol{\xi} + \boldsymbol{B}\boldsymbol{u} + \boldsymbol{L}\boldsymbol{y} . \tag{A.35}$$

For the error $\boldsymbol{\delta} = \boldsymbol{\xi} - \boldsymbol{x}$ it remains, cp. Eq. (A.32),

$$\dot{\boldsymbol{\delta}} = \boldsymbol{D}\boldsymbol{\delta} = (\boldsymbol{A} - \boldsymbol{L}\boldsymbol{C})\boldsymbol{\delta} . \tag{A.36}$$

The corresponding system setup is given in Fig. A.3. Obviously, the complete observer is basically a simulation of the original system. The error signal between original and simulated system is fed back and used for control.

A.6 Observer Design

For the design of an n-dimensional observer the gain matrix $\boldsymbol{L}$ is required, cp. Eq. (A.35). This is done by specifications of the observer matrix $\boldsymbol{D} = \boldsymbol{A} - \boldsymbol{L}\boldsymbol{C}$. Beside the requirement of asymptotical stability Eq. (A.33) the transient response of the observer shall be much faster than that of the dynamical system. With the transition to the transposed matrix $\boldsymbol{D}^{\mathrm{T}} = \boldsymbol{A}^{\mathrm{T}} - \boldsymbol{C}^{\mathrm{T}}\boldsymbol{L}^{\mathrm{T}}$ the similarity to a linear control loop is evident, cp. Eq. (A.9) with

$$\boldsymbol{A} \mathrel{\widehat{=}} \boldsymbol{A}^{\mathrm{T}}, \quad \boldsymbol{B} \mathrel{\widehat{=}} \boldsymbol{C}^{\mathrm{T}}, \quad \boldsymbol{K}_x \mathrel{\widehat{=}} \boldsymbol{L}^{\mathrm{T}} . \tag{A.37}$$

Therefore, the methods described in Sect. A.4 for designing linear state controller are adaptable to the design of observer, too.

A.6.1 Observer Design with Pole Assignment

If the system Σ of order n with $m \leq n$ measured variables is complete observable, there exist always matrices $\boldsymbol{S}_1$, $\boldsymbol{S}_2$, $\boldsymbol{T}$, $\boldsymbol{D}$ and $\boldsymbol{L}$, which fulfill the relations of Eq. (A.30)-(A.32). The order s of the observer can be chosen arbitrarily in the range $n - m \leq s \leq n$. The eigenvalues $\lambda_i\left(\boldsymbol{D}\right), i = 1\,(1)\,s$, can also be set arbitrarily regarding Eq. (A.33). For complete observable systems with only one output, $m = 1$,

$$\dot{\boldsymbol{x}}(t) = \boldsymbol{A}\boldsymbol{x}(t) + \boldsymbol{B}\boldsymbol{u}(t) , \quad y(t) = \boldsymbol{c}^{\mathrm{T}}\boldsymbol{x}(t) , \tag{A.38}$$

the dimension s of the observer is either $s = n$ or $s = n - m = n - 1$. For an n-dimensional observer the gain matrix $\boldsymbol{L}$ results in a $n \times 1$ feedback vector $\boldsymbol{l}$. The matrix

$$\boldsymbol{D} = \boldsymbol{A} - \boldsymbol{l}\boldsymbol{c}^{\mathrm{T}} \tag{A.39}$$

has arbitrarily prescribed eigenvalues $\bar{\lambda}_i, \mathrm{Re}\bar{\lambda}_i < 0$, $i = 1\,(1)\,n$, or the corresponding characteristic polynomial

$$\bar{p}(\lambda) = (\lambda - \bar{\lambda}_1)(\lambda - \bar{\lambda}_2)\ldots(\lambda - \bar{\lambda}_n) = \lambda^n + d_1\lambda^{n-1} + \ldots + d_{n-1}\lambda + d_n \,. \tag{A.40}$$

The feedback vector $\boldsymbol{l}$ is determined uniquely as

$$\boldsymbol{l} = \bar{p}(\boldsymbol{A})(\boldsymbol{Q}_B^{\mathrm{T}})^{-1}\boldsymbol{e}_n \,. \tag{A.41}$$

Here $\boldsymbol{e}_n$ is the n-th unit vector, $\boldsymbol{Q}_B$ the observability matrix and $\bar{p}\,(\boldsymbol{A})$ the matrix-polynomial built of the system matrix $\boldsymbol{A}$:

$$\boldsymbol{e}_n = [0, 0, 0 \ldots 0, 1]^{\mathrm{T}} \,, \tag{A.42}$$

$$\boldsymbol{Q}_B = [\boldsymbol{c} \vdots \boldsymbol{A}^{\mathrm{T}}\boldsymbol{c} \vdots \cdots \vdots \boldsymbol{A}^{Tn-1}\boldsymbol{c}] \,, \tag{A.43}$$

$$\bar{p}(\boldsymbol{A}) = \boldsymbol{A}^n + d_1\boldsymbol{A}^{n-1} + \ldots + d_{n-1}\boldsymbol{A} + d_n\boldsymbol{E} \,. \tag{A.44}$$

The matrices $\boldsymbol{S}_1$ and $\boldsymbol{S}_2$ can be chosen as

$$\boldsymbol{S}_1 = \boldsymbol{E}_n - \frac{\boldsymbol{c}\boldsymbol{c}^{\mathrm{T}}}{\boldsymbol{c}^{\mathrm{T}}\boldsymbol{c}} \,, \quad \boldsymbol{S}_2 = \frac{\boldsymbol{c}}{\boldsymbol{c}^{\mathrm{T}}\boldsymbol{c}} \tag{A.45}$$

or

$$\boldsymbol{S}_1 = \boldsymbol{E}_n \,, \quad \boldsymbol{S}_2 = \boldsymbol{0} \,. \tag{A.46}$$

A.6.2 Optimal Observer Due to a Quadratic Integral Criterion

For an n-dimensional observer the equation for the error $\boldsymbol{\delta} = \boldsymbol{\xi} - \boldsymbol{x}$ is given by Eq. (A.36). The transposed error differential equation

$$\dot{\boldsymbol{\eta}}(t) = (\boldsymbol{A}^{\mathrm{T}} - \boldsymbol{C}^{\mathrm{T}}\boldsymbol{L}^{\mathrm{T}})\boldsymbol{\eta}(t) \tag{A.47}$$

has the structure of a control loop with state feedback, cp. Eq. (A.9),

$$\dot{\boldsymbol{\eta}}(t) = \boldsymbol{A}^{\mathrm{T}}\boldsymbol{\eta}(t) + \boldsymbol{C}^{\mathrm{T}}\boldsymbol{\nu}(t) \,, \quad \boldsymbol{\eta}(0) = \boldsymbol{\eta}_0 \,, \tag{A.48}$$

$$\boldsymbol{\nu}(t) = -\boldsymbol{L}^{\mathrm{T}}\boldsymbol{\eta}(t) \,. \tag{A.49}$$

The complete controllability of the system of Eq. (A.48) corresponds to the complete observability of the system given by Eqs. (A.1) and (A.2). In analogy to Sect. A.4.2 the following can be stated: For the complete observable system or the complete controllable system (A.48), respectively, both of order n, and the cost function

$$J[\boldsymbol{\eta}(t), \boldsymbol{\nu}(t)] = \frac{1}{2}\int_0^\infty [\boldsymbol{\eta}^{\mathrm{T}}(t)\boldsymbol{Q}\boldsymbol{\eta}(t) + \boldsymbol{\nu}^{\mathrm{T}}(t)\boldsymbol{R}\boldsymbol{\nu}(t)]\mathrm{d}t \rightarrow \mathrm{Min} \tag{A.50}$$

with the weighting matrices $\boldsymbol{Q} = \boldsymbol{Q}^{\mathrm{T}} \geq \mathbf{0}, \boldsymbol{R} = \boldsymbol{R}^{T} > \mathbf{0}$ and $\left(\boldsymbol{A}^{T}\boldsymbol{Q}\right)$ completely observable, there exists one unique optimal control

$$\boldsymbol{\nu}^{*}(t) = -\boldsymbol{R}^{-1}\boldsymbol{C}\boldsymbol{P}\boldsymbol{\eta}(t) = -\boldsymbol{L}^{\mathrm{T}}\boldsymbol{\eta}(t) \,. \tag{A.51}$$

Herein $\boldsymbol{P} = \boldsymbol{P}^{\mathrm{T}} > \mathbf{0}$ is the unique, positive definite solution of the Riccati equation

$$\boldsymbol{A}\boldsymbol{P} + \boldsymbol{P}\boldsymbol{A}^{\mathrm{T}} - \boldsymbol{P}\boldsymbol{C}^{\mathrm{T}}\boldsymbol{R}^{-1}\boldsymbol{C}\boldsymbol{P} + \boldsymbol{Q} = \mathbf{0} \,. \tag{A.52}$$

The optimal value for the criterion results in

$$J^{*} = \boldsymbol{\eta}_0^{\mathrm{T}}\boldsymbol{P}\boldsymbol{\eta}_0 \,. \tag{A.53}$$

The closed-loop control

$$\dot{\boldsymbol{\eta}}(t) = \boldsymbol{D}^{\mathrm{T}}\boldsymbol{\eta}(t) \,, \quad \boldsymbol{D}^{\mathrm{T}} = \boldsymbol{A}^{\mathrm{T}} - \boldsymbol{C}^{\mathrm{T}}\boldsymbol{L}^{\mathrm{T}} = \boldsymbol{A}^{\mathrm{T}} - \boldsymbol{C}^{\mathrm{T}}\boldsymbol{R}^{-1}\boldsymbol{C}\boldsymbol{P} \tag{A.54}$$

is asymptotically stable, and, therefore, the n-dimensional observer of Eq. (A.36), too. The results of Sect. A.4.3 for the choice of the poles and the weighting matrices can also be used for designing the observer.

A.7 Structure of (Optimal) Controlled Multivariable Systems

For the complete controllable and complete observable system Σ a state feedback according to Sect. A.4 and a state observer according to Sect. A.6 can be found, each with the specified properties. The structure of the complete system is depicted in Fig. A.2 c).

The complete multivariable control system is described mathematically as:

Dynamical system:

$$\dot{\boldsymbol{x}}(t) = \boldsymbol{A}\boldsymbol{x}(t) + \boldsymbol{B}\boldsymbol{u}(t) \,, \tag{A.55}$$

Measurement:

$$\boldsymbol{y}(t) = \boldsymbol{C}\boldsymbol{x}(t) \,,$$

Observer:

$$\dot{\boldsymbol{\xi}}(t) = \boldsymbol{D}\boldsymbol{\xi}(t) + \boldsymbol{T}\boldsymbol{B}\boldsymbol{u}(t) + \boldsymbol{L}\boldsymbol{y}(t) \,, \tag{A.56}$$

$$\hat{\boldsymbol{x}}(t) = \boldsymbol{S}_1\boldsymbol{\xi}(t) + \boldsymbol{S}_2\boldsymbol{y}(t) \,,$$

Controller:

$$\boldsymbol{u}(t) = -\boldsymbol{K}_x\hat{\boldsymbol{x}}(t) \,. \tag{A.57}$$

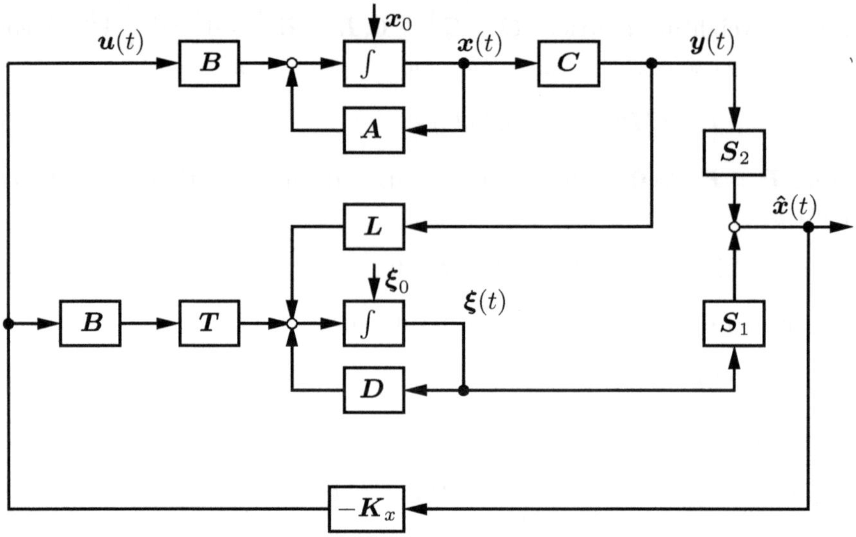

Fig. A.4. Block diagram of the complete control system

The eigenvalues of the linear, time invariant complete system, Fig. A.4, are composed of the n eigenvalues of the control loop with ideal state feedback and the s eigenvalues of the observer. For the characteristic polynomial and the eigenvalues holds:

$$p_{ges}(\lambda) = \hat{p}_{A-BK_x}(\lambda) \cdot \bar{p}_D(\lambda) \ , \tag{A.58}$$

$$\lambda_{i,ges} = \begin{cases} \hat{\lambda}_i(\boldsymbol{A} - \boldsymbol{B}\boldsymbol{K}_x) & \quad i = 1(1)n \\ & \text{for} \\ \hat{\lambda}_{i-n}(\boldsymbol{D}) & \quad i = (n+1)(1)(n+s) \ . \end{cases}$$

The realizable complete control loop is extended by an observer compared to the ideal state feedback. With respect to the eigenvalues of the complete system, the controller can be designed separately from the design of the observer.

B

Appendix: Key Words

B.1 English - German

A	
acceleration	Beschleunigung
aligning torque stiffness	Bohrmomentbeiwert
B	
beam	Balken
continuously bedded	kontinuierlich gebetteter
double-span	Zweifeldträger
pillared	periodisch gestützter
single-span	Einfeldträger
bicycle model	Riekert-Schunck Modell
body slip angle	Schwimmwinkel
boundary value problem	Randwertproblem
braking force	Bremskraft
brush model	Bürstenmodell
C	
Cardano angle	Kardanwinkel
connecting element	Bindungselement
constraint	Bindung, Zwangsbedingung
contact force	Kontaktkraft
law	-gesetz
tire-road	Reifen-Straße
wheel-rail	Rad-Schiene
control gain	Reglerverstärkung
control vector	Steuervektor
controllability	Steuerbarkeit

controller	Regler
design	-entwurf
gain	-verstärkung
cornering	Kurvenfahrt
stiffness	Seitenkraftbeiwert
cost function	Gütekriterium
coupling element	Koppelelement
creepage	Radsatzschlupf

D

Damper	Dämpfer
characteristic	-Kennlinie
parallel combination	-Parallelschaltung
series combination	-Reihenschaltung
damping	Dämpfung
normalized coefficient	-smaß, Lehrsches
degree of efficiency	Wirkungsgrad
degree of freedom	Freiheitsgrad
degree of unevenness	Unebenheitsgrad
derailment	Entgleisung
differential	Differentialgetriebe
driving comfort	Fahrkomfort
driving performance	Fahrleistung
driving performance diagram	Fahrzustandsschaubild
driving safety	Fahrsicherheit
driving stability	Fahrstabilität
durability	Lebensdauer

E

eigenmode	Schwingungsform
elementary rotation	Elementardrehung
equation of motion	Bewegungsgleichung
equation of reaction	Reaktionsgleichung
excitation	Erregung
bump	Bodenwellen-
ramp	Rampen-
random	Zufalls-
time delay	zeitverzögerte
unbalance	Unwucht-
exposure time	Einwirkungsdauer

F

feedback matrix	Rückführmatrix
force	Kraft
applied	eingeprägte

constraint	Reaktions-
contact	Kontakt-
dissipative	dissipative
cornerning	Seiten-
friction	Reibungs-
generalized	verallgemeinerte
lateral	Seiten-
reaction	Reaktions-
travelling	bewegte
force actuator	Kraftstellglied
foundation	Bettung
frame	Koordinatensystem
frequency decoupling	Frequenzentkopplung
frequency response	Frequenzgang
friction	Reibung
sliding	Gleit-
sticking	Haft-
friction coefficient	Reibungsbeiwert

G

gradability	Steigfähigkeit
gravitational stiffness	Gravitationssteifigkeit
guidance system	Führsystem
guideway	Fahrweg
gyro matrix	Kreiselmatrix

H

handling	Kurshaltung
hunting motion	Sinuslauf

I

inertia matrix	Massenmatrix
inertia properties	Trägheitseigenschaften
initial value problem	Anfangswertproblem
input matrix	Eingangsmatrix

L

lateral motion	Querbewegung

M

maglev vehicle	Magnetschwebefahrzeug
magnetic actuator	Magnetstellglied
magnetic wheel	Magnetisches Rad
maximum rolling contact coefficient	Kraftschlußbeiwert
mean value	Mittelwert

mileage	Kraftstoffverbrauch
model	Ersatzsystem
modeling	Modellbildung
moment of inertia	Trägheitsmoment
motion	Bewegung
hunting	Sinuslauf
plane	ebene
moving load	bewegte Last
multibody system	Mehrkörpersystem
multivariable system	Mehrgrößensystem
N	
node displacement	Knotenverschiebung
noise process	Rauschprozeß
O	
observability	Beobachtbarkeit
oversteer	übersteuern
P	
perception	Wahrnehmung
pitch	nicken
pole assignment	Polvorgabe
position vector	Lagevektor
power spectral density	spektrale Leistungsdichte
probability density	Wahrscheinlichkeitsdichte
product of inertia	Devitationsmoment
R	
railway wheelset	Eisenbahnradsatz
random vibration	Zufallsschwingungen
reliability interval	Vertrauensintervall
resistance	Widerstand
climbing	Steigungs-
rolling	Roll-
ride comfort	Fahrkomfort
rolling condition	Rollbedingung
rolling contact coefficient	Kraftschlußbeanspruchung
normalized	normierte
maximum	maximale
rolling elastic contact	rollender elastischer Kontakt
rolling radius	Rollradius
roll	rollen
rotation matrix	Drehmatrix

S

safety margin	Sicherheitsreserve
saturation	Sättigung
shape filter	Formfilter
slip	Schlupf
aligning	Bohr-
braking	Brems-
driving	Antriebs-
lateral	Quer-
longitudinal	Längs-
micro	Mikro-
rigid body	Starrkörper-
slip angle	Schräglaufwinkel
speed	Fahrgeschwindigkeit
spring	Feder
characteristic	-kennlinie
leaf	Blatt-
parallel connection	-Parallelschaltung
series connection	-Reihenschaltung
state equation	Zustandsgleichung
stiffness matrix	Steifigkeitsmatrix
subsystem	Teilsystem
surface pressure	Flächenpressung
suspension	Federung
suspension travel	Federweg
system boundary	Systemgrenze

T

time integration	Zeitintegration
tire	Reifen
carcass	-karkasse
torque	Moment
aligning	Bohr-
braking	Brems-
driving	Antriebs-
track model	Schienenmodell
trail	Nachlauf
suspension	konstruktiver
trailer	Anhänger
transmission ratio	Übersetzung
twist	Bewegungswinder

U

understeer	untersteuern
unevenness	Unebenheit

V

vehicle-guideway-system	Fahrzeug-Fahrweg-System
velocity	Geschwindigkeit
vibration	Schwingung

W

waviness	Welligkeit
wheel	Rad
braked	gebremstes
conical	konisches
cornering	schräglaufendes
deformable	deformierbar
driven	angetriebenes
elastic	elastisches
rigid	starres
wheel load	Radlast
wheelset	Radsatz
wrench	Kraftwinder

Y

yaw	gieren, schleudern

B.2 Deutsch - Englisch

A	
Anfangswertproblem	initial value problem
Anhänger	trailer
B	
Balken	beam
Einfeldträger	single-span
kontinuierlich gebetteter	continously bedded
periodisch gestützter	pillared
Zweifeldträger	double-span
Beiwert	coefficient
Beobachtbarkeit	observability
Beschleunigung	acceleration
Bettung	foundation
bewegte Last	moving load
Bewegung	motion
ebene	plane
Sinuslauf	hunting
Bewegungsgleichung	equation of motion
Bewegungswinder	twist
Bremskraft	braking force
Bindung	constraint
Bindungselement	connecting element
Bohrmomentbeiwert	aligning torque stiffness
Bürstenmodell	brush model
D	
Dämpfer	damper
-Parallelschaltung	parallel combination
-Reihenschaltung	series combination
-kennlinie	characteristic
Dämpfung	damping
-smaß, Lehrsches	normalized coefficient
Devitationsmoment	product of inertia
Differentialgetriebe	differential
Drehmatrix	rotation matrix
E	
Eingangsmatrix	input matrix
Einwirkungsdauer	exposure time
Eisenbahnradsatz	railway wheelset

Elementardrehung	elementary rotation
Entgleisung	derailment
Erregung	excitation
Bodenwellen-	bump
Rampen-	ramp
stochastische	stochastic
Unwucht-	unbalance
zeitverzögerte	time delay
Zufalls-	random
Ersatzsystem	model

F

Fahrgeschwindigkeit	speed
Fahrkomfort	driving comfort, ride comfort
Fahrleistung	driving performance
Fahrsicherheit	driving safety
Fahrstabilität	driving stability
Fahrweg	guideway
Fahrzeug-Fahrweg-System	vehicle-guideway-system
Fahrzustandsschaubild	driving performance diagram
Feder	spring
-kennlinie	characteristic
-Parallelschaltung	parallel connection
-Reihenschaltung	series connection
Blatt-	leaf
Federung	suspension
Federweg	suspension travel
Flächenpressung	surface pressure
Formfilter	shape filter
Freiheitsgrad	degree of freedom
Frequenzentkopplung	frequency decoupling
Frequenzgang	frequency response
Führsystem	guidance system

G

Geschwindigkeit	velocity
gieren	yaw
Gravitationssteifigkeit	gravitational stiffness
Gütekriterium	cost function

K

Kardanwinkel	Cardano angle
Knotenverschiebung	node displacement
Kontaktkraft	contact force
-gesetz	law

Rad-Schiene	wheel-rail
Reifen-Straße	tire-road
Koordinatensystem	frame
Koppelelement	coupling element
Kraft	force
bewegte	travelling
dissipative	dissipative
eingeprägte	applied
Kontakt-	contact
Reaktions-	constraint
Reibungs-	friction
Seiten-	cornerning
verallgemeinerte	generalized
Kraftschlußbeanspruchung	rolling contact coefficient
normierte	normalized
Kraftschlußbeiwert	maximum rolling contact coefficient
Kraftstellglied	force actuator
Kraftstoffverbrauch	mileage
Kraftwinder	wrench
Kreiselmatrix	gyro matrix
Kurshaltung	handling
Kurvenfahrt	cornering

L

Lagevektor	position vector
Lebensdauer	durability

M

Magnetisches Rad	magnetic wheel
Magnetschwebefahrzeug	maglev vehicle
Magnetstellglied	magnetic actuator
Massenmatrix	inertia matrix
Mehrgrößensystem	multivariable system
Mehrkörpersystem	multibody system
Mittelwert	mean value
Modellbildung	modeling
Moment	torque
Antriebs-	driving
Bohr-	aligning
Brems-	braking

N

Nachlauf	trail
konstruktiver	suspension
nicken	pitch

P

Polvorgabe	pole assignment

Q

Querbewegung	lateral motion

R

Rad	wheel
angetriebenes	driven
deformierbar	deformable
elastisches	elastic
gebremstes	braked
konisches	conical
schräglaufendes	cornering
starres	rigid
Radlast	wheel load
Radsatz	wheelset
Radsatzschlupf	creepage
Randwertproblem	boundary value problem
Rauschprozeß	noise process
Reaktionsgleichung	equation of reaction
Regler	controller
-entwurf	design
-verstärkung	gain
Reibung	friction
Gleit-	sliding
Haft-	sticking
Reibungsbeiwert	friction coefficient
Reifen	tire
-karkasse	carcass
Riekert-Schunck Modell	bicycle model
Rollbedingung	rolling condition
rollen	roll
rollender elastischer Kontakt	rolling elastic contact
Rollradius	rolling radius
Rückführmatrix	feedback matrix

S

Sättigung	saturation
Schienenmodell	track model
schleudern	yaw
Schlupf	slip
Antriebs-	driving
Bohr-	aligning

Brems-	braking
Längs-	longitudinal
Mikro-	micro
Quer-	lateral
Starrkörper-	rigid body
Schräglaufwinkel	slip angle
Schwimmwinkel	body slip angle
Schwingung	vibration
Schwingungsform	eigenmode
Seitenkraftbeiwert	cornering stiffness coefficient, lateral force coefficient
Sicherheitsreserve	safety margin
Sinuslauf	hunting motion
spektrale Leistungsdichte	power spectral density
Steigfähigkeit	gradability
Steuervektor	control vector
Steuerbarkeit	controllability
Systemgrenze	system boundary

T

Teilsystem	subsystem
Trägheitseigenschaften	inertia properties
Trägheitsmoment	moment of inertia

U

Übersetzung	transmission ratio
übersteuern	oversteer
Unebenheit	unevenness
Unebenheitsgrad	degree of unevenness
untersteuern	Understeer

V

Vertrauensintervall	reliability interval

W

Wahrnehmung	perception
Wahrscheinlichkeitsdichte	probability density
Welligkeit	waviness
Widerstand	resistance
Roll-	rolling
Steigungs-	climbing
Wirkungsgrad	degree of efficiency

Z

Zeitintegration	time integration

Zufallsschwingungen	random vibration
Zustandsgleichung	state equation
Zwangsbedingung	constraint

References

Adams Multibody Dynamics (2009), `http://www.msc.software.com`

ATP Automotive Testing Papenburg (2009), `http://www.atp.papenburg.com`

Bae, D., Haug, E.: A recursive formulation for constrained mechanical system dynamics: Part i, open loop systems. Mechanics of Structures and Machines 15, 359–382 (1987a)

Bae, D., Haug, E.: A recursive formulation for constrained mechanical system dynamics: Part ii, closed loop systems. Mechanics of Structures and Machines 15, 481–506 (1987b)

Baumal, A., McPhee, J., Calamai, P.: Application of genetic algorithms to the design optimization of an active vehicle suspension system. Computer Methods in Applied Mechanics and Engineering 163, 87–94 (1998)

Bestle, D.: Analyse und Optimierung von Mehrkörpersystemen. Springer, Berlin (1994) (in German)

Boehm, F.: Theorie schnell veraenderlicher Rollzustaende fuer Guertelreifen. Ing.-Arch. 55, 30–44 (1985) (in German)

Bogacz, R., Krzyzynski, T., Popp, K.: On the generalization of Mathews' problem of the vibrations of a beam of elastic foundation. Z. angew. Math. Mech. 69, 243–252 (1989)

Bormann, V.: Messungen von Fahrbahnunebenheiten paralleler Fahrspuren und Anwendung der Ergebnisse. Vehicle System Dynamics 7, 65–81 (1978) (in German)

Bosch, R.: Kraftfahrtechnisches Taschenbuch. Vieweg, Wiesbaden (2007) (in German)

Brabie, D.: On Derailment-Worthiness in Rail Vehicle Design: Analysis of vehicle features influencing derailment processes and consequences. PhD thesis, KTH, Aeronautics and Vehicle Engineering, Stockholm (2007)

Brandl, H., Johanni, R., Otter, M.: A very efficient algorithm for the simulation of robots and similar multibody systems without inversion of the mass matrix. In: Kopacek, P., Troch, I., Desoyer, K. (eds.) Theory of Robots, pp. 95–100. Pergamon, Oxford (1988)

Braun, H.: Untersuchungen von Fahrbahnunebenheiten und Anwendungen der Ergebnisse. Dr.-Ing. Diss., TU Braunschweig (1969) (in German)

Bryson, A.E.: Applied Linear Optimal Control. Cambridge University Press, Cambridge (2002)

Burckhardt, M.: Die Mercedes-Benz/Bosch-Antriebs-Schlupf-Regelung (ASR). Verkehrsunfall und Fahrzeugtechnik 10, 285–290 (1986) (in German)

Carter, F.W.: On the action of a locomotive driving wheel. Proc. Roy. Soc. A, 112–151 (1926)

Cebon, D.: Simulation of the response of leaf springs to broad band random excitation. Vehicle System Dynamics 15, 375–392 (1986)

Claus, H.: Dynamikanalyse eines Eisenbahnwagens mit radialelastischen Raedern. Fort. Ber. VDI. Reihe 12, Nr. 568. VDI-Verlag, Duesseldorf (2004)

Claus, H., Schiehlen, W.: Symbolic-numeric analysis of flexible multibody systems. Mechanics of Structures and Machines 30(1), 1–30 (2002)

Crandall, S.H., Mark, W.D.: Random Vibration in Mechanical Systems. Academic Press, New York (1963)

Czerny, L.: Analyse stationaerer Zufallsschwingungen. Fort. Ber. VDI-Reihe 11, Nr. 99. VDI-Verlag, Duesseldorf (1987) (in German)

De Pater, A.: Optimal design of railway vehicles. Ing.-Arch. 57, 25–38 (1987)

Dodds, C., Robson, J.: The description of road surface roughness. J. Sound Vibrations 31, 175–183 (1973)

Doerr, J.: Der unendliche, federnd gebettete Balken unter dem Einfluss einer gleichfoermig bewegten Last. Ing.-Arch. 14, 167–192 (1943) (in German)

Fleissner, F., Lehnart, A., Eberhard, P.: Dynamic simulation of sloshing fluid and granular cargo in transport vehicles. Vehicle System Dynamics 48, 3–15 (2010)

Foellinger, O.: Regelungstechnik, vol. 6. Aufl. Huethig Buch Verlag, Heidelberg (2008) (in German)

Fromm, H.: Berechnung des Schlupfes beim Rollen deformierbarer Scheiben. Z. angew. Math. Mech. 7, 27–58 (1927) (in German)

Fryba, L.: Vibration of Solids and Structures under Moving Loads, 3rd edn. Thomas Telford Ltd., London (1999)

Gaebel, G., Moldenhauer, P., Kroeger, M.: Local effects between tire and road. ATZ Autotechnology 4, 48–53 (2008)

Garg, V.K., Dukkipati, R.V.: Dynamics of Railway Systems. Academic Press, Toronto (1984)

Gillespie, T.: Fundamentals of Vehicle Dynamics. Society of Automotive Eng., Troy (1992)

Goodall, R.M.: Dynamics and control requirements for EMS maglev suspensions. In: Proceedings 18th Int. Conf. on Magnetically Levitated Systems and Linear Drives (MAGLEV 2004), Shanghai, October 26-28, pp. 926–934 (2004)

Gottzein, E.: Das 'Magnetische Rad' als autonome Funktionseinheit modularer Trag- und Fuehrsysteme fuer Magnetbahnen. Fort. Ber. VDI. Reihe 8, Nr. 68. VDI-Verlag, Duesseldorf (1984) (in German)

Gutzeit, F., Sextro, W., Kroeger, M.: Unsteady rolling contact of rubber wheels. analysis and simulation of contact problems. Lecture Notes in Applied and Computational Mechanics 27, 261–270 (2006)

Hairer, E., Norset, S., Wanner, G.: Solving Ordinary Differential Equations I, Nonstiff Problems. Springer, Berlin (2008)

Haug, E., Arora, J.: Applied Optimal Design. Wiley Interscience, New York (1979)

Heimann, B., Gerth, W., Popp, K.: Mechatronik. Hanser Fachbuchverlag, Leipzig (2007) (in German)

Heinrich, W., Henning, K.: Zufallsschwingungen mechanischer Systeme. Vieweg, Braunschweig (1978) (in German)

Helms, H., Strothmann, W.: Lateral rail irregularities - measurement and applications. In: Slibar, A., Springer, H. (eds.) Proc. 5th VSD-2nd IUTAM Symposium on the Dynamics of Vehicles on Roads and on Tracks, pp. 430–459. Swets and Zeitlinger, Amsterdam (1977)

Hertz, H.: Ueber die Beruehrung fester elastischer Koerper. In: Gesammelte Werke, vol. 1. Barth, Leipzig (1895) (in German)

Hirschberg, W.: Vergleich verschiedener Fahrzeugmodelle. Studienarbeit STUD-1. Inst. Eng. Comp. Mechanics, University of Stuttgart, Stuttgart (1981) (in German)

Hollerbach, J.: A recursive lagrangian formulation of manipulator dynamics and comparative study of dynamics formulation complexity. IEEE Trans. Syst. Man. Cybern. 11, 730–736 (1980)

ISO 2631. Mechanical Vibration and Shock-Evaluation of Human Exposure to Whole-Body Vibrations. Beuth, Berlin (1997-2004)

ISO International Standard 2631. Guide for the Human Exposure to Whole-Body Vibrations. Beuth, Berlin (1974)

Iwnicki, S.: Handbook of Railway Vehicle Dynamics. CRC Press, Boca Raton (2006)

Johnson, K.L., Vermeulen, P.J.: Contact of non-spherical bodies transmitting tangential forces. J. Appl. Mech. 31, 338–340 (1964)

Kalker, J.J.: Introduction to the Fortran Program CONTACT for the Solution of 3D Elastostatic Half-Space Contact Problems With and Without Friction. Technical report, TH Delft (1967a)

Kalker, J.J.: On the Rolling Contact of Two Elastic Bodies in the Presence of Dry Friction. PhD thesis, TH Delft (1967b)

Kalker, J.J.: Survey on wheel-rail rolling contact theory. Vehicle System Dynamics 5, 317–358 (1979)

Kane, T., Levinson, D.: Dynamics: Theory and Applications. McGraw-Hill, New York (1985)

Kiencke, U., Nielsen, L.: Automotive Control Systems. Springer, Berlin (2005)

Klingel, J.: Ueber den Lauf der Eisenbahnwagen auf gerader Bahn. Organ Fortschr. des Eisenbahnwesens 20, 113–123 (1883) (in German)

Knothe, K.: Vergleichende Darstellung verschiedener Verfahren zur Berechnung der Eigenschwingungen von Rahmentragwerken. In: Fort. Ber. VDI Reihe 11, Nr. 9, VDI-Verlag, Duesseldorf (1971) (in German)

Knothe, K., Wessels, H.: Finite Elemente, 4th edn. Springer, Berlin (2008)

Kolousek, V., McLean, R.F., Fleming, J.S.: Dynamics in Engineering Structures. Newnes-Butterworth, London (1973)

Kortuem, W., Sharp, R.: Multibody Computer Codes in Vehicle System Dynamics. Swets and Zeitlinger, Lisse (1993)

Kortuem, W., Wormley, D.N.: Dynamic interactions between travelling vehicles and guideway systems. Vehicle System Dynamics 10, 285–317 (1981)

Kraft, F.: Der Einfluss der Fahrgeschwindigkeit auf den Haftwert zwischen Rad und Schiene. Arch. Eisenbahntechnik 22, 58–78 (1976) (in German)

Kranz, A.: Beitrag zur Beschreibung der Eigenschaften geschichteter Trapez- und Parabelfedern (in German). Dr.-Ing. Diss., Universitaet Hannover (1983)

Krause, H., Poll, G.: Mechanik der Festkoerperreibung. VDI-Verlag, Duesseldorf (1980)

Kreisselmeier, G.: A solution of the bilinear matrix equation AY + YB = -Q. SIAM J. Appl. Math. 23, 334–338 (1972)

Kreuzer, E., Leister, G.: Programmsystem NEWEUL - Beispielsammlung. Anleitung AN-20. Inst. Eng. Comp. Mechanic, Universitaet Stuttgart (1988) (in German)

Kreuzer, E., Rill, G.: Vergleichende Untersuchung von Fahrzeugschwingungen an raeumlichen Ersatzmodellen. Ing.-Arch. 52, 205–219 (1982) (in German)

Kreuzer, E.J.: Numerische Untersuchung nichtlinearer dynamischer Systeme. Springer, Berlin (1987) (in German)

Kropac, O., Mucka, P.: Be careful when using the international roughness index as an indicator of road unevenness. Journal of Sound and Vibration 287, 989–1003 (2005)

Krugmann, H.L.: Lauf der Schienenfahrzeuge im Gleis. Oldenbourg, Muenchen (1982) (in German)

Kurz, T., Eberhard, P.: Symbolic modeling and analysis of elastic multibody systems. In: Terzc, Z., Lacor, C. (eds.) Proceedings of International Symposium on Coupled Methods in Numerical Dynamics, pp. 187–201 (2009); Fac. Mech. Eng. Naval Arch., Zagreb

Law, E.H., Cooperrider, N.K.: A survey of railway vehicle dynamics research. J. Dyn. Syst. Meas. Control, Trans. ASME 96, 132–146 (1974)

Leiber, H., Czinczel, A.: Antiblockiersystem fuer Personenwagen mit digitaler Elektronik, Aufbau und Funktion. Automobilt. Zeitschrift 81, 569–583 (1979) (in German)

Lewis, P.R., Gagg, C.: Aesthetics versus function: the fall of the dee bridge, 1847. Interdisciplinary Science Reviews 29(2), 177–191 (2004)

Litz, L., Preuss, H.P.: Ein Vorschlag zur Wahl der Pole beim Verfahren der Polvorgabe. Regelungstechnik 25(10), 318–323 (1977) (in German)

Ljung, L.: System Identification - Theory of the User. Prentice Hall, Upper Saddle River (1999)

Luenberger, D.G.: The state of a linear system. IEEE Trans. Military Electronics MIL-8, 74–80 (1964)

Lugner, P., Ploechl, M.: Tyre Models for Vehicle Dynamics Analysis. Taylor & Francis, Abington (2005)

Magnus, K., Mueller-Slany, H.H.: Grundlagen der Technischen Mechanik, 7th edn. B.G. Teubner, Wiesbaden (2005) (in German)

Magnus, K., Popp, K., Sextro, W.: Schwingungen. 8. Aufl. B.G. Teubner, Wiesbaden (2008) (in German)

Matthews, P.M.: Vibration of a beam on elastic foundation. Z. angew. Math. Mech. 38, 105–115 (1958)

Matthews, P.M.: Vibrations of a beam on elastic foundation II. Z. angew. Math. Mech. 39, 13–19 (1959)

Meinders, T., Meinke, P.: Rotordynamics and irregular wear of elastic wheelsets. In: Popp, K., Schiehlen, W. (eds.) System Dynamics and Long-Term Behaviour of Railway-Vehicles, Track and Subgrade, pp. 133–152. Springer, Berlin (2003)

Melzer, F.: Symbolisch-numerische Modellierung elastischer Mehrkoerpersysteme mit Anwendung auf rechnerische Lebensdauervorhersagen. In: Fort. Ber. VDI. Reihe 20, Nr. 139. VDI-Verlag, Duesseldorf (1994) (in German)

Mitschke, M., Wallentowitz, H.: Dynamik der Kraftfahrzeuge. Springer, Berlin (2004) (in German)

Moon, F.C.: Chaotic vibrations. Wiley, New York (1987)

Mueller, P., Popp, K.: Kovarianzanalyse von linearen Zufallsschwingungen mit zeitlich verschobenen Erregerprozessen (in German). Z. angew. Math. Mech. 59, T144–T146 (1979)

Mueller, P., Popp, K., Schiehlen, W.: Berechnungsverfahren fuer stochastische Fahrzeugschwingungen. Ing.-Arch. 49, 235–254 (1980) (in German)

Mueller, P., Schiehlen, W.: Linear Vibrations. Martinus Nijhoff Publ. (Springer), Dorbrecht (1985)

Newland, D.E.: Random Vibrations and Spectral Analysis. Longman, London (1975)

Nishimura, T., Ohoha, M.: Cal Desk for adaptative cruise control. dSPACE News 1, 6–7 (2007)

ORE. Wechselwirkungen zwischen Fahrzeugen und Gleis. In: Bericht Nr. 1: Spektrale Dichte der Unregelmaeßigkeiten in der Gleislage, ORE-Report, C116/RP1/D, Utrecht (1971)

Pacejka, H.B.: Principles of plane motions of automobiles. In: Pacejka, H. (ed.) Proc. of IUTAM Symp. on the Dynamics of Vehicle on Roads and on Railway Tracks, pp. 33–59. Swets and Zeitlinger, Amsterdam (1975)

Pacejka, H.B.: Modelling of the pneumatic tire and its impact on vehicle dynamic behaviour. In: CCG-Course Vehicle and Transport Technology, Oberpfaffenhofen (1986)

Pacejka, H.B.: Tire and Vehicle Dynamics. Butterworth-Heinemann, Oxford (2002)

Pasimodo (2009), `http://www.itm.uni-stuttgart.de/research/pasimodo/pasimodo_en.php`

Pintelon, R., Schoukens, J.: System Identification: A Frequency Domain Approach. IEEE, New York (2001)

Popp, K.: Beitraege zur Dynamik von Magnetschwebefahrzeugen auf gestaenderten Fahrwegen. In: Fort. Ber. VDI Reihe 12, Nr. 35, Duesseldorf (1978)

Popp, K.: Dynamik von Fahrweg-Strukturen unter wandernden Lasten. VDI-Berichte 419, 153–161 (1981) (in German)

Popp, K., Bremer, H.: Modalanalyse von unverzweigten Balkentragwerken. Z. angew. Math. Mech. 63, T86–T88 (1983) (in German)

Popp, K., Mueller, P.C.: Ein Beitrag zur Gleisdynamik. Z. angew. Math. Mech. 62, T65–T67 (1982) (in German)

Popp, K., Schiehlen, W.: System Dynamics and Long-Term Behaviour of Vehicles, Track and Subgrade. Springer, Berlin (2003)

Rajamani, R.: Vehicle Dynamics and Control. Springer, Berlin (2006)

RecurDyn (2009), `http://www.functionbay.co.kr`

Riekert, P., Schunck, T.E.: Zur Fahrdynamik des gummibereiften Kraftfahrzeugs. Ing.-Arch. 11, 210–224 (1940) (in German)

Rill, G.: Auswahl eines geeigneten Integrationsverfahren fuer nichtlineare Bewegungsgleichungen bei Erregung durch beliebige Zeitfunktionen. Research report FB-4, Inst. Eng. Comp. Mechanics, Stuttgart (1981) (in German)

Rill, G.: Instationaere Fahrzeugschwingungen bei stochastischer Erregung. Dr.-Ing Diss. (1983) (in German)

Rill, G.: Simulation von Kraftfahrzeugen. Vieweg, Braunschweig (1994) (in German)

Rill, G.: Beurteilungskriterien. In: Rill, G., Schiehlen, W. (eds.) Handbuch Seminar 320811 Fahrzeugdynamik, pp. 6.0–6.9. VDI-Wiessensforum, Duesseldorf (2007) (in German)

Rill, G., Schiehlen, W.: Performance assessment of time integration methods for vehicle dynamics simulations. In: Arczewski, K., Frazek, J., Wojtyra, M. (eds.) Multibody Dynamics 2009, ECCOMAS Thematic Conference, Warsaw, Poland (2009)

Saha, S., Schiehlen, W.: Recursive kinematics and dynamics for parallel structural closed-loop multibody systems. Mechanics of Structures and Machines 29, 143–175 (2001)

Schiehlen, W. (ed.): Multibody Systems Handbook. Springer, Berlin (1990)

Schiehlen, W.: Computational aspects in multibody system dynamics. Comp. Meth. appl. Mech. Eng. 90, 569–582 (1991)

Schiehlen, W.: Computational dynamics: theory and applications of multibody systems. European J. of Mechanics A/Solids 25, 566–594 (2006)
Schiehlen, W., Eberhard, P.: Technische Dynamik, 2nd edn. Teubner, Wiesbaden (2004) (in German)
Schiehlen, W., Fehr, J., Kim, Y.: Covariance analysis for active vehicle suspensions. PAMM Proceedings in Applied Mathematics and Mechanics 7, 4120001–4120002 (2007)
Schramm, D.: Ein Beitrag zur Dynamik reibungsbehafteter Mehrkoerpersysteme. In: Fort. Ber. VDI. Reihe 18, Nr. 32. VDI-Verlag, Duesseldorf (1986) (in German)
Simpack (2009), http://www.simpack.de
Smith, R.A.: Matrix equation XA + BX = C. SIAM J. Appl. Math. 16, 198–201 (1968)
Snyder III, J.E., Wormley, N.D.: Dynamic interactions between vehicles and elevated, flexible randomly irregular guideways. J. Dyn. Syst. Meas. Control. Trans. ASME 99, 23–33 (1977)
Sperling, E.: Zur Kinematik und Kinetik elastischer Raeder aus der Sicht verschiedener Theorien. Dr.-Ing. Diss., Techn. Universitaet Muenchen (1977)
Stadler, W.: Multicriteria Optimization in Engineering and in the Sciences. Plenum Press, New York (1988)
Sussmann, N.E.: Statistical ground excitation models for high speed vehicles dynamic analysis. High Speed Ground Transportation Journal 8, 145–154 (1974)
Transrapid International. Driving Without Wheels (2009), http://www.transrapid.de
True, H.: Dynamics of railway vehicles and rail/wheel contact. In: Schiehlen, W. (ed.) Dynamical Analysis of Vehicle Systems. CISM Courses and Lectures, vol. 497, pp. 75–128. Springer, Vienna (2007)
VDI - Richtlinie 2057. Beurteilung der Einwirkung mechanischer Schwingungen auf den Menschen (in German). VDI, Duesseldorf (1975-1979) (in German)
VDI 2057. Einwirkung mechanischer Schwingungen auf den Menschen - Ganzkoerper-Schwingungen. VDI, Duesseldorf (2002) (in English and German)
VORtech Computing (2009), http://www.vortech.nl
Voy, C.: Die Simulation vertikaler Fahrzeugschwingungen. In: Fort. Ber. VDI Reihe 12, Nr. 30. VDI-Verlag, Duesseldorf (1977) (in German)
Weber, R.: Reifenfuehrungskraefte bei schnellen Aenderungen von Schraeglauf und Schlupf. Habilitationsschrift, Universitaet Karlsruhe (1981) (in German)
Wickens, A.H.: Stability criteria for articulated railway vehicles possessing perfect steering. Vehicle Systems Dynamics 7, 165–182 (1987)

Williams II, R.L., Lawrence, D.A.: Linear State-Space Control Systems. John Wiley & Sons, Hoboken (2007)

Wimmer, J., Rauh, J.: Multicriteria optimization as a tool in the vehicle's design process. In: Bestle, D., Schiehlen, W. (eds.) IUTAM Symp Optimization Mech Systems, pp. 333–340. Kluwer, Dordrecht (1996)

Wong, J.Y.: Theory of Ground Vehicles. Wiley, New York (2001)

Index

MIX
Papier aus verantwortungsvollen Quellen
Paper from responsible sources
FSC® C105338

If you have any concerns about our products,
you can contact us on
ProductSafety@springernature.com

In case Publisher is established outside the EU,
the EU authorized representative is:
Springer Nature Customer Service Center GmbH
Europaplatz 3, 69115 Heidelberg, Germany

Printed by Libri Plureos GmbH
in Hamburg, Germany